CW00921384

Advanced Sciences and Technologies for Security Applications

The series Advanced Sciences and Technologies for Security Applications comprises interdisciplinary research covering the theory, foundations and domain-specific topics pertaining to security. Publications within the series are peer-reviewed monographs and edited works in the areas of:

- biological and chemical threat recognition and detection (e.g., biosensors, aerosols, forensics)
- crisis and disaster management
- terrorism
- cyber security and secure information systems (e.g., encryption, optical and photonic systems)
- traditional and non-traditional security
- energy, food and resource security
- economic security and securitization (including associated infrastructures)
- transnational crime
- human security and health security
- social, political and psychological aspects of security
- recognition and identification (e.g., optical imaging, biometrics, authentication and verification)
- smart surveillance systems
- applications of theoretical frameworks and methodologies (e.g., grounded theory, complexity, network sciences, modelling and simulation)

Together, the high-quality contributions to this series provide a cross-disciplinary overview of forefront research endeavours aiming to make the world a safer place.

The editors encourage prospective authors to correspond with them in advance of submitting a manuscript. Submission of manuscripts should be made to the Editor-in-Chief or one of the Editors.

More information about this series at http://www.springer.com/series/5540

Babak Akhgar · P. Saskia Bayerl
Fraser Sampson
Editors

Open Source Intelligence Investigation

From Strategy to Implementation

Editors
Babak Akhgar
School of Computing and Management Sciences
Sheffield Hallam University
Sheffield
UK

Fraser Sampson
Office of the Police and Crime Commissioner for West Yorkshire
Wakefield
UK

P. Saskia Bayerl
Rotterdam School of Management
Erasmus University
Rotterdam
The Netherlands

ISSN 1613-5113 ISSN 2363-9466 (electronic)
Advanced Sciences and Technologies for Security Applications
ISBN 978-3-319-47670-4 ISBN 978-3-319-47671-1 (eBook)
DOI 10.1007/978-3-319-47671-1

Library of Congress Control Number: 2016955064

Printed on acid-free paper

This Springer imprint is published by Springer Nature
The registered company is Springer International Publishing AG
The registered company address is: Gewerbestrasse 11, 6330 Cham, Switzerland

Preface

It is our great privilege to welcome you to our book *Open Source Intelligence—From Strategy to Implementation*. In this collection, we offer an authoritative and accessible guide on how to conduct open-source intelligence (OSINT) investigations from data collection to analysis to the design and vetting of OSINT tools. It further highlights the broad range of challenges and complexities faced by law enforcement and other security agencies utilizing OSINT to increase our communities' security as well as to combat terrorism and organized crime.

One of the most important aspects for a successful police operation is the ability for the police to obtain timely, reliable, and actionable intelligence related to the investigation or incident at hand. OSINT provides an invaluable avenue to access and collect such information in addition to traditional investigative techniques and information sources. Examples of OSINT covered in this volume range from information posted on social media as one of the most openly available means of accessing and gathering open-source intelligence to location data, OSINT obtained from the darkweb to combinations of OSINT with real-time analytical capabilities and closed sources. And while OSINT by its nature is not generally gathered as 'evidence', it can be powerful when deployed in proceedings against criminals. The book therefore concludes with some consideration of the legal and procedural issues that will need to be addressed if OSINT is to be used in this way.

This book thus provides readers with an in-depth understanding to OSINT from a theoretical, practical, and legal perspective. It describes strategies for the design and deployment of OSINT for LEAs as well as other entities needing to capitalize on open-source data. The book offers a wide range of case examples and application scenarios from LEAs to defense and security agencies to industry, as well as hands-on guidance on the OSINT investigation process. The book outlines methods and illustrates benefits and challenges using real-life cases and (best) practices used by LEAs, security agencies, as well as industry. Another important aspect is the inclusion of legal and ethical considerations in the planning and conducting of OSINT investigations.

We would like to take the opportunity to recognize the work of our contributors to allow us to draw upon their expertise for this book—a process that has enabled us to highlight many of the important aspects of OSINT-related needs and requirements of LEAs and other security actors within its chapters. This interdisciplinary approach has helped us to bring together a wide range of domain knowledge from law enforcement, academia and industry to present our readers with an operational focused aspect of OSINT-based investigations and related strategic narratives from planning to deployment. We hope that this book will serve as a compendium for practitioners, academics, teachers, and students for state-of-the art knowledge ranging from conceptual considerations to hands-on practical information to legal and ethical guidance.

Sheffield, UK — Babak Akhgar
Rotterdam, The Netherlands — P. Saskia Bayerl
Wakefield, UK — Fraser Sampson

Acknowledgements

The editors wish to thank the multidisciplinary team of experts who have contributed to this book, sharing their knowledge, experience, and latest research. Our gratitude is also extended to the following organizations and projects:

- CENTRIC (Centre of Excellence in Terrorism, Resilience, Intelligence and Organised Crime Research), UK
- Rotterdam School of Management, Erasmus University, Netherland
- Information Technologies Institute, Centre for Research and Technology Hellas (CERTH-ITI), Thessaloniki, Greece
- National University of Public Service, Budapest, Hungary
- National Academy of Sciences, Institute for Computer Science and Control, Hungary
- Hungarian Competition Authority
- Police Services of Northern Ireland
- Home Office CAST, UK
- Serco Plc.
- EU-FP7 Project ATHENA (313220)
- EU-H2020 Project TENSOR (700024)
- EU-FP7 Project HOMER (312388)
- DG Home Project UNIFC2 (HOME/2013/ISEC/AG/INT/4000005215)

Contents

Editors and Contributors

About the Editors

Babak Akhgar is Professor of Informatics and Director of Center of Excellence in Terrorism, Resilience, Intelligence and Organised Crime Research (CENTRIC) at Sheffield Hallam University (UK) and Fellow of the British Computer Society. He has more than 100 refereed publications in international journals and conferences on strategic information systems with aspecific focus on knowledge management (KM) and intelligence management. He is member of editorial boards of several international journals and has acted as Chair and Program Committee Member for numerous international conferences. He has extensive and hands-on experience in the development, management, and execution of KM projects and large international security initiatives (e.g., the application of social media in crisis management, intelligence-based combating of terrorism and organized crime, gun crime, cyber-crime and cyber-terrorism, and cross cultural ideology polarization). In addition to this, he acts as technical lead in EU Security projects (e.g., "Courage" on Cyber-Crime and Cyber-Terrorism and "Athena" on the Application of Social Media and Mobile Devices in Crisis Management). Currently, he is the technical lead on EU H2020-project TENSOR on dark web. He has co-edited several books on Intelligence Management. His recent books are titled *Strategic Intelligence Management (National Security Imperatives and Information and Communications Technologies), Knowledge Driven Frameworks for Combating Terrorism and Organised Crime, Emerging Trends in ICT Security, and Application of Big Data for National Security.* Professor Akhgar is a board member of the European Organisation for Security and member of the academic advisory board of SAS UK.

P. Saskia Bayerl is Associate Dean of Diversity and Associate Professor of Technology and Organizational Behavior at Rotterdam School of Management, Erasmus University, the Netherlands. She further is Co-Director of the Centre of Excellence in Public Safety Management (CESAM, Erasmus University) and Visiting Research Fellow at CENTRIC (Center of Excellence in Terrorism, Resilience, Intelligence and Organised Crime Research, Sheffield Hallam University, UK). She is a regular speaker at police and security conferences and workshops and member of advisory boards of EU projects, as well as program committee member for international conferences. Her current research interests lie at the intersection of human–computer interaction, organizational communication, and organizational change with a special focus on the impact of technological innovations and public safety. Her research has been published in journals such as *MIS Quarterly, Communications of the ACM, New Media and Society*, and *Journal of Organizational Behavior* as well as international conferences in psychology, management, computational linguistics, and computer sciences and books. Most recently, she co-edited the book *Application of Big Data for National Security* (Elsevier).

Fraser Sampson, LL.B. (Hons), LL.M., MBA Solicitor has over 30 years experience in the criminal justice sector. A former police officer, he is the Chief Executive and Solicitor for the Office of the Police and Crime Commissioner for West Yorkshire. While practicing with national law firms, he represented police officers and the Police Federation in a number of high profile disciplinary cases and inquiries. A graduate of the Top Management Programme at the National School of Government, he is the founding author of *Blackstone's Police Manuals*, has written other key policing books published by Oxford University Press and is the editor of *Blackstone's Police Operational Handbook and the Routledge Companion to UK Counter Terrorism* (by Andrew Staniforth). Having published over 90 articles Fraser is on the editorial board of the Oxford Journal *Policing: A journal of strategy and practice*, is a member of the board of the Centre of Excellence in Terrorism, Resilience, Intelligence, and Organised Crime Research at Sheffield Hallam University and is an Associate Member of the Scottish Institute for Policing Research. Recent publications include chapters in *The Cyber Crime and Terrorism Investigators' Handbook* (Akhgar et al., Elsevier), *Big Data for National Security—A Practitioner's Guide to Emerging Technologies* (Akhgar et al., Elsevier) and *Policing in Northern Ireland—A New Beginning? It Can Be Done* (Rea & Masefield, Liverpool University Press).

Contributors

Babak Akhgar CENTRIC/Sheffield Hallam University, Sheffield, UK

András Benczúr Institute for Computer Science and Control of the Hungarian Academy of Sciences (MTA SZTAKI), Budapest, Hungary

Ben Brewster CENTRIC/Sheffield Hallam University, Sheffield, UK

Elisavet Charalambous Advanced Integrated Technology Solutions & Services Ltd, Egkomi, Cyprus

Neil Cunningham Police Service Northern Ireland, Belfast, Ireland

Tony Day CENTRIC/Sheffield Hallam University, Sheffield, UK

Géza Füzesi Hungarian Competition Authority, Budapest, Hungary

Helen Gibson CENTRIC/Sheffield Hallam University, Sheffield, UK

Eliot Higgins Bellingcat, Leicester, UK

Christos Iliou Centre for Research and Technology Hellas, Information Technologies Institute (CERTH-ITI), Thermi-Thessaloniki, Greece

George Kalpakis Centre for Research and Technology Hellas, Information Technologies Institute (CERTH-ITI), Thermi-Thessaloniki, Greece

Dimitrios Kavallieros Center for Security Studies (KEMEA), Hellenic Ministry of Interior and Administrative Reconstruction, Athens, Greece

Ioannis Kompatsiaris Centre for Research and Technology Hellas, Information Technologies Institute (CERTH-ITI), Thermi-Thessaloniki, Greece

Nikolaos Koutras Advanced Integrated Technology Solutions & Services Ltd, Egkomi, Cyprus

George Leventakis Center for Security Studies (KEMEA), Hellenic Ministry of Interior and Administrative Reconstruction, Athens, Greece

Alison Lyle Wakefield, UK

Laurence Marzell SERCO, Hook, UK

Jonathan Middleton Police Service Northern Ireland, Belfast, Ireland

Sándor Munk National University of Public Service, Budapest, Hungary

George Papalexandratos Center for Security Studies (KEMEA), Hellenic Ministry of Interior and Administrative Reconstruction, Athens, Greece

Steve Ramwell CENTRIC/Sheffield Hallam University, Sheffield, UK

Quentin Revell Centre for Applied Science and Technology, Home Office, St Albans, UK

Fraser Sampson Office of the Police and Crime Commissioner for West Yorkshire, West Yorkshire, UK

Tom Smith Centre for Applied Science and Technology, Home Office, St Albans, UK

Robert Stacey Centre for Applied Science and Technology, Home Office, St Albans, UK

Andrew Staniforth Trends Institution, Abu Dhabi, United Arab Emirates

Fahimeh Tabatabaei Mehr Alborz University, Tehran, Iran

Theodora Tsikrika Centre for Research and Technology Hellas, Information Technologies Institute (CERTH-ITI), Thermi-Thessaloniki, Greece

Pál Vadász National University of Public Service, Budapest, Hungary

Stefanos Vrochidis Centre for Research and Technology Hellas, Information Technologies Institute (CERTH-ITI), Thermi-Thessaloniki, Greece

Douglas Wells CENTRIC/Sheffield Hallam University, Sheffield, UK

B.L. William Wong Interaction Design Centre, Middlesex University, London, UK

Part I
Introduction

Chapter 1
OSINT as an Integral Part of the National Security Apparatus

Babak Akhgar

Abstract The roles of law enforcement agencies include maintaining law and order, protecting citizens and preventing, detecting and investigating crime. OSINT can provide critical capability for LEAs and security services to complement and enhance their intelligence capability, as the ability to rapidly gather and accurately process and analyze open source data can be a significant help during investigations and used for national level strategic planning to combat crime. Thus, purposeful and legal monitoring, analyzing and visualizing data from open data sources should be considered as mandatory requirement of any national security strategy. This chapter showcases the breadth of current and potential uses of OSINT based on UK's CONTEST strategy which provides the underlying basis of measures to prevent, pursue, protect and prepare against terror. It further proposes that to achieve efficient and innovative solutions, LEAs may be well advised to consider collaborations with private and public partners including academia using the successful implementation of the CENTRIC OSINT Hub is an example of how academia and LEAs can collaborate within the OSINT sphere in order to bring research into reality for the security and protection of citizens.

1.1 Introduction

A rise in the prevalence of Open Source Intelligence (OSINT) and its application by law enforcement and security agencies is set against a background of conflict, insecurity and the resurgence of violence in troubled regions across the world. For the United Kingdom, like many other nations, we remain under the constant threat of actual and potential attacks from all manner of hazards including terrorism, organized crime and cyber-related threats that—if left unchecked—can cause untold harm to citizens, communities, public services, businesses and the wider economy.

B. Akhgar (✉)
CENTRIC/Sheffield Hallam University, Sheffield, UK
e-mail: B.Akhgar@shu.ac.uk

B. Akhgar et al. (eds.), *Open Source Intelligence Investigation*,
Advanced Sciences and Technologies for Security Applications,
DOI 10.1007/978-3-319-47671-1_1

The scale and level of atrocities of recent terrorist attacks such as those in Paris, Brussels, Nice and Munich provide a cold and tangible reminder of the very real threat different nations across the world face. One of the common factors seen in the way these threats are realized is in the use of internet based communication platforms by terrorist individuals or formal groups such as the self-proclaimed Islamic State. For example, social media has increasingly become the dominant platform for projection onto overseas individuals, primarily through the dissemination of propaganda, complex indoctrination methodologies and recruitment campaigns,[1] creating a theatre of manipulation, with unprecedented ease of usage as well as access to the vulnerable. Indeed, this is reflected in both the record number of foreign nationals fighting in areas such as Iraq and Syria[2] as well as the controversial arrests and imprisonment of UK children between the ages of 14–17 for encouraging and masterminding terror attack plots.[3,4]

1.2 OSINT and Counter Terrorism Strategy

OSINT has, over the last five to ten years, been increasingly utilized by private sector organizations as a means to measure customer loyalty, track public opinion and assess product reception. Similarly, law enforcement and security agencies are acknowledging the requirement to apply similar techniques in order to enhance their investigative capability and improve their ability to identify and respond to criminal threats (see Chaps. 2, 3 and 13). The criminal entities perpetrating these threats are exploiting the internet for purposes such as recruitment (see Chap. 5), formation of illegal cartels (see Chap. 16) and the transfer of information and money to finance and co-ordinate their illicit activities.

The expansion of the internet has interwoven continents, cultures and communities, in addition to integrating with the majority of contemporary technologies. Whilst social media remains the dominant online platform for criminal and extremist psychological operations, there is an increasing potential for it to follow

[1]Helmus, T. C., York, E., Chalk, P. (2013). Promoting Online Voices for Countering Violent Extremism. (Rand Corporation, Santa Monica, California). Available at: http://www.rand.org/pubs/research_reports/RR130.html (Accessed: 03/08/2016).

[2]Bartlett, E. (2014). Record number of foreign nationals fighting in Iraq and Syria. The Independent Online. Available at: http://indy100.independent.co.uk/article/record-number-of-foreign-nationals-fighting-in-iraq-and-syria–gk2635auox (Accessed: 03/08/2016).

[3]BBC. (2015) "Anzac Day terror plot: Blackburn boy sentenced to life". Available at: http://www.bbc.co.uk/news/uk-34423984 (Accessed Online: 03/08/2016).

[4]Dodd, V. (2016). "Counter-terrorism detectives arrest east London teenager". The Guardian Online, Available at: https://www.theguardian.com/uk-news/2016/jun/16/counter-terrorism-detectives-arrest-east-london-teenager (Accessed Online: 03/08/2016).

the path of the internet, branching out, utilizing the likes of gaming consoles,[5] mobile applications,[6] cloud storage[7] and P2P services. Whilst social media and the surface web are used fundamentally for psychological, moral and emotional tactics, the dark web is used to a greater degree for the physical and tactical side of operations, focusing on arms and munitions,[8] false documents,[9] explosive making guides, crypto currency funding[10] and encrypted anonymous strategic communications.

The ubiquity of the internet has vastly increased the quantity, value and accessibility of OSINT sources. By definition, OSINT is intelligence based upon information that is freely available from public sources such as newspaper reports, journals, radio and television broadcasts, and more commonly in the current environment; social media and the internet.

When dealing with intelligence derived from the public domain, and specifically social media, there is a requirement to manage the public's privacy expectations appropriately, as although often freely available, much of the information posted to sites such as Facebook and Twitter is considered to be personal by registered users. When dealing with OSINT as opposed to more traditional closed intelligence sources, the concerns of the intelligence community turn from the availability of information to the identification of pertinent and accurate information. For these reasons it is increasingly necessary to validate intelligence derived from open sources with that from robust, closed source intelligence and the domain expertise of security professionals (see Chap. 9). Intelligence validation in this way is a particularly poignant topic when addressing social media content, as users often choose not to disclose or to falsify the personal information that they provide on these platforms (Bayerl and Akhgar 2015).

---

[5]Tassi, P. (2015). "How ISIS terrorists may have used PlayStation 4 to discuss and plan attacks". Forbes Online. Available at: http://www.forbes.com/sites/insertcoin/2015/11/14/why-the-paris-isis-terrorists-used-ps4-to-plan-attacks/#39d5c755731a (Accessed Online: 03/08/2016).

[6]Billington, J. (2015). "Paris terrorists used WhatsApp and Telegram to plot attacks according to investigators". International Business Times. Available at: http://www.ibtimes.co.uk/paris-terrorists-used-whatsapp-telegram-plot-attacks-according-investigators-1533880 (Accessed: 03/08/2016).

[7]Hall, K. (2011). "Cyber terrorism set to increase after al-Qaeda calls for more cyber-attacks, says government". Computer Weekly Online. Available at: http://www.computerweekly.com/news/2240105012/Cyber-terrorism-set-to-increase-after-al-Qaeda-calls-for-more-cyber-attacks-says-government (Accessed Online: 03/08/2016).

[8]See listed Dark Web onion gun site vendors: https://www.deepdotweb.com/tag/guns/.

[9]Charlton, A. (2015). "Dark web vendors sell blank British passport and entry to database for just £2000". Available at: http://www.ibtimes.co.uk/dark-web-vendors-sell-blank-british-passports-entry-passport-database-just-2000-1509564 (Accessed Online: 03/08/2016).

[10]Smith, M. (2015). "Hacktivists claim ISIS terrorists linked to Paris attacks had bitcoin funding". Network World. Available at: http://www.networkworld.com/article/3005308/security/hacktivists-claim-isis-terrorists-linked-to-paris-attacks-had-bitcoin-funding.html (Accessed Online: 03/08/2016).

Table 1.1 Overview of the CONTEST strategy principles and their application for OSINT

	Contest components	OSINT proposition
Prevent strategy	The Prevent strategy is concerned with tackling the radicalisation of people who sustain the international terrorist and organised crime threat	Identification of terrorist narratives, influencers and propaganda over the surface web, particularly in countering attempts to turn people to terrorism by 'incitement and recruitment', thus tackling the factors or root causes which can lead to radicalisation and recruitment, in Europe and internationally More effective development of counter-extremists narratives and to encourage inter-cultural dialogue promoting good governance, democracy, education and economic prosperity Understanding of communities and areas of concern
Pursue strategy	The Pursue strategy is concerned with reducing the terrorist threat by disrupting terrorists and organised criminal groups and their operations	Gathering intelligence from the dark web (see also Chap. 8) Legal and ethical collection of evidence for securing convictions International corporation helping to pursue and investigate terrorist threats inside and outside of national borders, to impede the travel and communication of terrorists and criminals, to disrupt their support networks and to cut off funding and access to attack materials, and to bring individuals to justice
Protect strategy	The Protect strategy is concerned with reducing vulnerability to terrorist and organised crime attacks for European Member States	Proactive threat assessment of vulnerabilities (e.g., border security) and social areas of risk such as Child Sexual exploitation (CSE) (see Chap. 15) Proactive and live assessment of threats to mass gatherings Protection of National Critical Infrastructure and reduction of their vulnerability to attacks, including through increased security of borders and transport. (see Chap. 4)
Prepare strategy	The Prepare strategy is concerned with ensuring that the population and European Member States are as ready as they can be for the consequences of a terrorist attack and organised criminal event.	Building of communities' resilience (see Chap 4) Preparation for potential CBRNE (Chemical, Biological Radioactive, Nuclear and Explosive) attacks Modelling of emerging organised crime Early warning for health hazards

From the national security perspective, OSINT-based solutions should enhance the capabilities of law enforcement agencies and security services, providing access to more actionable intelligence that can support existing decision making, tasking and coordination activities. The core of any OSINT solution should focus on internet-centric data gathering and exploitation. The latter includes development of enhanced capabilities and services to gather, analyze, visualize and combine relevant data from which dynamic and real time hypothesizes can be generated.

Measures to combat crime and terrorism—online and offline—are an increasingly important element of any national security strategy. Looking at current approaches to counter terrorism in the UK, the 4 Ps of the CONTEST strategy[11] provide the underlying basis of measures to prevent, pursue, protect and prepare against terror. An overview of the four principles and how OSINT may be employed in their support can be found in Table 1.1. Later chapters in the book discuss how these examples manifest within a real operational context.

It should be noted that although the goals of terrorist and organised crime groups (OCGs) are different, the connections between terrorist and organised criminal activities appear to be growing. For example, in the recent attack in Munich (22nd July 2016) the perpetrator of the attack is believed to have procured his weapon through the dark web. Criminal activities that terrorist groups are involved in, either through affiliation with individual criminals and criminal groups or through their own operations, can include the trafficking of illegal goods and substances such as weapons and drugs, trafficking in human beings, financial fraud, money laundering and extortion (see Chap. 16)

OSINT is already being utilized as one of the key intelligence sources for national security and its importance is only increasing. And as Table 1.1 demonstrates, the breadth of current and potential uses is enormous. However, OSINT cannot be the only source that LEAs and security agencies rely on. OSINT is at its most powerful, when it is able to augment existing closed source intelligence by providing additional information and direction to where further intelligence may be required (see Chap. 9). The combined effect of OSINT and traditional intelligence sources reflect national security intelligence apparatus of a nation.

A promising approach to ensure efficient and innovative solutions and processes are collaborations between various public and private actors and organizations such as LEAs, industry and academia, amongst others. In the following we illustrate this approach by describing the setup and functioning of the OSINT Hub at CENTRIC (Centre of Excellence for Terrorism, Resilience, Intelligence and Organized Crime Research).

[11]Contest Strategy. The UK's Strategy for Countering Terrorism, Presented to Parliament by the Secretary of State for the Home Department by Command of Her Majesty, July 2011.

1.3 The CENTRIC OSINT Hub

Since 2012, the Centre of Excellence in Terrorism, Resilience, Intelligence & Organised Crime Research (CENTRIC) has built a strong research and development capability focused on the operational utilization of OSINT in regards to its application in relation to counter terrorism, cybercrime, crisis management and in the identification and modeling of organized crime.

At the beginning of 2016, CENTRIC launched its Open Source Intelligence Hub, or OSINT Hub, which has been gaining momentum as a physical and virtual space for the operational exploitation, dissemination and development of CENTRIC capabilities. Such capabilities are constantly being acquired, developed and improved in technical and non-technical expertise and tooling. This is happening in close collaboration with national, pan-European, and international partners in academia, public and private sectors. The OSINT Hub is ultimately the sum of CENTRIC's and its partners' experiences and is quickly setting a benchmark in research and development around:

- Counter-terrorism
- Major investigations
- Cybercrime
- Crisis management
- Public order
- Child sexual exploitation
- Identification and modeling of Organized Crime

Domain expertise has been ingrained into the Hub through direct collaboration with a number of law enforcement agencies and investigatory teams to directly influence and increase the capabilities of the hub.

The foundation of the OSINT Hub's situational awareness and data processing capabilities were born out of CENTRIC's participation in major EU projects, in collaboration with law enforcement, as a major technical partner responsible for the delivery of the projects situational awareness dashboard, web crawling, entity extraction, content categorization, social media and data aggregation functionalities—all of which are built on state-of-the-art tools offered by leading providers, open source communities and existing academic research. Harnessing these capabilities has enabled CENTRIC to more efficiently deliver data processing and command and control capabilities to its partners. To date, the OSINT Hub provided support to various live investigations ranging from child sexual exploitation to terrorism.

The secure physical environment in the OSINT Hub enables investigators to work directly with the CENTRIC team and their tools and to provide direct input into the development of future capabilities. Many of the investigatory capabilities of the OSINT Hub have been developed through such collaboration and can clearly benefit law enforcement through cost reductions in both targeted investigations and strategic situational awareness. They may also explore the potential of OSINT making use of the Hub's tools, ensuring compatibility with existing workflows and processes and compliance with existing governance and legal requirements such as RIPA and the Data Protection Act (see Chaps. 17 and 18).

1.4 Concluding Remarks

The roles of law enforcement agencies include maintaining law and order, protecting citizens and preventing, detecting and investigating crime. In achieving these goals, LEAs will fulfill the purpose of protecting the security of society and the citizens they serve. OSINT can potentially provide critical capability for LEAs and security services to complement and enhance their intelligence capability. Purposeful and legal monitoring, analysing and visualizing public open source data source should be considered as mandatory requirements of any national security strategy. The ability to rapidly gather and accurately process and analyse open source data can be a significant help during investigations, whilst it can also be used for national level strategic planning to combat crime. However, to achieve efficient and innovative solutions, LEAs may be well advised to consider collaborations with private and public partners including academia. The successful implementation of the CENTRIC OSINT Hub is an example of how academia and LEAs can collaborate within the OSINT sphere in order to bring research into reality for the security and protection of citizens.

References

Bayerl PS, Akhgar B (2015) Surveillance and falsification implications for open source intelligence investigations. Commun ACM 58(8):62–69

Billington J (2015) Paris terrorists used WhatsApp and Telegram to plot attacks according to investigators. International Business Times. Available at: http://www.ibtimes.co.uk/paris-terrorists-used-whatsapp-telegram-plot-attacks-according-investigators-1533880 Accessed Aug 03 2016

Charlton A (2015). Dark web vendors sell blank British passport and entry to database for just £2000. Available at: http://www.ibtimes.co.uk/dark-web-vendors-sell-blank-british-passports-entry-passport-database-just-2000-1509564 Accessed Online Aug 03 2016

Contest Strategy (2011) The UK's strategy for countering terrorism, Presented to Parliament by the Secretary of State for the Home Department by Command of Her Majesty, July 2011

Hall K (2011) Cyber terrorism set to increase after al-Qaeda calls for more cyber attacks, says government. Computer Weekly Online. Available at: http://www.computerweekly.com/news/2240105012/Cyber-terrorism-set-to-increase-after-al-Qaeda-calls-for-more-cyber-attacks-says-government Accessed Online Aug 03 2016

Tassi P (2015) How ISIS terrorists may have used PlayStation 4 to discuss and plan attacks. Forbes Online. Available at: http://www.forbes.com/sites/insertcoin/2015/11/14/why-the-paris-isis-terrorists-used-ps4-to-plan-attacks/#39d5c755731a Accessed Online Aug 03 2016

Smith M (2015) Hacktivists claim ISIS terrorists linked to Paris attacks had bitcoin funding. Network world. Available at: http://www.networkworld.com/article/3005308/security/hacktivists-claim-isis-terrorists-linked-to-paris-attacks-had-bitcoin-funding.html Accessed Online 3 Aug 2016

Chapter 2
Open Source Intelligence and the Protection of National Security

Andrew Staniforth

Abstract Given the scale and complexity of the threats from international terrorism, intelligence agencies must continue to advance counter-terrorism measures to keep us all safe; and most importantly, seek new ways in which to embed progressive developments to ensure that the primary driver for change in counter-terrorism practice is not simply the next successful attack. Harnessing the power of OSINT via Big Data continues to be a game-changer for counter-terrorism policy-makers, professionals and practitioners. The purpose of this chapter is to explain the importance of OSINT within the context of national security and the role of intelligence agencies to prevent and protect citizens from the threat of international terrorism. To outline the operational requirements for intelligence agencies use of OSINT, this chapter also outlines key components of the modern terrorist threat, which includes explanations of terrorist radicalization development processes and how OSINT and the power of Big Data analytics is increasingly being used to combat terrorism and prevent violent extremism.

2.1 Introduction

The first duty of government remains the security of its citizens. The range of threats to national security is becoming increasingly complex and diverse. Terrorism, cyber-attack, unconventional attacks using chemical, nuclear, or biological weapons, as well as large-scale accidents or natural hazards—anyone could put citizen's safety in danger while inflicting grave damage to a nation's interests and economic well-being. When faced with a combination of current levels of economic uncertainly and political instability, governments must be able to act quickly and effectively to address new and evolving threats to their security. Tough security measures are needed to keep citizens, communities and commerce safe

A. Staniforth (✉)
Trends Institution, Abu Dhabi, The United Arab Emirates
e-mail: info@trendsinstitution.org

B. Akhgar et al. (eds.), *Open Source Intelligence Investigation*,
Advanced Sciences and Technologies for Security Applications,
DOI 10.1007/978-3-319-47671-1_2

from contemporary security hazards, the most pressing of which remains the enduring threat from international terrorism.

Threats from terrorism are matters of intense public and political concern but they also raise acute challenges for the security apparatus of the State. These challenges arise because terrorism can inflict significant loss of life, yet it is not the scale of the atrocities committed in its name that gives terrorism its special status; it is the threat it poses to the State, for it undermines the basis of State legitimacy—the capacity to protect its citizens. Therefore, measures known as Counter-Terrorism (CT), as a major aspect of national security, attract high-profile political and public attention and correlatively failures in CT lead to significant outcries followed by stringent scrutiny from a variety of quarters including the media, public opinion, police investigation, government inquiry, parliamentary questioning and academic study.

The purpose of this chapter is to explain the importance of OSINT within the context of national security and the role of intelligence agencies to prevent and protect citizens from the threat of international terrorism (see Chap. 1). To outline the operational requirement for intelligence agencies use of OSINT, this chapter also outlines key components of the modern terrorist threat, which includes explanations of terrorist radicalization development processes and how OSINT and the power of Big Data analytics is increasingly being used to combat terrorism and prevent violent extremism.

2.2 From Threat to Threat

To understand the importance of OSINT in protecting national security from international terrorism we must first examine the nature of the changing threat from terrorists in this post-9/11 era of global terrorism. Over recent years, following a decade of tackling the terrorist atrocities committed by Al Qaeda and their global network of affiliates and inspired lone-actors, a new global terrorist phenomenon has risen from the conflict in Syria. The rise of Daesh—the self-proclaimed 'Islamic State'—provides further evidence that many nations across the world face a continuing threat from extremists who believe they can advance their aims through acts of terrorism. This threat is both serious and enduring, being international in scope and involving a variety of groups, networks and individuals who are driven by violent and extremist beliefs (see Chap. 12).

The violent progress of Daesh through towns and villages in Iraq has been swift —aided by foreign fighters from across the Middle East and Europe. Daesh have now taken control of large swathes of Iraq leading the British Prime Minister David Cameron to warn his Cabinet that violent Daesh jihadists were planning attacks on British soil. The warning came amid growing concerns amongst senior security officials that the number of Britons leaving the UK to fight alongside extremist groups abroad was rising. The export of British born violent jihadists is nothing new but the call to arms in Iraq this time had been amplified by a slick online recruitment campaign, urging Muslims from across the world to join their fight and to post messages of support for Daesh. In a chilling online recruitment video designed to lure jihadists to Iraq, 20 year-old Nasser Muthana, a medical student from Cardiff in Wales, and his 17 year old brother Aseel, declared their support for Daesh while sitting alongside each other holding their semi-automatic assault rifles. In the video, Nasser states: 'We understand no borders. We have participated in battles in Syria and in a few days we will go to Iraq and will fight with them.' (Dassanayake 2014). Despite Nasser attaining 12 GSCE's at grade A, studying for his A-levels and being offered places to enroll on medical degrees at four UK universities, he instead volunteered to join the ranks of Daesh. Unbeknown to his parents or authorities, the former school council member and his younger brother, who was studying A-levels at the time, travelled to Syria via Turkey to fight the Assad regime. The father of the brothers-in-arms fighting for Daesh, Mr Muthana, declared no knowledge of their intended travel plans to Syria and had reported them missing to the police during November 2013. Mr Muthana remained devastated that his sons had turned to violent extremism, stating that: 'Both my sons have been influenced by outsiders, I don't know by whom. Nasser is a calm boy, very bright and a high achiever', going on to say that: 'He loved rugby, playing football and going camping with friends. But he has been got at and has left his home and everyone who loves him' (Dassanayake 2014).

The online propaganda of Daesh has proved ruthlessly effective. On 13 June 2015, a group of Daesh suicide bombers delivered a deadly attack near the city of Baiji in the Salahuddin province of Iraq. A four-strong terrorist cell killed eleven people in two separate explosions as they detonated bombs placed in their vehicles at an oil refinery. One of the bombers was 17 year old Talha Asmal, who had travelled thousands of miles from his home town in Dewsbury, West Yorkshire in England to join and fight alongside Daesh. Talha was described by his school teacher as a "conscientious student" (BBC 2015). His family, utterly devastated and heartbroken by the loss of their son, said: "Talha was a loving, kind, caring and affable teenager", going on to suggest that he had been: "exploited by persons unknown who were hiding behind the anonymity of the World Wide Web." (Grierson 2015) In committing his act of martyrdom in support of the Daesh cause, Talha became the youngest British suicide bomber. UK security forces are rightly concerned by British citizens fighting in Syria and Iraq. The numbers reported by various research institutes are certainly shocking. The Foreign Policy Research Institute evaluates that between 6000 and 12,000 volunteers have passed through Syrian territory, arriving from 70 different countries (Helfont 2012). Among these are approximately 2000 European citizens, creating a new dimension to terror threats across Europe, which has resulted in deadly and determined Daesh-inspired attacks in Paris and Brussels, leading to heightened levels of security across European Member States (Helfont 2012). Cecilia Malmstromm, the European Commissioner of Home Affairs, raised the alarm regarding foreign fighters to her counterparts in EU Member States saying that: "Often set on the path of radicalization in Europe by extremist propaganda or by recruiters, Europeans travel abroad to train and to fight in combat zones, becoming yet more radicalized in the process.

Armed with newly acquired skills, many of these European 'foreign fighters' could pose a threat to our security on their return from a conflict zone. And as the number of European foreign fighters rises, so does the threat to our security" (European Commission 2014). Of critical concern for the security of Western nations, is the way in which their citizens are being influenced by the extreme single narratives of religious or political ideologies promoted by terrorist and extremist groups and individuals such as Daesh. This narrative, when combined with a complex malaise of social and economic factors, serves to manipulate individuals towards adopting extremist perspectives, cultivating a terrorist threat which presents a clear and present danger to the free and democratic way of life in the West. This is a threat which is propagated via open-source information on the internet and the online radicalization of citizens remains a major source of support and new recruits for the Daesh cause.

2.3 Online Radicalisation

Understanding why young men and women from our communities have travelled to take up arms in conflict overseas is the question which most troubles intelligence agencies across Europe. The obsession with finding the answer has come to dominate the whole debate over the causes of terrorism. In the post-9/11 world, understanding how people become terrorists has come to be discussed in terms of 'radicalisation', a rather exotic term which presumably describes a similarly exotic process (Silke 2014). What is called radicalisation today, in the past was referred too much more mundanely as 'joining' a terrorist group or being 'recruited'. No one talked of the individuals joining the Irish Republican Army (IRA) or Euskadi Ta Askatasuna (ETA) as being 'radicalised', though they all certainly were according to our modern understanding arising from the Al Qaeda-inspired genre of international terrorism. After 9/11 it became awkward to talk about people 'becoming' terrorists, 'joining' terrorist groups or being 'recruited'. Those terms were too banal, too ordinary for the dawn of new-era global terrorism. Ordinary terms might imply ordinary processes and worse still, ordinary solutions. So these simple terms had to make way for something more exotic, more extreme and 'radicalisation' fitted the bill nicely, especially with the rise of online radicalisation via the internet which has changed—and continues to change—the very nature of terrorism. The internet is well suited to the nature of terrorism and the psyche of the terrorist. In particular, the ability to remain anonymous makes the internet attractive to the terrorist plotter. Terrorists use the internet to propagate their ideologies, motives, grievances and most importantly communicate and execute their plan. The most powerful and alarming change for modern terrorism, however, has been its effectiveness for attracting new terrorist recruits, very often the young and most vulnerable and impressionable in our societies. Modern terrorism has rapidly evolved, becoming increasingly non-physical, with vulnerable 'home grown' citizens being recruited, radicalized, trained and tasked online in the virtual and ungoverned domain of cyber space. With an increasing number of citizens putting more of their lives online, the interconnected and globalized world in which we now live provides an extremely large pool of potential candidates to draw into the clutches of disparate terrorists groups and networks.

The openness and freedom of the internet unfortunately supports 'self-radicalization'—the radicalization of individuals without direct input or encouragement from others. The role of the internet in both radicalization and the recruitment into terrorist organizations is a growing source of concern for security authorities. The internet allows individuals to find people with shared views and values and to access information to support their radical beliefs and ideas. The unregulated and ungoverned expanse of the internet knows no geographical boundaries, thus creating a space for radical activists to connect across the globe. This is especially problematic as the easy access to like-minded people helps to normalize radical ideas such as the use of violence to solve grievances. Yet, solving the complex issue of radicalization by simple processes—such as the suggestion to 'clean up' the internet—is impracticable and well beyond the scope of any single government (see Chap. 5).

Understanding how and why people in our communities move towards extremist perspectives, and creating an alternative to allow them to resist adopting such views, remains the key challenge in countering the contemporary threat from terrorism. The intelligence agencies have a central role to both identify those individuals and groups who are radicalising and recruiting others to their extremist cause, as well as identifying individuals who are adopting violent views and putting in place preventative mechanisms and interventions to stop and deter the development of radicalism at source. Identifying radicalised individuals is a complex task and continues to present the greatest challenge to intelligence agencies across the world. It is widely acknowledged that nobody suddenly wakes up in the morning and decides that they are going to make a bomb. Likewise no one is born a terrorist. Conceptualisations of radicalisation have increasingly recognized that becoming involved in violent extremism is a process: it does not happen all at once. Similarly, the idea that extremists adhere to a specific psychological profile has been abandoned, as has the view that there may be clear profiles to predict who will follow the entire trajectory of radicalisation development (Hubbard 2015). Instead, empirical work has identified a wide range of potential 'push' and 'pull' factors leading to (or away from) radicalisation. There are many potential factors which may influence an individual towards adopting extremist perspectives. These include not only politics, religion, race and ideology, the very core motivations of terrorism, but may also include elements of a sense of grievance or injustice. It is important to recognize that terrorist groups can fulfill important needs for an individual: they give a clear sense of identity, a strong sense of belonging to a group, the belief that the person is doing something important and meaningful, and also a sense of danger and excitement. For some individuals, and particularly young men, these are very attractive factors. Individuals deemed to be vulnerable and potentially at risk of radicalisation share a widely held sense of injustice. The exact nature of this perception of injustice varies with respect to the underlying motivation for violence, but the effects are highly similar. Personal attitudes such as strong political views against government foreign policies regarding conflicts overseas can also play an important role in creating initial vulnerabilities.

Intelligence agencies have come to recognize that terrorism is a minority-group phenomenon, not the work of a radicalized mass of people following a twisted version of faith. People are often socialized into this activity leading to a gradual deepening of their involvement over time. Radicalization is thus a social process, which requires an environment that enables and supports a growing commitment (Silke 2014). The process of radicalization begins when these enabling environments intersect with personal 'trajectories', allowing the causes of radicalism to resonate with the individual's personal experience. Some of the key elements in the radicalization process are thus related to the social network of the individual, for example, who is the person spending time with, and who are his or her friends, whether this activity and interaction is in the physical or cyber world. Intelligence agencies also recognize that no single radicalisation 'push' or 'pull' factor predominates. The catalyst for any given individual developing extremist views will more likely be a combination of different factors, which makes prediction with any certainty a challenging task. The manifestation of individual radicalisation factors may be subtle, resulting in very weak signs and signals of radicalisation development, while other factors may be more visible. Identifying these factors remains the key to the early prevention and intervention of radicalisation, ensuring intelligence agencies and their partners can act appropriately to stop terrorism at its source (Silk et al. 2013).

2.4 Counter Measures

The contemporary phase of counter-terrorism has evolved important new preventative trends, alongside palpable moves towards expansion and localism (Masferrer and Walker 2013). Many governments now seek to ensure that mechanisms are in place to be able to draw upon the valuable information and goodwill of communities from which aberrant extremists are recruited and radicalised. This role has fallen to both intelligence agencies and Law Enforcement Agencies (LEAs) who not only find specialist their counter-terrorism units tackling the extremist threat but a requirement for police officers engaged in community policing activities and their local authority partners (Spalek 2012). The working relationship between intelligence agencies and LEAs across Europe is developing and information to counter terrorism is no longer handled on a 'need to know' basis but rather on a 'need to share' approach (Silk et al. 2013). Although the security services and police forces retain their individual roles and responsibilities, there is currently greater sharing between them and between individual police forces within the arena of counter-terrorism. This has forged not only a greater readiness for intelligence sharing but also a sharing of equipment and human assets and a more jointly co-ordinated response to counter-terrorism activities such as surveillance operations. Collaboration in counter-terrorism, given the immediacy and severity of the terrorist threat, is now absolutely essential, being fuelled by the relentless pursuit to gather intelligence to prevent attacks and pursue terrorists.

To kerb the terrorists' use of the internet, authorities are seeking to harness the full power of new technologies to keep communities safe from terrorist threats and intelligence agencies are using OSINT to inform and provide a richer picture to their covert and clandestine operations. An important aspect of intelligence agency use of OSINT is social media, which represents an increasing and fundamental part of the online environment in which the users are authors of the content who do not passively receive information, but they create, reshape and share it (see Chaps. 6 and 9). In some cases, the interaction among users based on social media creates communities and virtual worlds providing an excellent source of information for intelligence agencies. Although there are significant differences in the nature of these outputs, two aspects are always present and are relevant to the work of intelligence agencies: large amounts of information and user generated content. The social media platforms aggregate huge amounts of data generated by users which are in many cases identified or identifiable. When combined with other online and stand-alone datasets, this contributes to create a peculiar technological landscape in which the predictive ability that is Big Data analytics, has relevant impact for the implementation of social surveillance systems by States. Big Data is nothing new, but it is currently at the final stage of a long evolution of the capability to analyze data using computer resources which for the intelligence agencies of government provides an excellent opportunity to tackle terror and keep communities safe.

Big Data represents the convergence of different existing technologies that permit enormous data centers to be built, create high-speed electronic highways and have ubiquitous and on-demand network access to computing resources, more commonly referred to as 'cloud computing'(Akhgar et al. 2015). These technologies offer substantially unlimited storage, allow the transfer of huge amounts of data from one place to another, and allow the same data to be spread in different places and re-aggregated in a matter of seconds. All these resources permit a large amount of information from different sources to be collected and the pet bytes of data generated by social media represent the ideal context in which Big Data analytics can be used. The whole dataset can be continuously monitored by analytics, in order to identify the emerging trends in the flows of data and obtaining real-time or nearly real time results in a way that is revolutionary. Within the context of counter-terrorism, the availability of these new technologies and large datasets provides a competitive advantage, representing the greatest opportunity to increase the effective delivery of counter-terrorism. Big Data can help the identification of terrorist networks and their associations using OSINT and provide valuable corroboration of other intelligence sources to support the holistic development of intelligence. It can also rapidly support the identification of radical roots within online communities providing significantly increased capabilities and opportunities not just to prevent terrorist attacks, but to identify attack planning activity and most importantly, spot the early signs and signals of radicalization and recruitment to stop violent and extremist development at source.

2.5 Conclusions

Counter-terrorism is no longer the hidden dimension of statecraft. It has over recent years moved out of the shadows due in part to the understanding of intelligence agencies that not all counter-terrorism measures need to be cloaked in secrecy in order for them to be effective. Harnessing the power of OSINT via Big Data analytics capabilities presents a unique opportunity for governments to address the increasing threats from international terrorism at relatively low cost. But the handling of such large data-sets raises acute concerns for existing storage capacity, together with the ability to share and analyze large volumes of data. The accessibility of OSINT and the introduction of Big Data capabilities will no doubt require the rigorous review and overhaul of existing intelligence models and associated processes to ensure all in authority are ready to exploit Big Data OSINT.

While OSINT and Big Data analytics present many opportunities for national security, any developments in this arena will have to be guided by the State's relationship with its citizenry and the law. Citizens remain rightly cautious and suspicious of the access to and sharing of their online data—especially by agents of the state. As citizens put more of their lives online, the safety and security of their information matters more and more. Any damage to public trust is counter-productive to contemporary national security practices and just because the state may have developed the technology and techniques to exploit OSINT and harness Big Data does not necessarily mean that it should. The legal, moral and ethical approach to OSINT via Big Data analytics must be fully explored alongside civil liberties and human rights, yet balanced with protecting the public from security threats. Big Data analytics must not be introduced by stealth, but through informed dialogue, passing though the due democratic process of governments. Citizens are more likely to support robust measures against terrorists that are necessary, appropriate and proportionate but many citizens, and politicians for that matter, will need to be convinced that extensive use and increased reliability upon publicly-available information harnessed through the power of Big Data is an essential part of keeping communities safe from terrorism and violent extremism.

All in authority must also avoid at all costs the increased use of Big Data to maximize the potential of OSINT as a knee-jerk reaction to placate the public and the press following a terrorist attack. Experience over recent years shows that in the aftermath of terrorist events political stakes are high: politicians and legislators fear being seen as lenient or indifferent and often grant the executive broader authorities without thorough debate. New special provisions intended to be temporary turn out to be permanent. Although governments may frame their new provisions in terms of a choice between security and liberty, sometimes the loss of liberty is not necessarily balanced by the gain in safety and the measures introduced become counter-productive. The application of Big Data OSINT for national security should be carefully considered and not quickly introduced as any misuse of its power may result in long term damage of relations with citizens and communities due to the overextended and inappropriate use of Big Data capabilities.

In conclusion, it is important to remember that, when compared against other types of serious crime, terrorism, thankfully, remains a relatively rare occurrence but the cost is high when attacks succeed. Terrorism therefore continues to demand a determined proportionate response. The history of terrorism reveals with alarming regularity that terrorist plotters achieve their intended objectives, defeating all of the State's security measures put in place at the time. Unfortunately, this pattern is not set to change and Daesh have proved their expertise in carrying out deadly attacks in major cities across Europe. The intelligence agencies will prevent further terrorist atrocities, but there is a very strong likelihood that they will not stop them all. In the light of that conclusion all in authority must dedicate themselves to increasing counter-terrorism capabilities and developing new approaches to better protect the public. To ignore or dismiss the positive benefits of OSINT would be both misplaced and unwise as all citizens expect the intelligence agencies of their governments to take the necessary steps to keep them safe. Harnessing the power of OSINT via Big Data continues to be a game-changer for counter-terrorism policy-makers, professionals and practitioners. Given the scale and complexity of the threats from international terrorism, intelligence agencies must continue to advance counter-terrorism measures to keep us all safe; and most importantly, seek new ways in which to embed progressive developments to ensure that the primary driver for change in counter-terrorism practice is not simply the next successful attack.

References

Akhgar B, Saathoff GB, Arabnia HR, Hill R, Staniforth A, Bayerl PS (2015) Application of big data for national security: a practitioner's guide to emerging technologies. Elsevier/Butterworth, Heinemann

BBC (2015) Dewsbury teenager is UKs youngest ever suicide bomber. Retrieved 26 Jul 2016 from http://www.bbc.co.uk/news/uk-england-leeds-33126132

Dassanayake D (2014) Why is he doing this? Devastated father of UK jihadists in Isis video speaks out. Retrieved 26 Jul 2016 from http://www.express.co.uk/news/world/483929/Isis-Father-of-UK-jihadist-Nasser-Muthana-speaks-out

European Commission (2014) Preventing radicalisation to terrorism and violent extremism: strengthening the EU's response

Grierson J (2015) Dewsbury in 'utter shock' over Talha Asmal's death in Iraq suicide bombing. The Guardian. Retrieved from https://www.theguardian.com/world/2015/jun/15/dewsbury-in-utter-shock-at-over-talha-asmals-death-in-iraq-suicide-bombing

Helfont T (2012) The foreign fighter problem: a conference report. Retrieved from http://www.fpri.org/article/2012/10/the-foreign-fighter-problem-a-conference-report/

Hubbard EM (2015) Hostile intent and counter terrorism: human factors theory and application. Ergonomics, 1–2

Masferrer A, Walker C (eds) (2013) Counter-terrorism. Crossing Legal Boundaries in Defence of the State. Edward Elgar Publishing, Human Rights and the Rule of Law

Silk P, Spalek B, O'Rawe M (eds) (2013) Preventing ideological violence: communities, police and case studies of "success". Springer

Silke A (2014) Terrorism: all that matters. Hachette, UK

Spalek B (2012) Counter-Terrorism: community-based approaches to preventing terror crime. Palgrave Macmillan

Chapter 3
Police Use of Open Source Intelligence: The Longer Arm of Law

Andrew Staniforth

Abstract While the internet and online social networks have positively enriched societal communications and economic opportunities, these technological advancements have changed—and continue to change—the very nature of crime, serving to breed a new sophisticated and technically capable criminal. Furthermore, the borderless nature of the phenomenon of cybercrime and the transnational dimensions of human trafficking, drugs importation and the illegal movement of firearms, cash and stolen goods means that criminals can plan their crimes from jurisdictions across the world, making law enforcement particularly challenging, the very reason why LEAs must maximise the potential of OSINT and seek new and innovative ways to prevent crime. Hence, it is essential for all practitioners, policy-makers and policing professionals to understand what OSINT is and what it is not, how it can be used and the limitations or conditions on it, as well as understanding more about the scale, scope and complexity of the threats from criminals whose methods of operating are becoming increasingly sophisticated. The purpose of this chapter is to explain the role and function of OSINT within the context of policing and existing intelligence collection disciplines, as well as to define OSINT from an LEA perspective and describe its position within the intelligence profession of policing.

3.1 Introduction

Since before the advent of advanced technological means of gathering information, Law Enforcement Agencies (LEAs) have planned, prepared, collected, and produced intelligence from publicly available information and open sources to gain knowledge and understanding in support of preventing crime and pursuing criminals. While traditional threats from crime have historically begun at the most local level, in today's increasingly interconnected and interdependent world, many new

---

A. Staniforth (✉)
Trends Institution, Abu Dhabi, United Arab Emirates
e-mail: info@trendsinstitution.org

B. Akhgar et al. (eds.), *Open Source Intelligence Investigation*,
Advanced Sciences and Technologies for Security Applications,
DOI 10.1007/978-3-319-47671-1_3

hazards have a cross-border, trans-national dimension, being amplified by the Internet, online social networks and smarter mobile communications. Social and technical innovations are now occurring at an ever-increasing speed, causing fast and drastic changes to society. These changes, driven by the possibilities offered by new and emerging technologies, affect citizens, their wider communities, the private sector, the government, and of course, the police.

To tackle all-manner of contemporary security hazards effectively, LEAs have tapped into an increasingly rich source of intelligence that is gathered from publicly available information. The relentless pursuit of intelligence by police officers to keep communities safe via the use of open sources of information has produced the fastest-growing policing discipline of the 21st century—a discipline which adds significant value and increasing efficiency to combatting contemporary crime called Open Source Intelligence (OSINT).

The purpose of this chapter is to explain the role and function of OSINT within the context of policing and existing intelligence collection disciplines, as well as to define OSINT from an LEA perspective and describe its position within the intelligence profession of policing. It is essential for all practitioners, policy-makers and policing professionals to understand what OSINT is and what it is not, how it can be used and the limitations or conditions on it, as well as understanding more about the scale, scope and complexity of the threats from criminals whose methods of operating are becoming increasingly sophisticated. These other areas are covered in later chapters; here we are concerned with the nature of OSINT and its value to LEAs.

3.2 Understanding Intelligence in Policing

The effectiveness of policing can be judged in part by the capacity and capability of LEAs to harness and effectively utilize their intelligence collection opportunities to prevent crime and to pursue criminals. An essential element in the creation and management of this knowledge collection is the police workforce—the most valuable asset of any LEA—including police officers and especially support staff who now largely populate the increasingly specialist intelligence functions of indexing, research and analysis. Throughout the history of policing, the intelligence gathering efforts of the men and women who enforce the law to keep us safe has been recognised as an integral part of their duties. The value of intelligence to the mission of policing and the challenges in sharing it effectively have long been understood by police officers. During 1881, Assistant Commissioner Howard Vincent of the Metropolitan Police stated: "Police work is impossible without information, and every good officer will do his best to obtain reliable intelligence, taking care at the same time not to be led away on false issues. Information must not be treasured up, until opportunity offers for action by the officer who obtains it, but should be promptly communicated to a superior, and those who are in a position to act upon it. Not only is this the proper course of action to task in the public interest, but it will be certainly recognised by authorities and comrades, promoting esteem and confidence, which will bring their own reward" (Cook et al. 2013).

Academic studies differ on how intelligence might be defined but broadly they fall into two distinct categories. The first describes intelligence as a process, defined as "the systematic and purposeful acquisition, sorting, retrieval, analysis, interpretation and protection of information" (Harfield and Harfield 2008). The second category describes intelligence in terms of information rather than a process defined as "information developed to direct police action" (Cope 2004) and "a mode of information that has been interpreted and analysed in order to inform future actions of social control against an identified target" (Innes et al. 2005). For police officers, a rather more practical working description of intelligence is provided described as "information that has been given some added value after being collated and assessed" (Kleiven 2007). General consensus amongst academic and LEA practitioners seems to view intelligence as "being created from raw material information that has been evaluated and analysed" (Herfield and Herfield 2008). But for many police officers across the world, the English word 'intelligence' has negative connotations, associated with the work of the security services and clandestine and secretive intelligence agencies. Indeed, the fact that some native speakers of English also associate the word exclusively with secretly obtained information adds to the confusion (Brown 2008). This can have cultural implications and, as a consequence, lead to the lack of a consistent method for assessing crime threats.

Despite the differences in defining 'intelligence', police officers understand that if they are to have any kind of impact upon combatting contemporary crime, it is essential for their efforts to be intelligence-led. At its simplest, 'intelligence-led policing' conveys the relatively obvious notion that LEA activity—whether focused upon local community aspects of policing, the investigation of crime or public order —should be informed and directed rather than undertaken randomly (see Chap. 6). An intelligence-led approach to policing also serves to support the evidence-gathering of police investigators and detectives to achieve successful prosecutions. As criminals are rendered vulnerable by the fact that their criminality is rarely random but patterned, the structure of this pattern from the movement of criminals, their lifestyle, communications, capability and their intent may be inferred and so intervention options identified. For the police practitioner, and especially those engaged in specialist intelligence-gathering and intelligence development roles, intelligence is not so much a way of working as a way of thinking. Such is the importance of intelligence to LEAs success is that it is has become a holistic discipline to which the whole of an LEA contributes and in which all officers and staff have a role to play directly or indirectly. An intelligence-led approach to policing is therefore not just something that the intelligence unit or the proactive investigation unit within an LEA undertake. Intelligence-led policing demands that the whole organisation undertakes or supports this vital function which now includes a suite of complimentary intelligence collection disciplines which can be used by LEAs adding great value to police investigations.

3.3 Intelligence Collection Disciplines

Various kinds of intelligence—military, political, economic, social, environmental, health, and cultural—provide important information for decisions. Intelligence is very often incorrectly assumed to be just gathered through secret or covert means, and while some intelligence is indeed collected through clandestine operations and known only at the highest levels of government, other intelligence consists of information that is widely available. According to the Federal Bureau of Investigation (FBI) in the United States, and recognised by many LEAs across the world operating today, there are various ways of gathering intelligence that are collectively and commonly referred to as "intelligence collection disciplines"[1] which are shown in Table 3.1.

Within this suite of specialist intelligence collection disciplines, OSINT has now found its place, being extensively used by local and national LEAs, intelligence agencies and the military. Given the scale, accessibility and high yield of intelligence return for minimum resource, OSINT provides the glue which binds, compliments and increasingly corroborates and confirms other LEA intelligence functions, all of which are relevant operational reasons as to why OSINT has quickly become a rich source of information to disrupt and detect the modern criminal (see Chaps. 13 and 16).

3.4 Characteristics of Open Source Intelligence

According to the FBI, OSINT is the intelligence discipline that pertains to intelligence produced from 'publicly available information that is collected, exploited, and disseminated in a timely manner to an appropriate audience for the purpose of addressing a specific intelligence and information requirement'.[2] OSINT also applies to the intelligence produced by that discipline. OSINT is also intelligence developed from the overt collection and analysis of publicly available and open-source information which is not under the direct control of government authorities. OSINT is derived from the systematic collection, processing and analysis of publicly available information in response to intelligence requirements. Two important related terms are 'open source' and 'publicly available information' defined by the FBI as:

---

[1]Federal Bureau of Investigation, Intelligence Branch, Intelligence Collection Disciplines https://www.fbi.gov/about-us/intelligence/disciplines.

[2]Federal Bureau of Investigation, Intelligence Branch, Intelligence Collection Disciplines https://www.fbi.gov/about-us/intelligence/disciplines.

Table 3.1 Intelligence collection disciplines

Discipline	Description
Human intelligence (HUMINT)	The collection of information from human sources. The collection may be conducted openly, by police officers interviewing witnesses or suspects during the general course of their duties, or it may be collected through planned, targeted, clandestine or covert means. A person who provides information to the police as part of a covert relationship is known as an 'agent' or 'source' and in the UK is known as a Covert Human Intelligence Source (CHIS). LEAs who exploit covert sources of HUMINT have specialist and dedicated units with trained officers given the increasing levels of risk, security, complexity and legal and ethical considerations and compliance required.
Signals intelligence (SIGINT)	Refers to electronic transmissions that can be collected by ships, planes, ground sites or satellites. More commonly used in higher policing operations or investigations which pose potential threats to national security. SIGNIT is generally conducted by, and the responsibility of the military or intelligence services rather than domestic LEAs, although at federal and national levels SIGNIT may be used by LEAs in support of serious and organised crime and terrorism investigations which threaten public safety.
Communications intelligence (COMINT)	A type of SIGINT and refers to the interception of communications between two or more parties. This can be conducted in real-time or captured and stored for future interrogation. SIGNIT is primarily the responsibility of intelligence agencies such as the National Security Agency (NSA) in the United States or the Government Communications Headquarters (GCHQ) in the UK. As with SIGNIT, COMINT may be used by LEAs in support of organised crime and terrorism investigations which presents a serious threat to public safety.
Imagery intelligence (IMINT)	Also referred to as Photo Intelligence (PHOTINT), IMINT includes aerial and more increasingly satellite images associated with reconnaissance. Traditionally the preserve of military and intelligence agencies, IMINT is increasingly used by LEAs being captured by their aerial capabilities including helicopters and more recently Unmanned Aerial Vehicles (UAVs).
Measurement and signatures intelligence (MASINT)	A relatively little-known collection discipline that concerns weapons capabilities and industrial activities. MASINT includes the advanced processing and use of data gathered from overhead and airborne IMINT and SIGINT collection systems.
Telemetry intelligence (TELINT)	Used by the military and intelligence agencies, sometimes deployed to indicate data relayed by weapons during tests in the international monitoring to counter the proliferation of chemical, biological, nuclear or radiological weapons.
Electronic intelligence (ELINT)	Indicates electronic emissions picked up from modern weapons and tracking systems. Both TELINT and ELINT can be types of SIGINT and contribute to MASINT as part of a multi-pronged approach to intelligence gathering operations from the military and intelligence agencies.

- Open source is any person or group that provides information without the expectation of privacy—the information, the relationship, or both is not protected against public disclosure. Open-source information can be publicly available but not all publicly available information is open source.
- Publicly available information is data, facts, instructions, or other material published or broadcast for general public consumption; available on request to a member of the general public; lawfully seen or heard by any casual observer; or made available at a meeting open to the general public.[3]

OSINT collection by LEAs is usually accomplished through monitoring, data-mining and research. OSINT refers to a broad array of information and sources that are generally available, including information obtained from the media (newspapers, radio, television, etc.), professional and academic records (papers, conferences, professional associations, etc.), and public data (government reports, demographics, hearings, speeches, etc.). OSINT includes internet online communities and user-generated content such as social-networking sites, video and photo sharing sites, wikis and blogs. It also includes geospatial information such as hard and softcopy maps, atlases, gazetteers, port plans, gravity data, aeronautical data, navigation data, geodetic data, human terrain data, environmental data and commercial imagery (see Chap. 10).

One of the most important aspects of OSINT is the value it adds to police operations in the ability to obtain timely, reliable and actionable intelligence related to the investigation or incident at hand. In addition to traditional investigative techniques and information sources, OSINT is proving to have significant success in accessing and collecting such information. Examples range from information posted on social media as one of the most openly available means of accessing and gathering open-source intelligence to location data or browsing information. But the real value of OSINT for LEAs is when it is combined with real-time analytical capabilities, providing LEAs with vital situational awareness capability to prevent incidents, combat crime and respond to emergencies or larger-scale crisis events.

OSINT has key characteristics which provides the majority of the necessary background information for any LEA operation. OSINT has the availability, depth, and range of publicly available information which enables LEAs to satisfy intelligence and information requirements without the use or support of specialized human or technical means of collection at low cost. It also enhances collection as open source information supports surveillance and reconnaissance activities by answering intelligence and information requirements. It also provides information that enhances and uses more technical means of collection. OSINT also has the capability to enhance intelligence production. As part of a multi-disciplinary intelligence effort, the use and integration of publicly available information and open sources ensures that officers have the benefit of all sources of publicly available information to make informative decisions.

[3]Federal Bureau of Investigation, Intelligence Branch, Intelligence Collection Disciplines https://www.fbi.gov/about-us/intelligence/disciplines.

Making sound judgements is a core role and important attribute of any successful police officer, particularly those leaders who are charged with the responsibility of managing and directing large scale investigations or managing emerging crisis. Effective decision making, particularly at the very outset of major incidents or investigations, will ensure opportunities are not missed and potential lines of enquiry are identified and rigorously pursued (Cook et al. 2013). In reality, law enforcement officers who have been engaged in the early developments of an investigation have had to cope with a lack of sufficient information to begin with, at a time when important decisions are needed to be made quickly and intuitively, but the rise and availability of OSINT has served to inform and dramatically improve the very beginning of police action and response which is vital to save lives, prevent disasters and respond effectively to unfolding events.

While there are significant opportunities for LEAs to exploit the potential of OSINT, all law enforcement officers must, when handling OSINT, especially those progressing complex investigations, ensure that nothing is taken for granted and it cannot be assumed that things are what they seem or that the correct intelligence processes and protocols have been followed correctly. There is no room for complacency in any investigation. Looking for corroboration, rechecking, reviewing and confirming all aspects of OSINT which can direct investigations, are hallmarks of an effective police officer, and shall ensure that no potential line of enquiry is overlooked during a dynamic and fast-moving investigation, including operations which have relied heavily upon OSINT (see Chaps. 17 and 18).

In order to progress an investigation effectively, it may have to be based on or guided by a hypothesis as a starting point. For all investigations a hypothesis is a proposition made as a basis for reasoning without the assumption of its truth and supposition made as a starting point for further investigation of known facts (Cook et al. 2013). Developing and using hypothesis is a widely recognised technique amongst criminal investigators which can be used as a means to establish the most logical or likely explanation, theory, or inference for why and how a crime has been committed. Ideally, before investigators develop hypothesis there should be sufficient reliable material available on which to base the hypothesis which traditionally has included the details of the victim, precise details of the incident or occurrence, national or international dimensions of the offence, motives of the crime and the precise modus operandi. But the use of OSINT now provides a new, richer and detailed data set to inform the investigators hypothesis, and simply by lawfully accessing social media networking sites reveals personal information, images and links to friends and associates that accelerates and informs LEA responses.

Unlike other intelligence collection disciplines, OSINT is not the sole responsibility of any one single agency but instead is collected by the entire LEA community. One advantage of OSINT is its accessibility, although the sheer amount of available information can make it difficult to know what is of value. Determining the data's source and its reliability can also be complicated. OSINT data therefore still requires review and analysis to be of use to LEA decision-makers and must be appropriately processed alongside other intelligence from other collection methods.

3.5 Modelling Open Source Intelligence

Since intelligence is comprised of information and data, when it is held by public authorities it is subject to the various laws on data protection and information disclosure that apply to public authorities. When using OSINT, LEAs must comply with their statutory duties to disclose information, protect the information and fulfil their discretionary power and responsibility to disclose and share the information. If the intelligence is ever to be utilised as evidence in criminal proceedings there will be other considerations (see Chap. 18) and there are legal requirements around its collation, processing and storage (see Chap. 17). The governance of intelligence management does not simply reside within statute law. Senior police officers and executives in LEAs utilising intelligence management have a responsibility for ensuring the appropriate and professional use of information gathering and analysis. To establish, consolidate and professionalise the intelligence process, especially where the use if OSINT is concerned, some basic principles are required to be adopted as follows:

- Intelligence work must be lawful, for a legitimate purpose, necessary and proportionate.
- Officers and staff throughout the LEA must be appropriately aware, trained and equipped for their role within the intelligence management process.
- Security and confidentiality are essential in establishing an effective and professional intelligence environment.
- Organisational and partnership activity should be based on knowledge of problems and their context, to which understanding intelligence contributes.
- Intelligence serves no purpose if it is not used.
- The effective use of intelligence as a tool requires objectivity and open-mindedness on both the part of the analyst and the subsequent user (Harfield and Harfield 2008).

All intelligence must be evaluated to check its reliability, and OSINT is no exception regardless of its free and open access. LEA investigators are responsible for evaluating any intelligence they submit, ensuring it is an accurate and unbiased evaluation based on their knowledge of the prevailing circumstances existing at that time (Cook et al. 2013).

Information-gathering and the subsequent analysis to generate intelligence does not operate within a vacuum and public and political expectations upon police practitioners is that their work will be intelligence-led, designed to prevent and reduce harm to the citizens and communities they serve. This is reinforced with the creation of prescribed business process models to embed intelligence-led approaches into the culture and operating frameworks of LEAs.

To embed the intelligence process into LEAs, many police forces have adopted an Intelligence Model (IM) which meets their specific requirements. The IM is a model for policing that provides intelligence which senior managers can use to help them formulate strategic direction, make tactical resourcing decisions and manage risk in any given investigation (Ratcliffe 2010). It is important to note that LEAs view their IM as not just being about crime or intelligence—but a model that can be used for many areas of the wider mission of policing. An effective IM offers LEAs the realisable goal of integrated intelligence in which all LEAs play a part in a system bigger than themselves.

Launched by the National Criminal Intelligence Service (NCIS) and adopted by the Association of Chief Police Officers (ACPO) in 2000, the British government placed their National Intelligence Model (NIM) at the centre of the UK Police Reform Agenda. Since its inception, NIM has become a cornerstone of policing which is now fully embedded in the operating culture of British policing, made possible by the issuing of a Code of Practice which ensures the principles and standards for implementation and continued improvement of the model (Harfield and Harfield 2008). The NIM has served to professionalise and standardise all manner of police and partner agency intelligence-related activities and the benefits of the NIM includes:

- Greater consistency of policing across the
- Allows operational strategies to focus on key priorities
- Allows more police officers to focus on solving priority problems and targeting the most active offenders
- Achieves greater compliance with human rights legislation
- Informs the management of risk
- Provides more informed business planning and a greater link to operational policing issues
- Improves direction and briefing of patrols
- Reduces rates of persistent offenders through targeting the most prolific
- Improves integration with partner agencies

The NIM model works at three levels as follows:

- Level 1—Local/Basic Command Unit (BCU)
- Level 2—Force and/or regional
- Level 3—Serious and organised crime that is usually national or international (Centrex 2005).

The NIM has three essential elements which includes 'Police Assets', 'Intelligence Products' and the 'Tasking and Co-ordinating Process' (Centrex 2005). These three elements are explained in Table 3.2.

An efficient IM is an integral aspect of all contemporary criminal investigations and LEAs have come to learn that the effective management of intelligence, including the gathering, assessment, prioritisation and dissemination provides a coherent and coordinated response to tackle all forms of criminality. But the scale and scope of OSINT, if not targeted correctly, can lead to acute challenges of intelligence overload where the sheer volume and velocity of OSINT from publicly available sources simply exceeds LEA capacity to process it effectively.

Table 3.2 Essential Elements of the National Intelligence Model

NIM essential element	Description
First: police assets	Certain key 'Assets' must be in place at each level to enable intelligence to be produced. Without them, the intelligence function will not operate efficiently. A sufficiently flexible 'Tactical Capability' must also be present at each level to deal with identified problems. Without it, too much intelligence will be produced with little or no capability to deal with it.
Second: intelligence products	Four key 'Intelligence Products' are produced to drive policing through the Tasking and co-ordinating process: • The strategic assessment • The tactical assessment • The problem profile • The target profile
Third: the tasking and co-ordinating process	The Tasking and co-ordinating Process takes place strategically and tactically at each level with information and intelligence flowing between levels and between neighbouring police forces and law enforcement agencies. At a strategic level, the NIM is strongly linked to all aspects of business planning, both in relation to local, regional and national plans for policing.

The over-extended use of OSINT can quickly overwhelm police operations, particularly when responding to real-time crisis events when verifying the validity and integrity of OSINT data sources can be complicated and time consuming.

3.6 Conclusions

The world is being reinvented by open sources. Publicly available information can be used by a variety of individuals and organisations to expand a broad spectrum of objectives and LEAs are increasingly making effective use of this free and accessible source of information. While the internet and online social networks have positively enriched societal communications and economic opportunities, these technological advancements have changed—and continue to change—the very nature of crime, serving to breed a new sophisticated and technically capable criminal. The nature of some 'traditional' crime types has been transformed by the use of computers and information communications technology (ICT) in terms of their scale and reach, with threats and risks extending to many aspects of social life. New forms of criminal activity have also been developed, targeting the integrity of computers and computer networks. Threats exist not just to individuals and businesses, but to national security and infrastructure.

Furthermore, the borderless nature of the phenomenon of cybercrime and the transnational dimensions of human trafficking, drugs importation and the illegal movement of firearms, cash and stolen goods means that criminals can plan their crimes from jurisdictions across the world, making law enforcement particularly challenging, the very reason why LEAs must maximise the potential of OSINT and seek new and innovative ways to prevent crime. For policing, progressive and forward-thinking LEAs who encourage, embrace and exploit the possibilities of OSINT will be better prepared to meet the security challenges of the future, and as a direct result, will provide themselves with a greater opportunity to keep the communities they serve safer and feeling safer.

The ability for LEAs to harness the power of OSINT will increasingly differentiate between those police forces who lead, with those who simply follow behind. Adopting a legal and ethical framework for the effective use of OSINT is therefore imperative, not only to realise the benefits from the opportunities of OSINT to prevent crime, pursue criminals and prepare for new and emerging challenges and threats, but also to improve the efficiency and effectiveness of the service LEAs deliver and to better safeguard the public information they use. But while OSINT remains largely unregulated amongst LEAs, with no one single agency having ownership or primacy, unnecessary duplications of access and unwelcome infringes on the privacy of citizens data are likely to be a continued feature of the increasing use of OSINT moving forward. LEAs must therefore never forget that inappropriate access and use of OSINT serves to strengthen the case to introduce rigid guidelines, tighten laws and to generally kerb their ready access and collection of public information. Moreover, LEAs would do well to ensure that their use of all OSINT follows formal intelligence management model processes. This approach will both retain and build the trust and confidence of the communities they serve that their publicly available information is being accessed legitimately to fight crime and to keep their fellow citizens safe.

References

Brown SD (ed) (2008) Combating international crime: the longer arm of the law. Routledge
Centrex ACPO (2005) Guidance on the national intelligence model. Wyboston: Centrex/NCPE (available online at www.acpo.police.uk/asp/policies/Data/nim2005.pdf)
Cook T, Hibbitt S, Hill M (2013) Blackstone's crime investigator's handbook. Oxford University Press, Oxford
Cope N (2004) Intelligence led policing or policing led intelligence? 'Integrating volume crime analysis into policing. B J Criminol 44(2):188–203
Harfield C, Harfield K (2008) Intelligence: investigation, community and partnership. OUP Oxford, Oxford
Innes M, Fielding N, Cope N (2005) The appliance of science? The theory and practice of crime intelligence analysis. Br J Criminol 45(1):39–57
Kleiven ME (2007) Where's the intelligence in the national intelligence model? Int J Police Sci Manage 9(3):257–273
Ratcliffe JH (2010) Intelligence-led policing: anticipating risk and influencing action intelligence

Chapter 4
OSINT as Part of the Strategic National Security Landscape

Laurence Marzell

Abstract This chapter looks at the context, application and benefits of OSINT for use in decision making, as an integrated part of the wider intelligence mix and, as an essential component within the overall Intelligence Cycle. OSINT is a growing and increasingly critical aspect in decision making by LEAs—and has been even before the burgeoning use of social media brought open source to the fore. But, its full integration into the wider intelligence mix, as well as into an overarching information governance framework, is essential to ensure efficient and effective contribution to usable intelligence able to support better informed decision making. Fundamentally, unless the system in which OSINT is used as interoperable as the system is in which decision-making is taking place, the application and value of OSINT will be far less effective, efficient and meaningful. This chapter addresses OSINT in the context of the Intelligence Process and the need to resolve the challenges and issues surrounding the integration and use of OSINT into the Intelligence Cycle. It further discusses how an overarching information governance framework may support OSINT for decision making within the wider Intelligence Mix.

4.1 Introduction

This chapter looks at the context, application and benefits of OSINT (Open Source Intelligence) for use in decision making, as an integrated part of the wider intelligence mix and, as an essential component within the overall Intelligence Cycle. Both of which are described in detail in the section on understanding the Intelligence Cycle in which OSINT must exist and the wider intelligence mix in which it must integrate. Then as part of this wider intelligence mix and cycle, how LEAs (Law Enforcement Agencies) need to enable their use of such an integrated OSINT, through a unified framework of information governance: able to bring both

L. Marzell (✉)
SERCO, Hook, UK
e-mail: Laurence.Marzell@serco.com

B. Akhgar et al. (eds.), *Open Source Intelligence Investigation*,
Advanced Sciences and Technologies for Security Applications,
DOI 10.1007/978-3-319-47671-1_4

open and closed source intelligence together in a meaningful and unified way to support better informed decision making.

So like all other data, OSINT must be understood, integrated and used in a unified way and as part of the mix with the myriad other data sources available to decision makers, to create the trusted intelligence they need to support their tasks. It is important to note that, in their use of OSINT for decision making—and indeed, not just OSINT but the entire mix of data across the spectrum of the intelligence cycle—both LEAs and the military face similar, shared challenges and issues.

In setting OSINT into context, this chapter uses the decision making needed in major crisis, humanitarian and emergency situations, and some areas of serious organised crime and counter terrorism as its basis. Here, while civil jurisdiction presides, LEA decision making often relies closely on the capability and expertise that underpins that of the military, in use of the wider intelligence mix and cycle, where data sources, information processing and analysis are heterogeneous and shared. It is here also that military capability and expertise can also often be used to support LEAs in an operational capacity.

4.2 Understanding the Strategic Landscape into Which OSINT Must Be Applied

Decision makers, in both the civilian and military domains, operating at Grand Strategic,[1] Strategic and Tactical levels, face a continuous need for faster, more trusted and better informed decision making. Analysts interpreting the data, turning it into meaningful intelligence, face a constant struggle to cope with the ever increasing Volume, Velocity, Variety and Validation of data (4Vs) (Akhgar et al. 2015), the plethora of formal and informal (OSINT) data sources and, the ambiguities surrounding the trustworthiness of that data.

Better informed decision making manifests itself in many ways. At the grand strategic level it can be seen in situations such as, for example, questioning whether sufficient and reliable intelligence is available on terrorist activity in Nigeria to warrant external intervention. Alternatively it can be at a lower but no less critical level, as might be found in peacekeeping, humanitarian relief and disaster response operations such as that in the Mediterranean supporting refugees. Or, in the Nepalese earthquake of April 25, 2015 where a massive 7.8 magnitude quake struck Nepal, just northwest of the capital of Kathmandu—the worst quake to strike the region in more than 80 years. Moreover, decision making in this context may arise when used by LEAs against serious organised crime or terrorism. In all of these

[1]Grand Strategic: An overarching concept that guides how nations employ all of the instruments of national power to shape world events and achieve specific national security objectives. Grand strategy is also called high strategy and comprises the "purposeful employment of all instruments of power available to a security community" (Military historian B. H. Liddell Hart).

examples a number of critical factors combine and compound themselves. This makes achieving the increasing 4V's of data being created, and therefore potentially that must be considered, for both short and long term analysis, a significant, increasing, and ever present challenge (see Chaps. 2 and 3).

For the military, this need for better informed decision making has seen the development and deployment of an integrated ability to collect and disseminate information through the use of extremely powerful sensors (Visual-optical and infra-red, Radar-active and Electronic Warfare—passive) fitted to a variety of high value air, land and naval platforms. This ability, known as C4ISTAR, addresses a spectrum of capabilities that covers Command, Control, Communications, Computers, Information/Intelligence, Surveillance, Targeting Acquisition and Reconnaissance.

This ability was supplemented in the early 2000s with constellations of geo-stationary and earth orbiting satellites that further enabled this and added time and positioning capability. The need for accurate and timely data for decision making is no less in the civilian world. While the technology and access to budgets might be somewhat different to those of the military, with the now relatively inexpensive access not only to open source space-based imagery and time and positioning or voice and data communications, the increase in technological advancement across a swathe of data generating sensors (both space-based and other) means that the same challenges of speed, volume, ambiguity and trustworthiness faced by the military (i.e., the 4Vs) in support of better informed decision making, affects civilian LEA end users too.

The Battlespace[2] is the term used to describe the domain in which the military now conduct their operations in time of hostility. In peacetime, where the military are operating in support of LEAs the term might best be described as the Security space; where all of the assets, people, information, networks and technology function together and are considered as a system of systems, when looked at and considered as a whole.

Of increasing significance for civilian authorities and LEAs—regardless of their geographic or territorial jurisdiction—is our increasingly complex and interconnected world in which decision making must now take place amidst greater complexity, ambiguity and interdependency; and, in which the supporting intelligence cycle must keep pace technologically as well as through all aspects of the human dimension.

---

[2]Battlespace: The effective combination, or integration, of all elements of a Joint force to form a coherent whole, clearly focused on the Joint Task Force Commander (JTFC)'s intent, is critical to successful Joint operations. Integration of individual Force Elements (FEs) enables their activities and capabilities to be coordinated and synchronised, in accordance with clear priorities, under unified command. On multinational and multi-agency operations, the contributions of other participating nations and non-military actors should also be harmonised wherever feasible. UK doctrine is, as far as practicable and sensible, consistent with that of NATO. The development of national doctrine addresses those areas not covered adequately by NATO; it also influences the evolution of NATO doctrine in accordance with national thinking and experience. Source: UK MoD Joint Doctrine Publication 3-70 (JDP 3-70), dated June 2008.

This can be seen in society's relationship with Critical Infrastructure (CI) upon which citizens, communities and society as a whole, depend for its health, well-being and very survival. Here society, and the many varied and different communities in which citizens live, has become wholly dependent upon the many sectors of CI and essential services such as transport, energy, health, finance, food and communications with which their health and well-being is intrinsically linked. CI and the network of interconnected and interdependent systems that support and underpin it, like the Security Space of decision making mentioned above, can also be described as a system of systems.

Critical infrastructure is often described as a 'system of systems', which functions with the support of large, complex, widely distributed and mutually supportive supply chains and networks. Such systems are intimately linked with the economic and social wellbeing and security of the communities they serve. They include not just infrastructure but also networks and supply chain that support the delivery of an essential product or service (Pitt 2008).

And sitting within, alongside or straddling these CI System of Systems, are the many varied and inextricably intertwined different communities, often referred to as ecosystems in how they evolve are structured and function. It is against this backdrop that the decision making must occur in relation to the planning, preparation, response and recovery to significant man-made, malicious or natural events.

This dependence of communities and society is exponentially increased to risks, hazards, threats and vulnerabilities by the very nature of how these System of Systems interoperate in order to function; with many unseen and indeed unknown, interdependencies and dependencies across and between them. When shocks to the system occur, whether man-made, malicious or natural, they can have unknown consequences or cascade effects, disproportionate to the event that may have caused them (see Chap. 1).

It is here where multiple decision makers from the many different civilian organisations and LEAs (both at local and regional levels within individual nations or in cross-border border cooperation) must come together and collaborate efficiently and effectively to achieve a set of common and shared outcomes. These outcomes manifest themselves in the planning and preparation for, response to and recovery from such shocks, as well as in the planning, preparation, conduct and often review or inquest, of such law enforcement activities as might be seen with serious organised crime or terrorism.

It is in these instances, where complexity is inherent in the system itself. Where with the intrinsic difficulty of multiple different organisations needing to work together efficiently and effectively to achieve a combined effect from their collective effort that the unseen consequences or cascade effects arising from the interdependencies of the system itself will mean that poorly informed decision making can be counted in lives lost as well as in terms of economic cost, physical damage or societal well-being. OSINT is a growing and increasingly critical aspect in decision making by LEAs in these situations—and has been even before the burgeoning use of social media has brought open source to the fore. The advent of social media as a source of open source intelligence has, for the most part, brought the extent and range of OSINT sources that are now available to a greater range of audiences, users and applications. Fundamentally, unless the system within which it sits and is used, in this case the Intelligence Cycle, is as interoperable as the system is in which decision making is taking place, then the application and value of OSINT will be far less effective, efficient and meaningful (see Chaps. 9 and 16).

One example which provides a useful historic context to this development and use, is taken from a 2002 case study in Australia as documented in a 2004 NATO report on OSINT (NATO 2002; see also Chaps. 12 and 16).

Case Study: The Heads of the Criminal Intelligence Agencies (HCIA) conference in Australia September 1998 directed that a business case be prepared for the establishment of a centralised open source unit. This was announced by Paul Roger, Director of Intelligence of the Queensland Criminal Justice Commission, in his paper presented at 'Optimising Open Source Information'. The open source unit will meet Australian law enforcement requirements for collection and dissemination of open source material. The clients of the open source unit will initially be limited to the agencies that comprise the Criminal Intelligence Agencies. After the unit is established, additional agencies may be included as clients of the open source unit. The establishment of an open source unit provides the criminal intelligence community with an ideal opportunity to market an information brokerage service to other agencies. There is also potential for the unit to become part of a larger unit and networking between other open source units including the Royal Canadian Mounted Police (RCMP), Europol and the UK's Metropolitan Police Service. The unit will initially concentrate on providing open source information rather than intelligence. When the unit has found its niche it can then concentrate on four other functions: current awareness briefs; rapid response to reference questions; contacting outside experts for primary research; and coordinating strategic forecasting projects. It will draw upon the full range of external open sources, software and services.

4.3 Understanding the Intelligence Cycle in Which OSINT Must Exist and the Wider Intelligence Mix in Which It Must Integrate

With the burgeoning use of and dependence on OSINT now firmly established, its full integration into the wider intelligence mix and with it, into an overarching information governance framework, is essential to ensure efficient and effective contribution to usable intelligence able to support better informed decision making. It is important at this stage to provide a context as to why information governance for the entirety of the intelligence mix, including that of OSINT, is so essential.

Different organisations view the world in which they exist and must operate very differently. These differing views are driven by many factors such as risk, history, culture, capability, economics and leadership. From these views flow how organisations conduct their business: their governance and policies, training, budgets, processes, systems, and so on. An organisation's view of the world, relative to any other organisation(s) with which it collaborates, is neither right nor wrong; it is just different. But these differences, especially in terms of governance and policy, where resulting information and decisions need to flow across organisational, operational or jurisdictional boundaries (both internally within organisations or nations as well as externally of those organisations or nations), are significant areas of risk. This is where failures can and often do occur, especially in our interconnected world where dependencies and interdependencies can often, in the wake of a major incident, lead to consequences and cascade effects way beyond the original cause.

Where different organisations need to come together in mission critical situations, to achieve a set of shared aims, objectives and outcomes, without a unified and common understanding and approach to how information is processed, analysed, understood and acted upon, as would be provided by good information governance, the likelihood of failure or less effective results in achieving those aims, objectives or outcomes is extremely high. This is based upon the risk of the different organisations and agencies involved, making conflicting or ineffective decisions, resulting from their different interpretations of the intelligence or their different responses to the intelligence, as influenced by their different organisational approaches and views of the world in which they exist and operate.

In simple terms, both strategic and tactical decision making, in military or civil domains, or in those of the shared space that occurs in significant crisis, emergency or serious organised crime or terrorism events or operations, requires the collection of data from all available, relevant sources—technical and human—and from both open source (e.g. social media, internet and commercially available) and closed (military or other specific). It then requires the critical processing where the analysis takes place to turn the disparate, ambiguous and multiple source information into usable, meaningful, trusted, accurate and timely intelligence. Then for onward dissemination out to appropriate end users. A feedback mechanism is required to verify and validate the accuracy along with a metric, if necessary or appropriate, to decide which is 'best'. All of which needs to sit within a unified and common information governance framework to ensure that regardless of each organisation's differing view of the world, everyone is viewing the resultant intelligence and decision making needs from the one unified and shared view. Better informed decisions can then result.

It is here, that it is important to describe the Intelligence Cycle or Process, how it supports decision making, and, the issues and challenges that OSINT, as an integrated part of that mix and cycle, faces. Owing to the inherent nature and sensitivity of much of that which surrounds intelligence gathering, analysis and use, with much of the topic area subject to security classification, this chapter uses freely available open source material as its basis. This material is of sufficient detail and accuracy to offer sound explanation—a fitting testament to the breadth and depth of previously classified information which is now freely available. Six phases make up the Intelligence Cycle: (1) Direction; (2) Collection; (3) Processing; (4) Analysis; (5) Dissemination and (6) Feedback. Figure 4.1 provides an overview of the Intelligence Process.

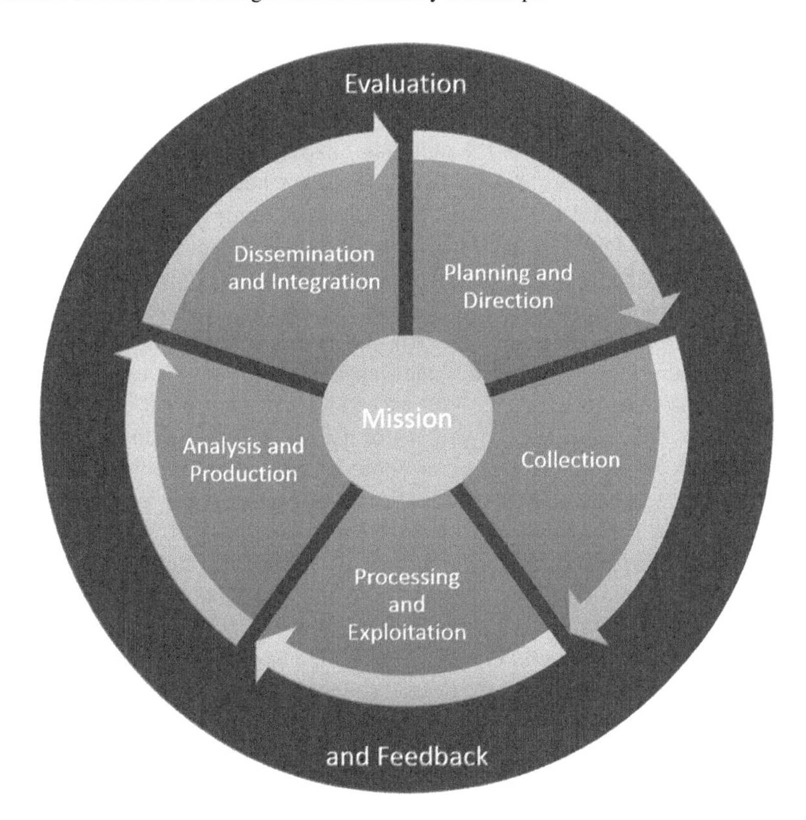

Fig. 4.1 The intelligence process/cycle (Source: "Joint Publication 2-0, Joint Intelligence". Defense Technical Information Center (DTIC). Department of Defense. February, 2013)

1. **Direction**: Using an example at the Grand Strategic or Strategic level, in the first instance, intelligence requirements and needs are determined by a decision maker to meet the objectives they seek to achieve. In NATO, a commander uses requirements (sometimes called 'Essential Elements of Intelligence (EEIs)) to initiate the intelligence cycle, whereas in the United States requirements can be issued from the White House or Congress. This is termed Direction.
2. **Collection**: In response to requirements, intelligence staff develop an intelligence collection plan applying available sources and methods and seeking intelligence from other agencies. Collection includes inputs from several intelligence gathering disciplines, such as HUMINT (human intelligence), IMINT (imagery intelligence), ELINT (electronic intelligence), SIGINT (Signals Intelligence), OSINT (open source, or publicly available intelligence), etc. This is termed Collection.
3. **Processing**: Once the collection plan is executed and information arrives, it is processed for exploitation. This involves the translation of raw intelligence materials, quite often from a foreign language, evaluation of relevance and

reliability, and collation of the raw intelligence in preparation for exploitation. This is termed Processing.
4. **Analysis**: Analysis establishes the significance and implications of processed intelligence, integrates it by combining disparate pieces of information to identify collateral information and patterns, then interprets the significance of any newly developed knowledge. This is termed Analysis.
5. **Dissemination**: Finished intelligence products take many forms depending on the needs of the decision maker and reporting requirements. The level of urgency of various types of intelligence is typically established by an intelligence organization or community. An indications and warning (I&W) bulletin would require higher precedence than an annual report, for example. This is termed Dissemination.
6. **Feedback**: The intelligence cycle is not a closed loop. Feedback is received from the decision maker and other sources and revised requirements issued. During the cold war, 90 % of the data used for intelligence based decision making came from military or other agencies and 10 % from open sources. Now, the figure is 90 % from open sources with the 10 % from military or other agencies. Defence assets are normally tasked to be on station, whereas civil Low earth orbit (LEO) may only come around every 6 days. Optimizing both however is critical as useful civil data harvesting has a high potential to advance warn, or, retrospectively, work out how something was arrived at as opposed to real time streaming of an area of interest. This is termed Feedback.

So in the context of the above, if we imagine the current state of the art for the collection, processing, analysis and dissemination of this data—the intelligence cycle—as an increasingly narrowing funnel, with the exponential increase in the sources and demand for OSINT as an essential part of the mix, then the critical processing and analysis, where information is turned into usable intelligence, has become a considerable bottleneck (see Chaps. 1 and 2).

This critical bottleneck can be seen marked in red in Fig. 4.2. This process has not kept pace with the ability to collect data in today's digital world; especially, where 90 % of all data now comes from open sources. Nor has the current state of the art for the integration together of this multisource data kept pace in any meaningful way.

The increased use and exploitation of OSINT into the wider intelligence mix, will also include that which is space-based. Here, along with the challenges presented by the 4Vs of data, the issues associated with spaced based technology (especially that of imagery) from long communication times and either poor slave rates, or invariant (fixed) views of the terrain, will add to the issues that befall all other data in the current state of the art. Therefore in processing and analysis—the essential functions where the data becomes usable intelligence—the critical bottleneck is exacerbated when spaced based OSINT, an aspect that is increasingly common-place, is included into the mix.

There are issues too in the crucial human needs and behavioural understanding of the end user decision makers for how OSINT, as part of an integrated and wider

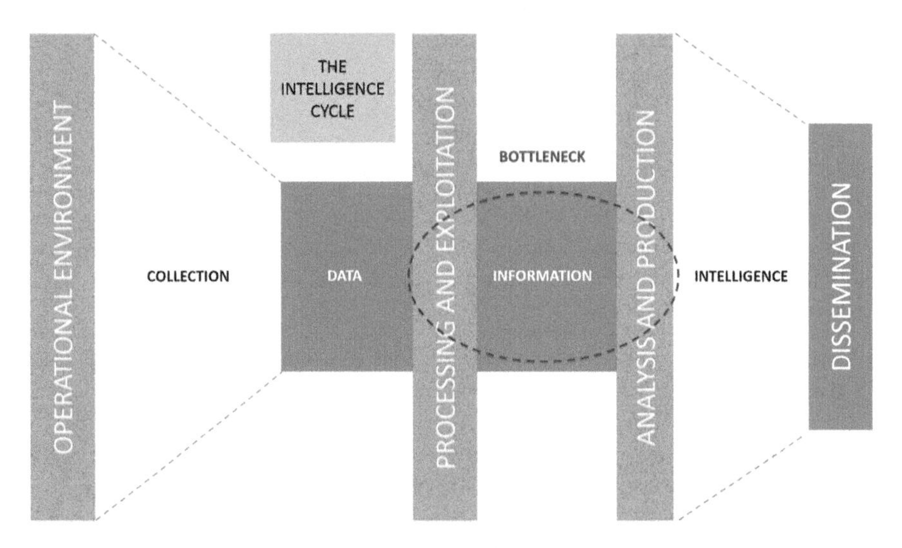

Fig. 4.2 Information bottleneck (Source: "Joint Publication 2-0, Joint Intelligence". Defense Technical Information Center (DTIC). Department of Defense. February, 2013)

intelligence mix, must be provided to them, in what manner it must be provided and through what means, in order to support better informed decision making. These can be summarised as the following:

- With the exponential increase in the availability and use of OSINT, the ability to collect data far exceeds techniques to analyse it and the 4V's of data requiring analysis is increasing logarithmically—thus the problem is only getting bigger. Thus more efficient (process) or faster (technology) approaches are required in Data Analytics—for those professional and expert individuals and analysts sifting through data, looking for themes and creating summaries.
- The pervasiveness of 24 h news and social media is leading to politicians needing an increasing confidence that intelligence has the highest probability of being correct and remaining time-stamped (i.e. valid) to enable an appropriate response. This is regardless of either civil or military context.
- The technology used by extremist groups/individuals equals, and may in some instances exceed, that available to either the military or civilian authorities and LEAs and often, their agility in how they apply such technologies, far surpasses that of the authorities.
- The application of Big Data and the benefits of Big Data analytics to the use of OSINT are much discussed and promoted but are little understood, let alone properly integrated into either the civilian or military decision making contexts (see Chap. 3).
- For the use of OSINT, the generation of data from space technology, generally in the form of imagery, Exocentric (god like) views of the earth, must now be recognised as a constant throughout the Intelligence Cycle and for this, there is

no exception to any other form of data. The 'views' need to make sense and their spatial orientation needs to be understood by the end user, trusted to be integrated into all the other data sources (ground based and aerial, human and technical, open source and specific) and when acted upon, some form of feedback needs to be provided to the end user that their actions are appropriate.

Taking all of the above into consideration, the need to resolve the challenges and issues surrounding the integration and use of OSINT into the Intelligence Cycle and wider mix is paramount, in order to enable decision makers to fully exploit its value and benefit. Such benefits as:

- The ability to ratify military intelligence, especially that from OSINT space-based imagery, which otherwise cannot be ratified
- A greater application and exploitation of OSINT as part of the wider mix for LEAs and emergency uses would have operational efficiency, effectiveness and economic benefits along-side those of better informed decision making
- The ability to speed up, make more accurate and increase the trustworthiness of OSINT that supports better decision making, would impact upon the quality of decisions made by politicians at the grand strategic level; as well as by strategic and tactical commanders operationally in times of stress, danger and need

One such example of this integration of OSINT into the wider intelligence mix and cycle is seen in the US model in moves by the Unites States intelligence community toward institutionalizing OSINT as seen in Fig. 4.3. It is taken from Open Source Intelligence: A Strategic Enabler of National Security produced by the Centre for Security Studies in Zurich, Switzerland in 2008.[3]

4.3.1 Understanding the Application of OSINT in Operational Decision Making

Gathering the data, processing and analysing it, then disseminating it as usable intelligence, is an international activity; as much as it is a local one. All dependent upon the task, need and outcomes sought. In many instances, local need translates and flows through into a national or international one. For the purposes of this chapter, UK decision making has been used as context and whilst structures and methods of working may differ from nation to nation, the principles and synergies to enable such international collaboration, especially where significant cross-sector cross border events are concerned, apply equally to one nation or LEA as they do to another.

At the highest level, the UK's National Security Council (NSC) and its supporting structures enable greater clarity of strategic direction, consolidated

[3]Pallaris, C. (2008). CSS Analysis in Security Policy. Available at: www.isn.ethz.ch.

Institutionalising OSINT: The US Model
Assistant Deputy Director of National Intelligence for Open Source • Establishes open source strategy, policy and program guidance • Makes sure that a single open source architecture is developed • Advises agencies and departments outside the National Intelligence Program regarding the acquisition of OSINT
National Open Source Committee • Provides guidance to the national open source enterprise • Members are senior executives from the Open Source Center, Office of the Under Secretary of Defense for Intelligence, department of Homeland Security, CIA, National Security Agency, National Geospatial-Intelligence Agency, Department of State's Bureau of and Research, Defense Intelligence Agency, Federal Bureau of Investigation, Office of the intelligence community's CIO
Open Source Center • Created in 2005 by the Director of National Intelligence, with the CIA as its executive agent • Several hundred full time personnel • Advances the intelligence community's exploitation of open source material; helps to develop mini open source centers within the respective agencies • Nutures acquisition, procurement, analysis, dissemination, and sharing of open source information, products, and services throughout the government • Makes reports, translations, and analytical products available online in a secure website available to government officials (www.opensource.gov)

Fig. 4.3 Institutionalising OSINT: The US model (Source: Best and Cumming 2007)

consideration of all national security risks and threats, and coordinated decision-making and responses to the threats faced. By way of providing context, the following, taken from the UK Government's National Intelligence Machinery, provides a useful overview to how all intelligence, whether OSINT or other, needs to be considered as a whole.[4]

4.3.2 UK Government Intelligence: Its Nature, Collection, Assessment and Use

Secret intelligence is information acquired against the wishes and (generally) without the knowledge of the originators or possessors. Sources are kept secret from readers, as are the many different techniques used. Intelligence provides privileged insights not usually available openly. Intelligence, when collected, may by its nature be fragmentary or incomplete. It needs to be analysed in order to identify significant facts, and then evaluated in respect of the reliability of the source and the

[4]National Intelligence Machinery: UK Government November 2010.

credibility of the information in order to allow a judgement to be made about the weight to be given to it before circulation either as single source reports or collated and integrated with other material as assessments.

SIS and GCHQ evaluate and circulate mainly single source intelligence. The Security Service also circulates single source intelligence although its primary product is assessed intelligence. Defence Intelligence produces mainly assessed reports on an all-source basis. The Joint Terrorism Analysis Centre produces assessments both on short-term terrorist threats and on longer term trends relating to terrorism. Assessment should put intelligence into a sensible real-world context and identify elements that can inform policy-making. Evaluation, analysis and assessment thus transform the raw material of intelligence so that it can be assimilated in the same way as other information provided to decision-makers at all levels of Government.

Joint Intelligence Committee (JIC) assessments, the collective product of the UK intelligence community, are primarily intelligence-based but also include relevant information from other sources. They are not policy documents. JIC products are circulated to No. 10, Ministers and senior policy makers. There are limitations, some inherent and some practical, on the scope of intelligence, which have to be recognised by its ultimate recipients if it is to be used wisely. The most important limitation is incompleteness. Much ingenuity and effort is spent on making secret information difficult to acquire and hard to analyse. Although the intelligence process may overcome such barriers, intelligence seldom acquires the full story. Even after analysis it may still be, at best, inferential.

Readers of intelligence need to bear these points in mind. They also need to recognise their own part in providing context. A picture that is drawn solely from secret intelligence will almost certainly be a more uncertain picture than one that incorporates other sources of information. Those undertaking assessments whether formally in a written piece or within their own minds when reading individual reports, need to put the intelligence in the context of wider knowledge available. That is why JIC assessments are "all source" assessments, drawing on both secret and overt sources of information. Those undertaking assessments also need to review past judgements and historic evidence. They need to try to understand, drawing on all the sources at their disposal, the motivations and thinking of the intelligence targets.

Where information is sparse or of questionable reliability, readers or those undertaking assessments, need to avoid falling into the trap of placing undue weight on that information and the need to be aware of the potential risk of being misled by deception or by sources intending to influence more than to inform. In addition readers and those undertaking assessments need to be careful not to give undue weight automatically to intelligence that reinforces earlier judgements or that conforms to others' expectations. If the intelligence machinery is to be optimally productive, readers should feedback their own comments on intelligence reports to the producers. In the case of human intelligence in particular, this is a crucial part of the evaluation process to which all sources continually need to be and are subjected.

The quality of the information underlying the decisions taken by the National Security Council is crucial. Piecemeal consideration of issues by too many different bodies risks leading to incoherent decision-making and a lack of overall prioritisation. An "all hazards" approach to national security ensures cohesion and includes:

- The creation of a new National Security Risk Assessment to be updated every other year
- Constant assessment of all sources of information concerning those priority risks, feeding directly into the National Security Council
- A coordinated early warning mechanism to identify and monitor emerging risks
- A cross-Government horizon-scanning system to look at risks and threats which might emerge in the longer term

Figure 4.4 illustrates the UK's National Security structures.

Sitting below the NSC is the Cabinet Office Briefing Room (COBR) or sometimes referred to as COBRA and refers to the location for a type of crisis response committee set up to coordinate the actions of bodies within the UK government in response to instances of national or regional crisis, or during events abroad with major implications for the UK. The constitution of a COBR meeting depends on the nature of the incident but it is usually chaired by the Prime Minister or another senior minister, with other key ministers as appropriate, and representatives of relevant external organizations. The following diagram illustrates the relationship between COBR and the local level Strategic Coordinating Groups which are set up across the UK and meet regularly for planning, training and exercising, as well as in times of actual need to respond to a major incident. Figure 4.5 shows the construct of a COBR meeting.

With all major incidents, whether from man-made, malicious or natural causes, as a general principle, the collective planning, response and recovery effort will have one or more strategic level commanders who ultimately, are in charge of the situation; take the decisions; and, are responsible for the consequences of their actions. These strategic commanders may operate at different levels: from strategic command of the collective effort on the ground, up to grand strategic command at a

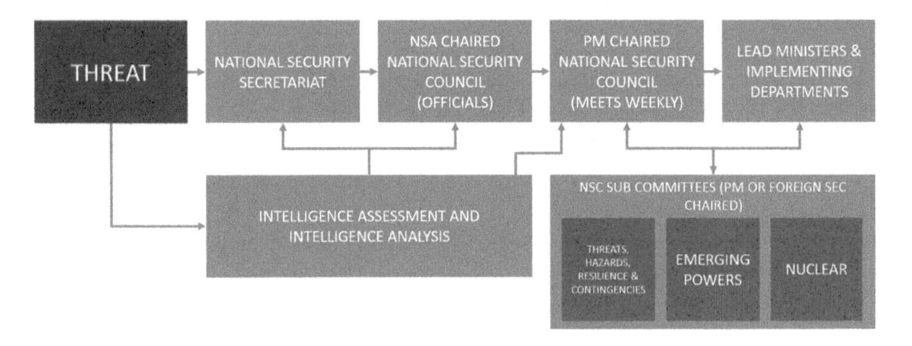

Fig. 4.4 UK National Security Council structure (Source: HM Government)

Fig. 4.5 Construct of a COBR meeting and Strategic Coordinating Group (Source: London Resilience, Strategic Coordination Protocol)

policy and/or political level. This is also the case for the strategic command of the individual multi-agency organisations involved in the event.

Their common, shared and defining criteria, is that they all need to clearly understand and have a full and shared situational awareness of the wide, strategic picture for the entirety of their remit, in which they need to make decisions. In the UK civilian context, these will be known as Gold Commanders. Whilst the command layers and decision making will differ in military only operations, where the military and civilian needs do converge, as is the case with MACA (Military Aid to the Civil Authority) operations, normally, the military will fall under this strategic command structure of the civilian authorities unless determined otherwise. Their existing very 'joined-up', interoperable and well-rehearsed decision making and command functions and structures, needing to work with and integrate with those of the civilian authorities. In instances of a serious and sustained terrorist attack for example, and against a political decision, this may well be reversed through the legislative ability to temporarily hand over command of an operation to the military.[5]

Sitting below the strategic commanders shown above, there generally sits two further levels of commanders that operate more closely to the front line. These are known in the UK context as Silver and Bronze. The Silver operating as the tactical command of the collective effort on the ground, focussing on achieving a less broad effect and outcomes from the Gold, with their effort directed into the incident itself and the immediate environs.

It should be noted that the terms Operational and Tactical are used in reverse between military and civilian organizations in the UK, including LEAs. Within the EU the military levels of command may be used in other countries. Bronze commanders will be responsible for the direct effort and effect of their organisations into the incident itself. Like the Golds, there may be multiple Silver and Bronze

[5]Military aid to the Civil Power (MACP).

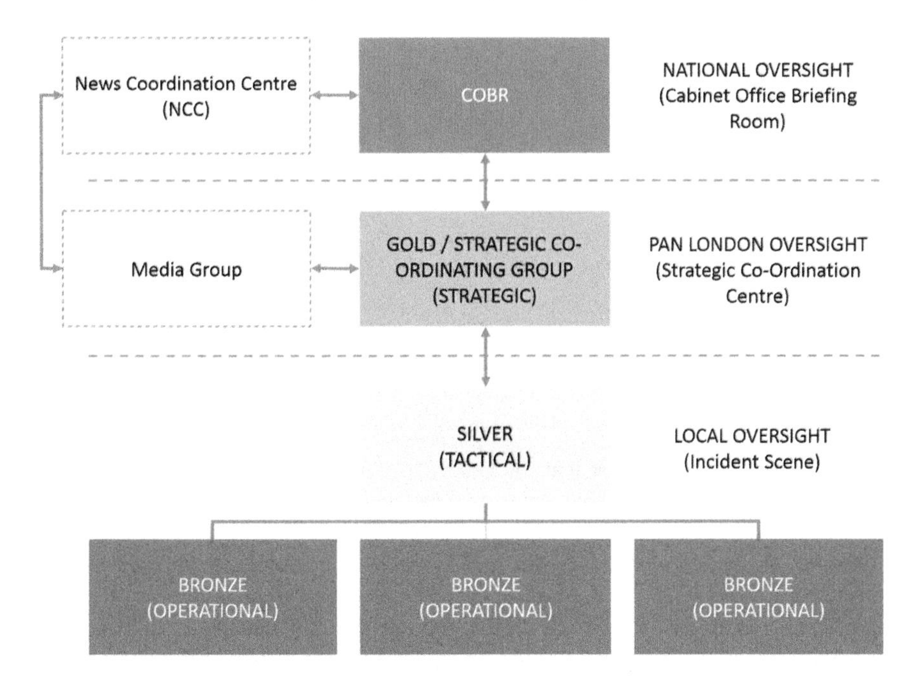

Fig. 4.6 Operational and tactical oversights (Source: HM Government)

commanders due to the multiple organisations and stakeholders involved. Aside from the establishment of clear lines of responsibility between all of the respective Gold, Silver and Bronze commanders, better informed decision making at all of the different levels from the application and exploitation of space-based imagery as part of the Integrated Intelligence mix, would be a much sought after outcome to benefit team and shared situational awareness. A recurring theme identified in many previous major incident inquests both in the UK as well as internationally. These structures can be seen in Fig. 4.6.

In support of the above structures and decision makers are the analysts and technicians working within the bottleneck of the data processing and analysis function of the Intelligence Cycle. This population, whether processing the data in order to create usable and meaningful intelligence for civil, military or converged operations, need to know and understand the end user needs and requirements emanating from the structures seen above; how decisions are informed by the intelligence and how, any greater exploited use of OSINT as part of the wider intelligence mix can be optimised for onward dissemination.

It is clear, that whether analysts are producing intelligence for politicians and diplomats to inform their decisions at a grand strategic or strategic level or, for operational commanders on the ground to use tactically, an understanding by those analysts of how the intelligence they provide needs to be received and used can only serve to help support better informed decision making. The following example taken from the Royal United Services Institute report in 2010 on interoperability in a crisis underlines the importance of this approach:

> Several recommendations made in RUSI 2010 report Interoperability in a Crisis 2: Human Factors and Organisational processes16 refer specifically to the need to improve mechanisms for building and disseminating situational awareness, in particular, Recommendation 23, which calls for the strengthening of joint training that increases organisations' understandings of one another; and Recommendation 24, which calls for stronger frameworks for sharing information and lessons identified from actual events and from exercises, so that planners and responders can learn from previous experience …Technology solutions need to suck in data, but there need to be trained and experienced analysts who look at all the information coming in and turn it into a 'so what?' that enables command decisions. Information and intelligence needs to be handled and disseminated so that it makes sense to the people who receive it … GIS is essential … a common risk information picture in particular is needed at Silver and Gold (Cole and Johnson 2010).

It is clear from the above that the ability to collect information, to amalgamate information from different sources, to process and analyse this information and to use it to produce a Commonly Recognised Information Picture that can inform the command decisions of Gold, Silver and Bronze commanders is far from mature. The growth in sources and demand of OSINT can only compound things. Creating situational awareness on the scale needed in many of the incidents faced in the 21st century and enabling the means for better informed decision making that must result, is beyond the Governance mechanisms that currently exist as well as beyond the experience and training of most incident commanders, at all levels of the command and decision making chain.

4.4 How Might an Overarching Information Governance Architecture Support OSINT for Decision Making Within the Wider Intelligence Mix and Cycle?

Previously mentioned is the complexity of the System of Systems and how this complexity compounds and effects the use of OSINT for decision making. Across these complex systems and their array of supporting networks, there is both a supply and demand side for the data and its disseminated usable intelligence. The complexity of this supply and demand side of data, the systems and networks in which it exists and the multiple different stakeholders across and throughout the Intelligence Cycle, can all be captured. Through the use of an Enterprise Architecture,[6] approach, such as that used by NATO in developing a model of a current or future state of an enterprise. An enterprise being an organisation, a system (including the human factors) or a project. The purpose of enterprise

[6] NATO Architecture Framework: http://nafdocs.org/introduction/.

architecture is to capture the complex dependencies that exist in large-scale systems of systems so as to aid with decision support. In using such an approach, an overarching Information Governance Architecture (IGA) can be created, as can a set of supporting business process flows that map the stages and progress of each and every component—supplier, demander and stakeholder. Such an IGA would provide all decision makers across the entire spectrum of the Intelligence Cycle and decision making process, with an enabling means and set of tools, by which to understand and manage the complexity of the systems in which they operate and in which they are asked to make often critical decisions to protect us (see Chap. 1).

The IGA could represent an integrated model of the System of Systems in which the supply and demand of information and intelligence exists. This would be from the operational and business aspects, to the technologies and systems that provide capability. By covering both the operational and technical aspects across such a system, the architecture enables all communities of interest to gain the essential common understanding needed to deliver benefits that are required from the application of OSINT as an integrated component of the wider intelligence mix. Such an IGA would enable a unified, end to end view of where changes and transformation to any stage of the Intelligence Cycle can take place, whether human or technical, process or procedure, Governance or application, to improve the efficiency and effectiveness of the intelligence Cycle and therefore support better informed decision making whether at a strategic or tactical level.

In doing so, one of the main focuses of the IGA would be to present a clear vision of the system in all of its dimensions and complexity in terms of its existing state, or Current Operating Model (COM) and its desired, future state(s) or Target Operating Model (TOM). The result of this would be to support all aspects of the requirement for the use of a fully integrated OSINT including: Governance and policy; Strategic planning; Tactical planning and operations (front line and logistics); Automation of processes; Capability/requirements capture. An IGA would manage and simplify the inherent complexity in a multi-stakeholder and dynamic environment, with its 'single source of truth' used to drive agile and iterative testing and governing rules and principals for the use of an integrated OSINT and wider intelligence mix whether strategic or tactical. It would enable the capture, interrogation and understanding of the following critical questions surrounding the use of OSINT:

1. **Capability Mapping**—capturing the human and technical capabilities and expertise, both civil and military, for OSINT, including that of space-based imagery, and where and how it must be applied and be useful as part of the wider intelligence mix within the Intelligence Cycle, allowing:

    - A mapping and analysis of OSINT capabilities, technology, expertise and best practice currently available to end users from both the civil and military sectors
    - A roadmap for where and how OSINT will provide the most utility and benefit to end users in the decision making tasks and where and how it integrates into the wider intelligence mix to support decision making needs

2. **Requirements Capture**—how will end users, both analysts and decision makers, need to request, be presented with, access and use OSINT as part of the wider intelligence mix within the Intelligence Cycle, allowing;
    - A comprehensive needs analysis across the spectrum of end user requirements throughout the collection, processing, analysis, disseminating and decision making Intelligence Cycle.
3. **Big Data Analytics**—data for intelligence use is already Big Data in that it is far in excess of end users ability to cope with it. Greater exploitation of OSINT, especially with the inclusion of space-based imagery, will add to that: how can Big data analytics be better applied to support better informed decision making given the limitations of the current state of the art in processing and analysis, allowing:
    - An audit, map and analysis of Big data applications and solutions that can ensure OSINT can be used efficiently and effectively as part of the wider mix across the Intelligence Cycle, including but not exclusive to
    - A review of how techniques such as natural language and image processing, geo-location extraction and other sophisticated querying techniques and/or map-based visualisations can be used to mine social media and other open sources and how Big data technologies such as Hadoop or NoSQL data-stores may be implemented for effective querying, perhaps in real-time
    - Integration of relevant findings into the Technology Blueprint, TOM and Concept of Operations.
4. **Social Media**—with 90 % of data for intelligence based decision making now coming from open sources, of which social media is a sizeable slice, what social media currently exists and what might exist in the future? How might it, as part of the wider intelligence mix, help validate other data sources and support better decisions;
    - Produce an audit, map and analysis of the Social Media applications and requirements that can contribute into, and complement the use of OSINT, for better informed decision making
    - A review and analysis for how these findings integrate with the tasks and outputs for the Big data activities seen above, social media having all the hallmarks of big data, i.e., large in size, changes quickly over time, comes from multiple data sources in multiple formats and, has a degree of uncertainty about the accuracy
    - Integration of relevant findings into the Technology Blueprint, TOM and Concept of Operations
5. **Human Factors and Behavioural Modelling**—how will end users need to understand, request and use OSINT within the context of the wider intelligence mix? What human cognitive and behavioural attributes need to be understood and designed for to support better informed decision making and how these might be measured:

- Incorporation of a model for the needs and behaviours of end users for shared and team situational awareness in the context of better informed decision making
- Use of a psychological model(s) of reasoning and decision making in crisis situations including effects of biases and heuristics; understanding the range of end user needs such as exocentric and FPV (First Person Views), the use of degraded imagery and mobile technologies in support of decision making where space-based OSINT is concerned
- Review of how control rooms and C3i centres need to deal with and display OSINT during times of uncertainty, with multiple inputs and where centres are remote and/or decision making is distributed
- Training needs analysis and development of curricula for civil, military and shared communication and situational awareness domains, including an analysis and scope for new immersive synthetic training that would develop new forensic approaches and skills for the end user analyst community
- Integration of relevant findings into the Technology Blueprint, TOM and Concept of Operations

6. **Technology Blueprint**—what technology is needed to 'glue' OSINT into the wider intelligence mix and within the Intelligence Cycle together and what technology will provide the interface(s) for how end users will want to use it? How will it need to work in both the separate civil and military domains as well as shared ones?

    - A review and analysis across the spectrum of technology requirements and needs for OSINT integration across the Intelligence Cycle of data, processing/analysis, dissemination and feedback, ensuring capture and focus of the technology priorities of the integration of OSINT
    - Scope and produce a technology blueprint to support a knowledge architecture, ensuring inclusion of the findings from the activity areas detailed above, focussing on the core technology outputs, namely the end user interface(s), supporting applications and the technology integration needs
    - The blueprint should include those aspects of data sources, communications and networks, e.g., the new 4th and 5th generation platforms, where relevant to OSINT outputs and especially, where being able to achieve and successfully share this level of intelligence will draw into the equation the critical Cyber and Crypto elements

7. **Target Operating Model (TOM)**—overall what does a fully integrated OSINT as part of the wider mix within the Intelligence Cycle look like at strategic and operational levels? Where and how do all of the moving parts, including end users and OSINT sit and fit together?

    - Building upon the Technology Blueprint and driven by the Human Factors and Behavioural Modelling outputs and design a TOM which takes the human and technical aspects and wraps around the required business and service delivery models that would support and enable OSINT capability for better informed decision making to be used operationally

- Use an overarching Enterprise Architecture framework such as that described earlier as used by NATO that enables a unified view to be defined and understood for the entire end to end process, in order to show the compatibility of OSINT across all end users in Law Enforcement, Government and the 5 Eyes[7] community
- Include within the TOM the necessary hooks into both the existing policy and governance arrangements for operational use of OSINT in LEA operations

8. **Concept of Operations (CONOPS)**—how will OSINT, including that of publicly available imagery such as Google maps or other space-based imagery, as an integral and integrated part of the wider Intelligence mix, be operated across the entirety of end users, both civil and military, across the Intelligence Cycle in both day to day use as well as in a crisis.

   - Working with end users across the entire spectrum of the Intelligence Cycle, scope and produce a detailed Concept of Operations for how OSINT will be needed and used in both a day to day role as well as in varying different crisis situations.
   - Include within the CONOPS scope a review for the use of OSINT in an integrated mix across the Intelligence Cycle that addresses the interoperability and interdependencies of the following.

     - *Training*: Do existing training methods for analysts and decision makers need to adapt or be re-written to accommodate, integrate and benefit from the use of OSINT?
     - *Equipment*: Are current or planned equipment, systems and technology fit for purpose for OSINT as an integrated part of the intelligence mix?
     - *Personnel*: Are the right people, skills and expertise in place to maximise the use and value of OSINT?
     - *Information*: Are existing information management approaches and outputs structured in the right way to accommodate a greater integration of OSINT?
     - *Concepts and Doctrine*: Are current and future methods of planning and implementation at a conceptual and doctrinal level affected by OSINT and in what way?
     - *Organisation*: Is the current structure of an organisation(s), it's governance, leadership, reporting and decision making enabled and in the right shape to benefit from OSINT?

[7]5 Eyes community: Canada, Australia, New Zealand, the United Kingdom (UK) and the United States (US) are members of the Five Eyes intelligence community. https://www.opencanada.org/features/canada-and-the-five-eyes-intelligence-community/.

  - *Infrastructure*: Is the existing infrastructure—the networks, systems and any physical infrastructure right for incorporating OSINT into the intelligence mix?
  - *Logistics*: Are the supporting logistics that enable and support the use of OSINT formed up correctly to exert maximum benefit and leverage from the use of OSINT?

9. **Operational Implementation—Enabling Functions**—An integrated model and part of the CONOPS embracing the spectrum of Policy and Governance, People and Process and Technology and Systems detailing what will be the functions used and needed by end users to implement an integrated OSINT; what are the service and business models needed to support the use of OSINT?

   - An integrated and integral outputs from the CONOPS but with inclusion of policy and governance frameworks and guidelines which determine the political, legislative and management frameworks within which OSINT must reside
   - Produce an audit, analysis and map of the policy, legislative and governance arrangements which currently surround the Intelligence Cycle, captured with the IGA
   - Articulate and align these with the TOM to analyse and understand the alignment of the COM with those of the TOM and whether a Delta exists and what Course of Action (COA) may be required to manage any misalignment prior to inclusion into the CONOPS
   - Scope, design, test and evaluate what the service delivery, commercial and business models might be to support and enable OSINT into an integrated mix for LEA use ensuring inclusion of the policy, legislative and governance needs
   - Scope, design, test and evaluate for inclusion into the CONOPS the people, processes and technology required for the service delivery and commercial/business models of OSINT where these are different from the COM

4.5 Summary

To summarise this chapter, it is useful to refer to the opening two paragraphs from the US Congressional Research Service report Open Source Intelligence (OSINT): Issues for Congress December 5, 2007 as follows.

Open source information (OSINT) is derived from newspapers, journals, radio and television, and the Internet. Intelligence analysts have long used such information to supplement classified data, but systematically collecting open source information has not been a priority of the U.S. Intelligence Community (IC). In recent years, given changes in the international environment, there have been calls, from Congress and the 9/11 Commission among others, for a more intense and focused investment in open source collection and analysis. However, some still emphasize that the primary business of intelligence continues to be obtaining and analysing secrets.

A consensus now exists that OSINT must be systematically collected and should constitute an essential component of analytical products. This has been recognized by various commissions and in statutes. Responding to legislative direction, the Intelligence Community has established the position of Assistant Director of National Intelligence for Open Source and created the National Open Source Centre. The goal is to perform specialized OSINT acquisition and analysis functions and create a centre of excellence that will support and encourage all intelligence agencies.

This statement, produced in 2007 provides a valuable reference to the direction of travel for the use of OSINT by LEAs in all aspects of their day to day use. However now, in 2016, there is a glaring absence in this statement of the terms Social media, Satellite and Drones, all of which have had an exponential increase in development, reduction of cost and use. This increase has led to a situation where during the Cold War, OSINT accounted for just 10 % of the information provision for intelligence with 90 % originating from closed source. Whereas today, this is reversed, with 90 % of information and data for intelligence use coming from OSINT.

The exponential increase in availability of OSINT and its use by LEAs, makes the need to ensure that full integration of OSINT, as it exists at present and might develop in the future, with that of closed source intelligence is essential; all within an overarching Information Governance framework. In so doing, a more accurate, timely and appropriate use by LEAs in their day to day decision making can be achieved.

Critically, this would provide a greater level of assurance in the use of such intelligence, the two sources providing mutual support, in order to assure both LEAs and the citizens whom they serve, with the knowledge that the decisions made and acted upon, have been based upon the most reliable, accurate and trusted information available at the time, and that better informed decisions have been the result. In achieving such an outcome, the perceived damage to the UK's political and intelligence communities, as indicated by the UK's inquest into the Iraqi war (BBC 2016), might be lessened or indeed, might never have occurred.

References

Akhgar B, Saathoff GB, Arabnia HR, Hill R, Staniforth A, Bayerl PS (2015) Application of big data for national security: a practitioner's guide to emerging technologies. Butterworth-Heinemann

BBC (2016) Chilcot report: findings at-a-glance. Retrieved 31 July 2016 from http://www.bbc.co.uk/news/uk-politics-36721645

Best RA, Cumming A (2007) Open source intelligence (OSINT): foreign affairs, defense, and trade division
Cole J, Johnson A (2010) Interoperability in a crisis 2 human factors and organisational processes. Retrieved from https://rusi.org/sites/default/files/201007_op_interoperability_in_a_crisis_ii.pdf
NATO (2002) NATO open source intelligence reader. Retrieved from http://www.oss.net/dynamaster/file_archive/030201/254633082e785f8fe44f546bf5c9f1ed/NATOOSINTReaderFINAL11OCT02.pdf
Pitt M (2008) The Pitt review: lessons learned from the 2007 floods. Cabinet Office, London, p 505

Chapter 5
Taking Stock of Subjective Narratives Surrounding Modern OSINT

Douglas Wells

Abstract This chapter highlights ongoing research towards improving current public perceptions of UK policing OSINT. The work aims to evaluate contemporary public misconceptions, exaggerations and under-acknowledgements of modern investigations and surveillance. In this sense the chapter is primarily qualitative building on existing literature that has focused specifically on the practicalities and various technical facets of modern OSINT usage. The chapter's positions contribute to the increasingly complex and diversified field of modern OSINT by highlighting public concerns and counter-narratives to the reactive and proactive benefits, in particular through concerns of disproportionality, transparency, misuse and accountability.

5.1 Introduction

This chapter highlights ongoing research towards improving current public perceptions of UK policing OSINT. The work aims to evaluate contemporary public misconceptions, exaggerations and under-acknowledgements of modern investigations and surveillance (Bayerl and Akhgar 2015). In this sense the chapter is primarily qualitative; building on existing literature (Carey 2015) that has focused specifically on the practicalities and various technical facets of modern OSINT usage. The chapter's positions may contribute to the increasingly complex and diversified field of modern OSINT by highlighting public concerns, and counter-narratives to the reactive and proactive benefits, in particular through concerns of disproportionality, transparency, misuse and accountability.

D. Wells (✉)
CENTRIC/Sheffield Hallam University, Sheffield, UK
e-mail: D.Wells@shu.ac.uk

B. Akhgar et al. (eds.), *Open Source Intelligence Investigation*,
Advanced Sciences and Technologies for Security Applications,
DOI 10.1007/978-3-319-47671-1_5

5.2 Contextual Background

Following the September 11th attacks, much of the Western-democratic world has increasingly struggled to balance traditional liberal values against modern security concerns (Moss 2011). In light of international terrorist killings having increased by fivefold since the year 2000 (Institute for Economics and Peace 2014) and the recent attacks in Brussels, Paris and Tunisia, reports suggest that the Conservative-led government has increasingly pushed surveillance and investigative technologies to new precedents (Elworthy 2015). Whilst the government defends such securitisation policies as essential, it has increasingly been met with controversy, both by opposition parties as well as by members of the public, political activists and even NGO's such as Amnesty International (2015). Whilst such concerns are primarily fuelled by terrorism, further criticisms have been caused through the recent ambiguity of potential domestic extremism and serious criminal definitions (Jones 2015) that are expected to come under surveillance. Often the media and activist groups have labelled such approaches as a general 'mass hysteria' of inappropriate and unjustified profiling (Evans and Bowcott 2014).

This chapter will examine three critical narratives that influence the public, political and policing sphere of internet-age OSINT: first, the lack of public clarity on the topic of UK police OSINT; secondly, the opposing narratives and contradictory arguments put forward against legitimacy and proportionality and thirdly, the chapter will address information and statements from independent and publically available reviews relating to OSINT practices.

The case of the UK is one worth studying amongst the wider European audience. The tragic increasing trends of successful terrorist attacks and casualties and the ongoing destruction, displacement and disruption throughout parts of the Middle East have created for many EU states; increasing security and surveillance challenges, in particular to fight terrorism, immigration and organised crime. Therefore, as one of the most geographically secure and heavily monitored (Barrett 2013) nations in Europe, the UK makes an excellent case study as the wider European audience may, considering the continuation of trends, follow suit. Indeed, regarding violent radicalisation, countries such as Spain, France, Belgium and Germany have stepped up security and surveillance (Ryan 2016; Youngs 2010).[1] Even temporary increases in security can cause large public "blowback", and so to consider the repercussions of UK OSINT and surveillance opinions makes for good resilience and continuity preparation.

[1]European Parliaments Privacy Platform, Policy Meeting Minutes. Available at: https://www.ceps.eu/sites/default/files/events/2011/10/Prog6299.pdf.

5.3 Lack of Public Clarity

Given the ongoing debate of security and liberty between political groups as well as members of the public and the current government, the relatively modern integration of internet-based OSINT capabilities for surveillance and investigation are suggested to be widely exaggerated (by for example the Snowden leaks, which relate more to the secret services) or misunderstood in terms of the actual practices and the legal frameworks they operate within (Carey 2015). For example, one such variation of OSINT software utilised by various LEAs, 'COSAIN', is a largely secretive branch of Capita (Cameron 2015) bought by company 'Barrachd Ltd', a big data solutions company based in Edinburgh, UK. COSAIN has very limited publicly available information (Cosain 2015), with no official individual website and cannot be located as part of the Barrachd Ltd. website domain. Although this does not imply that the service is necessarily controversial, the lack of transparency may undermine police-public relations and increase suspicion, mistrust and the spread of misinformation.

A similar argument can be made for services such as RepKnight, another intelligence solution used by "government, law enforcement, defence and security agencies worldwide."[2] Despite having a public website and even a Facebook page, there is very little information as to how the service operates, despite describing its capacity to analyse millions of messages every minute, to identify criminality and expose those behind it. RepKnight claims to do this through analysis of 'the widest range of open source digital, social media, and dark web sources.'[3] In particular, public concern may be drawn to the lack of clarity regarding the solution's ability to 'harvest millions of messages every minute' without proper explanation. This could be assumed to mean either the processing of open-format messages such as public tweets or Facebook status updates from open profiles. However, it could equally be assumed to involve the processing of private messages and content restricted to inside the privacy settings of social media accounts. Such assumptions cannot be argued to be completely irrational, particularly following recent allegations against the British government and GCHQ regarding the ongoing fallout of the Prism and Tempora projects (which is further discussed in the below Opposing Narratives section) (Pitt-Payne 2013).

Furthermore, there is a distinct lack of published literature and ratified knowledge on the subject, especially made easily accessible to the public. Current governmental, policing and legal literature is often most readily available through 'page-and-terminology-heavy' reports and documents such as the Regulation of Investigatory Powers Act 2000 (RIPA), the Protection of Freedom Act 2012 and the OSC independent surveillance reviews (Carey 2015; see Chaps. 17 and 18). It is of course important to point out that acts such as RIPA were written prior to, and not meant for, internet-based OSINT. At the time of writing the Home Office is able only to publish

[2]Taken from RepKnight Official Facebook. Available at: https://www.facebook.com/RepKnight/info?tab=page_info (Accessed online: 19/05/2016).

[3]Ibid.

recommendations for OSINT investigation guidelines instead of a coherent legally-approved framework (Home Office 2012). Additionally, the College of Policing has yet to upload any information to their OSINT website sub-division (College of Policing 2016). Such lack of transparency regarding ease of public access may exacerbate the continuing 'Big Brother Complex'. One final point of consideration is of deliberate misinformation used by anti-police and anti-governmental narratives; this will be considered and evaluated alongside the other subjective architectures.

Currently, it appears that the use of internet based OSINT, especially regarding big data analytics, are primarily used for intelligence gathering and investigations and not for general community-policing. As of 2015, it was noted that in general the majority of police forces and OSINT practitioners used "social media ... to inform strategies such as pre-emptive arrests, interceptions of activities, approaching particular individuals and groups, or change of tactics during events ... (The lack of identifying community needs) is not yet part of police practice and raises concerns within police about the level of overlap between intelligence and engagement" (Carey 2015). As with many aspects of policing, law enforcement and other relevant security practice, levels of engagement towards social media largely differ between police forces, with most mainly using it to engage with community regarding ad hoc notifications such as public information announcements and petty crime announcements. This in itself may concern members of the public who may feel that OSINT monitoring may be a 'two-way mirror', with intelligence practitioners able to observe and investigate with minimum community engagement and interaction.

Of course, it is essential to understand and respect the fact that many elements of OSINT investigations benefit from, and ought not to disclose the full nature of, tactics and solutions used. It should be made increasingly clear to the public that the tight rules and regulations warrant and authorise deployment as well as that public security and safety may benefit from the indiscretion and minimized disclosure of engagement, which may help to track down community threats, protect vulnerabilities and maximise order. It may also be beneficial to reassure the public that the police and other OSINT certified practitioners have to adhere to far stricter standards than the majority of private corporations and enterprises that utilise big data analytics and collect, store and correlate personal data. To the computer-literate generations, the loss of control and ownership over personal data to organisations and corporations is not a revolutionary or particularly terrifying revelation; however it may prove beneficial to reassure the collective that OSINT has to adhere to far stricter protocols than agencies such as Google, Facebook, Microsoft and Amazon (Comparing Business Product 2010).

5.4 Opposing Narratives

The aforementioned usage of big data solutions and social media analysis undoubtedly serves an instrumental and essential role amongst modern internet-based OSINT investigations and surveillance. However, this practice acts as double-edged sword in the hands of opposing narratives. In particular, concerns such as accountability, regarding big data investigation and information retention are leading causes of public debate. Additionally, relatively new questions are emerging over the ethical, legal, and social suitability of market-orientated software products for authoritative OSINT practice.

There are criticisms, particularly from privacy advocates (Schulz 2012) and surveillance critics (Crushing 2014), regarding the relationships between the state, law enforcement and security services, alongside private companies and organisations (Wallis Simons 2012). In recent years terrorists, criminals and suspects have increasingly used sophisticated technologies such as dark web and encryption services to preserve anonymity. In particular, the rise of encryption services offered by default by companies such as Apple and software services such as WhatsApp have provided a barrier to governmental and law enforcement duties. The December 2015 San Bernardino shooting, carried out by an Islamic extremist couple, went from a national tragedy to a complex international battle between the FBI and global technology giant Apple (Yadron 2016). A legal escalation between the two belligerents occurred with Apple's refusal to unlock an encrypted iPhone that was recovered from one of the assailants. The case arguably divided public and private opinion, with either side causing significant backlash to the government and law enforcement agencies. On the one hand, pro-surveillance commentators argued that 'the state has lost control' with democratic capitalism, represented through the legitimacy and authority of the high-government, being undermined by the technology industry (Morozov 2016). On the other hand, privacy advocates campaigned that the right to privacy was paramount and to obey the FBI would allow for weaknesses and vulnerabilities in all Apple products that may be further exploited for nefarious means.

These two points, although neither related to specifically open-source cases, nor UK based activity, are nevertheless an important contemporary reminder of the divided opinions that will regardless challenge modern OSINT capabilities. This situation is of course exacerbated by misinformation that often likens the powers and interests of national security to every-day OSINT operations. Regardless, the two opinions highlight a difficult middle-path for practitioners to tread; the reliance and involvement on social media, big data and other commercial services opens them up to speculation about the ethical and legitimate considerations of relying upon commercial services to such an extent they can effectively undermine the legitimate authority of power. External opinions may therefore need convincing that OSINT is a valuable asset and one that can coexist with modern technology organisations to provide valuable information in the interests of social resilience and community security.

In addition to such arguments are the campaigns of critical agents such as Amnesty International, Bytes for All, Liberty and Privacy International against pro-active surveillance and monitoring. In particular the steady increase of policing use of proactive surveillance, starting with (the not-so-contested) number plate recognition cameras through to the (more-controversial) facial recognition cameras mounted in high-streets, has been met with an equilibrium of slowly growing protest. The growing concern with Open Source Intelligence Investigations and surveillance is unlike the physical presence of cameras, airport security systems, police cars beside a motorway—there is no physical awareness of OSINT deployment.Therefore, proactive use of internet-based OSINT is likely to face even greater contest from opposition, and the cases for it ought to be transparent, proportionate and justifiable to the majority in order to minimise blowback.

The public should again be reassured of the effects of UK legislature as detailed in this book (see Chaps. 17 and 18) that help to limit practitioners from abusing the technology and authority available. Additionally, it is essential for practitioners to understand, stay up to date with and engage upon external grievances and misconceptions.

5.5 Independent Reviews

The third critical section that OSINT practitioners need to be actively aware of is the regular reports, publicly disseminated from independent bodies such as the Annual Report of the Chief Surveillance Commissioner (Rose 2015). It is worth noting that despite the age of the report, in terms of recent technological developments, governmental legislations (moving to ban end-to-end encryption) and the FBI vs. Apple San Bernardino case, a more recent review has not been published. This may be worrying to members of the public or members of groups that fight for internet freedom and anonymity, as the UK government (particularly GCHQ/NSA) has been seen, by some, to be using loopholes in outdated legal terminology and frameworks to justify and ratify the existence of PRISM; indeed, particularly so when acts such as RIPA have come under-fire for being outdated, not dedicated to the arrival of the modern internet and are being exploited by offshore government surveillance loopholes.

Indeed, this is especially concerning following excerpt from the most recent report, "*We are in an era of unprecedented scrutiny of the use of covert tactics by State bodies and how these are authorised and managed. The OSC has had, and will continue to have, a key role in ongoing oversight of these matters on behalf of parliament, the Government of the date, and the public. It concerns me that, after so many years of oversight under RUPA, RIP(S)A and the Police Act 1997, there remain some small, yet important, gaps in oversight*" (Rose 2015). Such publicly available comments, from an impartial review, are extremely important in ensuring maximum levels of authenticity and transparency. However, with similar comments like this being recorded in the previous reviews, practitioners must be aware of the ease that such high level content, may be deliberately or mistakenly misconstrued; "many commentators in the media do not accurately describe the extent and purpose to which these powers are used."[4] Therefore, it is essential to make sure

[4]Ibid., p. 3.

publicly available information regarding the usage of OSINT and other potentially controversial topics are not overly-technical or un-necessarily long - or alternatively it may be beneficial to include alongside them a condensed and simplified version that may be promoted and disseminated (see Chap. 4).

Furthermore, the report details how many local authorities have struggled to use OSINT as they should be regulated. On several occasions social networking sights "have been accessed … for an investigative approach without any corporate direction, oversight or regulation. This is a matter that every Senior Responsible Officer should ensure is addressed, lest activity is being undertaken that ought to be authorised.[5]" Of particular importance, is the publicised notion that legislature such as the Protection of Freedoms Act 2012 have "little evidence to be a move for the good", and that as a "political reaction to a media frenzy", they have placed far too much emphasis on actual OSINT practitioners and not towards better educating magistrates and other judicial staff. As a result of this, often OSINT and accompanying surveillance and investigation operations from the police struggle to get proper authorisation when necessary. Indeed, this reinforces the significant point that contemporary OSINT rarely functions as a stand-alone component, but as a complimentary tool for a wider framework of intelligence gathering. This wider framework is interlinked and co-dependent on a broad variety of factors: from hierarchical and legal authorisation through to ethical and political considerations, down to more indirect vectors such as potential public backlash and providing the media with misconceptions.

5.6 Conclusion

Overall, it is necessary to examine the subjective narratives of various parties that may be sensitive to, or potentially opposed to the practice of contemporary OSINT. An awareness of the ongoing friction between such schools of thought is necessary to minimise negative attention that may be detrimental to the reputation and resources of practitioners.

To evaluate modern challenges and perspectives that surround the sphere of Open Source Intelligence Investigation and surveillance. Public, private and state opinions on the topic have proven in previous years to be largely influential and authoritative over OSINT due to the highly sceptical and cautious socio-political stance that has emerged. Therefore it is in the mutual interest of the OSINT practitioners, security stakeholders and public and government representatives to ensure maximum transparency, awareness and education, with respect to the aspects of policing OSINT that require discretion such as specific tools, programs and tactics. Nonetheless, OSINT practitioners should make an effort to further publish the limitations of their capabilities, if feasible, to reassure the public as well as clearly documenting and

[5]Ibid., p. 28.

publishing the proportional rationales and authorisation that allow for OSINT to be deployed. Indeed, the lack of a physical police presence to be observed and monitored by the public likely creates anxiety for some individuals. Therefore greater dissemination via the OSINT practitioners of the strict guidelines of RIPA, the Protection of Freedom Act 2012 as well as information regarding data protection, officer monitoring and log-keeping standards should be publicised with reasonable dedication. In a similar manner that the wider public, media and private spheres feel both protected by and protected from law enforcement through common knowledge of the scope and limitations of police power, the digital spectrum of policing also needs to better roadmap the procedures and limitations of internet-based OSINT to reassure non-domain experts and avoid potential controversy and discontent.

Overall, by using the United Kingdom as a case study of high-level surveillance and security monitoring society, practitioners are able to anticipate potential scenarios, wherein the public, private, media, opposing and political narratives may cause problems, barriers and opportunities for discussions regarding OSINT investigations. In particular, it is important to look at examples such as the public availability of legal documents and independent reviews of OSINT investigations, as similar processes are carried out over Europe. It is through clearly disseminating such material, that narratives of misinformation and disinformation may be better dispelled.

References

Amnesty International (2015) Amnesty International takes UK to European Court over mass surveillance. Retrieved 20 July 2016, from https://www.amnesty.org/en/latest/news/2015/04/amnesty-international-takes-uk-government-to-european-court-of-human-rights-over-mass-surveillance/

Barrett D (2013) One surveillance camera for every 11 people in Britain, says CCTV survey. The Telegraph Online. Available at: http://www.telegraph.co.uk/technology/10172298/One-surveillance-camera-for-every-11-people-in-Britain-says-CCTV-survey.html

Bayerl PS, Akhgar B (2015) Surveillance and falsification implications for open source intelligence investigations. Commun ACM 58(8):62–69

Cameron G (2015) Capita buys Scottish analytics firm Barrachd (From Herald Scotland). Retrieved 20 July 2016, from http://www.heraldscotland.com/business/13416378.Capita_buys_Scottish_analytics_firm_Barrachd/

Carey HP (2015) Managing 'Threats': uses of social media for policing domestic extremism and disorder in the UK

College of Policing (Online) (2016) Intelligence and counter terrorism. Available at: http://www.college.police.uk/What-we-do/Learning/Curriculum/Intelligence/Pages/default.aspx

Compare Business Products (2010) Top 10 Big Brother Companies: ranking the worst consumer privacy infringers. Available at: http://www.comparebusinessproducts.com/fyi/top-10-big-brother-companies-privacy-infringement. Accessed online 20 May 2015

Cosain E (2015) Barrachd Edinburgh, UK. Available at: http://assets-production.govstore.service.gov.uk/G4/Barrachd-0879/523c2390354067df0ce14172/QD5/13_0125_BAR_Cosain_A4_flyer.pdf. Accessed 19 Jan 2016

Crushing T (2014) Police utilizing private companies, exploits to access data from suspects' smartphones. Wireless. Available at: https://www.techdirt.com/blog/wireless/articles/20140326/

08390126689/police-utilizing-private-companies-exploits-to-access-data-suspects-smartphones.shtml
Elworthy S (2015) Beyond deterrence: rethinking UK Security Doctrine|Oxford Research Group. Retrieved 20 July 2016, from http://www.oxfordresearchgroup.org.uk/publications/briefing_papers_and_reports/beyond_deterrence_rethinking_uk_security_doctrine
Evans R, Bowcott O (2014) Green Party peer put on database of 'extremists' after police surveillance. The Guardian. Online. Available at: http://www.theguardian.com/politics/2014/jun/15/green-party-peer-put-on-database-of-extremists-by-police. Accessed 19 Jan 2016
Home Office (2012) Open source software options for government: version 2.0, April 2012. Online. Available at: https://www.gov.uk/government/uploads/system/uploads/attachment_data/file/78964/Open_Source_Options_v2_0.pdf. Accessed 19 Jan 2015
Institute for Economics and Peace (2014) Global Terrorism Index 2014: measuring and understanding the impact of terrorism
Jones J (2015) The day I found out I'm a 'Domestic Extremist'. The Telegraph. Online. Available at: http://www.telegraph.co.uk/news/uknews/terrorism-in-the-uk/11357175/The-day-I-found-out-Im-a-Domestic-Extremist.html. Accessed 18 Jan 2016
Morozov E (2016) The state has lost control: tech firms now run western politics
Moss K (2011) Balancing liberty and security: human rights, human wrongs. Springer, Berlin
Pitt-Payne T (2013) PRISM and TEMPORA: ECtHR proceedings issues against UK
Rose C (2015) "Annual Report of the Chief Surveillance Commissioner". House of Commons Print, Presented to the Scottish Parliament June 2015. Available at: https://osc.independent.gov.uk/wp-content/uploads/2015/06/OSC-Annual-Report-2014-15-web-accessible-version.pdf. Accessed 10 May 2016
Ryan M (2016) Brussels attacks rekindle privacy vs. security debate in Europe. Washington post. Available at: https://www.washingtonpost.com/world/europe/brussels-attacks-rekindle-privacy-vs-security-debate-in-europe/2016/03/26/60a68558-f2dc-11e5-a2a3-d4e9697917d1_story.html
Schulz GW (2012) Private companies pitch Web surveillance tools to police. Reveal Online. Available at: https://www.revealnews.org/article/private-companies-pitch-web-surveillance-tools-to-police/. Accessed 26 May 2016
Wallis Simons J (2012) Who watches the private detectives? The Telegraph Online. Available at: http://www.telegraph.co.uk/culture/tvandradio/9344659/Who-watches-the-private-detectives.html. Accessed 20 May 2015
Yadron D (2016) FBI confirms it won't tell Apple how it Hacked San Bernardino shooter's iPhone. The Guardian Online. Available at: https://www.theguardian.com/technology/2016/apr/27/fbi-apple-iphone-secret-hack-san-bernardino. Accessed 20 May 2016
Youngs R (2010) The EU's role in world politics: a retreat from liberal internationalism. Routledge, London

Part II
Methods, Tools and Techiques

Chapter 6
Acquisition and Preparation of Data for OSINT Investigations

Helen Gibson

Abstract Underpinning all open-source intelligence investigations is data. Without data there is nothing to build upon, to combine, to analyse or draw conclusions from. This chapter outlines some of the processes an investigator can undertake to obtain data from open sources as well as methods for the preparation of this data into usable formats for further analysis. First, it discusses the reasons for needing to collect data from open sources. Secondly, it introduces different types of data that may be encountered including unstructured and structured data sources and where to obtain such data. Thirdly, it reviews methods for information extraction—the first step in preparing data for further analysis. Finally, it covers some of the privacy, legal and ethical good practices that should be adhered to when accessing, interrogating and using open source data.

6.1 Introduction

Underpinning all open-source intelligence investigations is data. Without data there is nothing to build upon, to combine, to analyse or draw conclusions from. Furthermore, as will be seen in Chap. 18 collecting data in the 'wrong' way may leave it inadmissible as evidence in court; therefore, it is imperative that the data collected is collected for the right reasons and that no more data is collected than is strictly necessary. This chapter will outline some of the processes an investigator can undertake to obtain data from open sources as well as methods for the preparation of this data into usable formats for further analysis. The following chapters will then build on this starting point by discussing techniques for analysis (Chap. 7), the deep and dark web (Chap. 8), integration with non-open source data (Chap. 9), and finally with an examination of the choice and design OSINT tools (Chaps. 10 and 11).

H. Gibson (✉)
CENTRIC/Sheffield Hallam University, Sheffield, UK
e-mail: H.Gibson@shu.ac.uk

B. Akhgar et al. (eds.), *Open Source Intelligence Investigation*,
Advanced Sciences and Technologies for Security Applications,
DOI 10.1007/978-3-319-47671-1_6

NATO (2001) splits open source information and intelligence into four categories: Open Source Data, Open Source Information, Open Source Intelligence (OSINT) and Validated Open Source Intelligence (Fig. 6.1). NATO defines each of these categories differently and through these definitions we can see that Open Source Intelligence is most appropriate to law enforcement investigations, in particular. NATO describes OSINT as "information that has been deliberately discovered, discriminated, distilled, and disseminated to a select audience … in order to address a specific question." Furthermore, we want to move towards validated OSINT, i.e., "information to which a high-degree of certainty can be attributed", which is, in part, what Chaps. 7 (validation) and 9 (fusion with non-OSINT data) will enable us to work towards achieving. The data collection and processing methods outlined in the chapter only take us, at best, as far as having open source information, and we leave it to the other chapters in the section to discuss how we move from information to intelligence.

With regards to the importance of OSINT in the future we need only look towards Mercado's (2009) article where he claims: "*Not only are open sources increasingly accessible, ubiquitous, and valuable, but they can shine in particular against the hardest of hard targets. OSINT is at times the "INT" of first resort, last resort, and every resort in between.*" That is, in effect, saying that data from open sources are perhaps the most useful weapon in the arsenal of an investigator above all other sources and all other intelligence. In fact, Pallaris (2008) estimates that between 80 and 95 % of all information used by the intelligence community comes from open sources. For law enforcement the above may be an exaggerated claim;

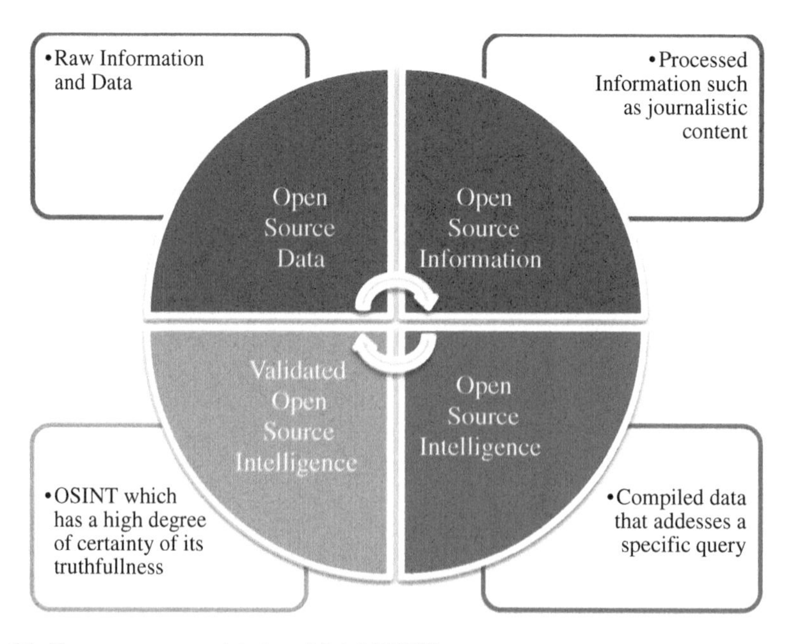

Fig. 6.1 From open source data to validated OSINT

however, with so much of people's lives being lived online it is becoming an ever more important resource in the fight against crime and terrorism.

There appear to be two clear ends of the spectrum with regards to the collection of open-source data. At one end, is the extremely specific and manual search of sources, such as the internet and social media, perhaps tracking down a particular individual or specific information about them, such as those conducted by the contributors to Bellingcat[1] (see Chap. 12) and the examples in Chap. 13 on LEA use cases. At the other end of the spectrum are much larger, although still targeted, investigations where investigators may be interested in the discussion happening around a specific topic or event directly relevant to their investigation but do not necessarily have a specific piece of information that they are searching for. Of course far beyond the end of this spectrum is the kind of mass surveillance we often hear about in the media, especially post Snowden's revelations (Greenwald et al. 2013) but this is outside the scope of this book and normal investigations.

The rest of this chapter proceeds as follows: First, we discuss the reasons for needing to collect data from open sources. Secondly, we introduce a number of different types of data that may be encountered including unstructured and structured data sources and where to obtain such data. Thirdly, we review methods for information extraction—the first step in preparing data for further analysis. Finally, we briefly cover some of the privacy, legal and ethical good practices that should be adhered to when accessing, interrogating and using open source data.

6.2 Reasons and Strategies for Data Collection

The reasons an investigator may wish to obtain information from open sources are wide and varied. It may be that such information is not available through normal closed-source intelligence channels, it may be to give direction in the search for closed-source intelligence or it may be that an investigator does not want to give up the source of their closed intelligence and so resorts to open sources to find the same information (Gibson 2004). The acquisition of intelligence is, in many organisations, governed by the intelligence cycle. This process, as shown in Fig. 6.2, moves from the identification of the intelligence required through the data collection and analysis phases to the feedback phase where the intelligence collected is measured against the initial requirements, and consequently new requirements are identified. The intelligence cycle is used by both the FBI (n.d.) and the UK College of Policing (2015).

Despite appearances and despite perhaps the lack of acknowledgement when the finished intelligence is presented, those working on data analysis estimate that between 50 and 80 % of their time can be spent on the data-wrangling process (Lohr 2014): that is, the effort to collect the right data, convert it into the required

[1] https://www.bellingcat.com/.

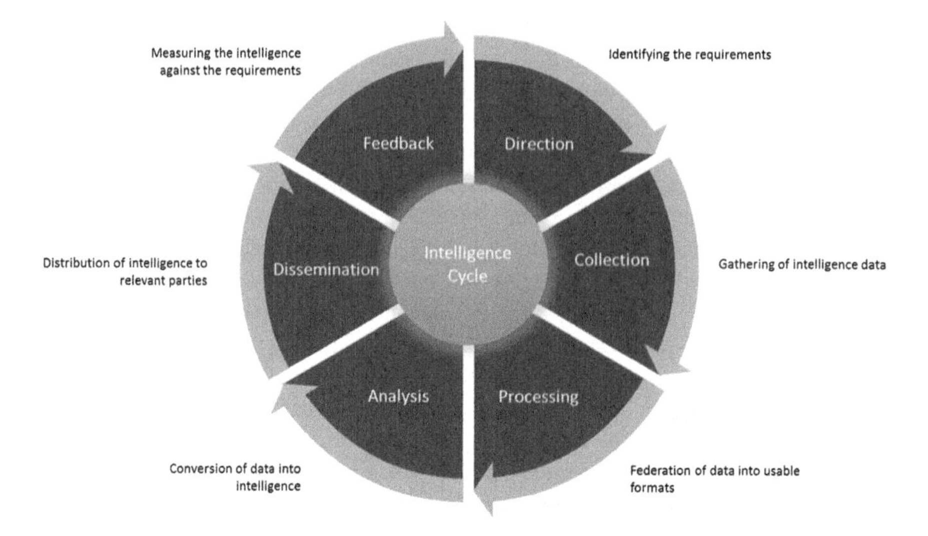

Fig. 6.2 The intelligence cycle

format for analysis, fuse it with other data sources (both open and closed), identify relevant data, and begin the extraction and aggregation processes. Thus carefully planning in identifying data relevant to the question to be answered and the processes for obtaining such data is an essential and also crucial first step in order to extract intelligence that is of the necessary quality and accuracy.

Following the identification of a crime, in a police investigation, investigators will take the necessary steps to identify the material that is relevant to their case (College of Policing 2013). This may include information from witnesses, victims and suspects as well as forensic information from the scene, prior intelligence on individuals and the areas involved as well as accessing data from passive data generators such as CCTV cameras and bank records, amongst others. More than ever data from open sources can augment that information and provide vital clues that may assist in the resolution of the case if identified, extracted, managed, analysed and presented correctly.

Investigations can be both reactive and proactive (Rogers and Lewis 2013). That is, sometimes a crime is committed and the investigation ensues; alternatively, an investigation that is proactive, often called intelligence-led policing, occurs when intelligence is received about crimes potentially being committed which then drives further investigation. Both of these concepts are compatible with open source intelligence; in the reactive case it may be that the investigator believes that information is being shared openly that relates to the case or those involved and wants to be privy to such information, whereas in the intelligence-led cases it may be searching for open sources that confirm, back-up or extend the intelligence that already exists.

In this chapter, we concentrate more on the acquisition of open source data compatible with the intelligence-led approach as we believe that currently this is where the automated methods of data collection can provide the most value at this point in time. Nevertheless, we recognise that for the individual investigator automated methods that can speed up their manual search would be exceptionally advantageous, and we will strive to highlight where manual techniques can be incorporated into automated ones. Furthermore, we refer the reader to Bazzell's (2016) book on open source intelligence techniques, which is a compendium of semi-automated techniques, tips and tricks for accessing open source data.

6.3 Data Types and Sources

6.3.1 Structured and Unstructured Data

A significant proportion of the data collected during an open source investigation will have an unstructured format or at best may be semi-structured. *Structured data* is data that is highly organised such as data held in typical relational databases with an underlying data model that describes each table, field and the relationships between them. *Unstructured data* is the exact opposite of this: It has no data model defined up-front and no prerequisite organisational structure. Unstructured data may typically include the content of web pages, books, audio, video and other files not easily read or interpreted by machines. Analysing unstructured data relies heavily on natural language processing (which is discussed in the next section) as well as image processing.

Between structured and unstructured data is *semi-structured data*, also sometimes called 'self-describing data'. This type of data is particularly representative of the type of information accessible through the web such as the type of data available through RESTful APIs (e.g. Twitter). However, even within these services we may find that they contain fields which allow the inclusion of free text, images, video and audio which are notoriously difficult to process automatically and extract information about the content. Table 6.1 contains some examples of how data from different and common open data sources is structured.

6.3.2 Where and How to Obtain Open Source Data

Just because open source data exists, it does not mean that is it necessarily straightforward to access. Identifying the data required to progress the investigation is the first step in determining, which is the best source and method for obtaining such data. Furthermore, accessing the right data but having it in an unusable format will significantly slow down the next phase of analysis and thus considering how

Table 6.1 Example sources of structured and unstructured Open Source Data

Structured data	Semi structured data	Unstructured data
Data model, database	Self-describing, XML/JSON formats	No structure, free text
Consented databases	Social media APIs	Webpages, blogs, forums, wikis
Electoral roll	Public body open data (e.g. police, government)	White papers, reports, academic publications
Statistical data (ONS, Eurostat)	Google/Bing search APIs	Books
Geonames, ordnance survey	RSS Feeds	Word documents
	Spreadsheets	Media (images/video/audio)
		Satellite imagery
		Street view
		Deep/Dark web sources

the data will be returned is also an important consideration. This section outlines some of the processes for obtaining different types of open source data.

6.3.2.1 Supporting Manual Searches

For the investigator at the coalface of an investigation access to open-source information may be an extremely manual process; that is, searching the web, social media sites, news and other open sources discussed below for specific mentions of a name, for an image, for a particular relationship, a phone number or other numerous data types. While nothing can replace the human-in-the-loop for the analytical and sense-making process of whether something is relevant or not and what it means in the context of an investigation, we can use automated searches to try to gather information faster, cast our net wider and bring back information more quickly. This section will discuss some avenues for obtaining data automatically and the following sections will explain some ideas on how to narrow down the data set once the initial search has been completed.

6.3.2.2 Web Crawling and Spiders

Sometimes a manual search is too time consuming, requires too many (human) resources and the result set is too wide for individuals to trawl through checking one site, the next and then the next until they find the information they are looking for following from link to link to link (see example in Fig. 6.3). Setting up a web crawler, sometimes called spider, automates this process by following these links, either indiscriminately or according to some pre-determined rules.

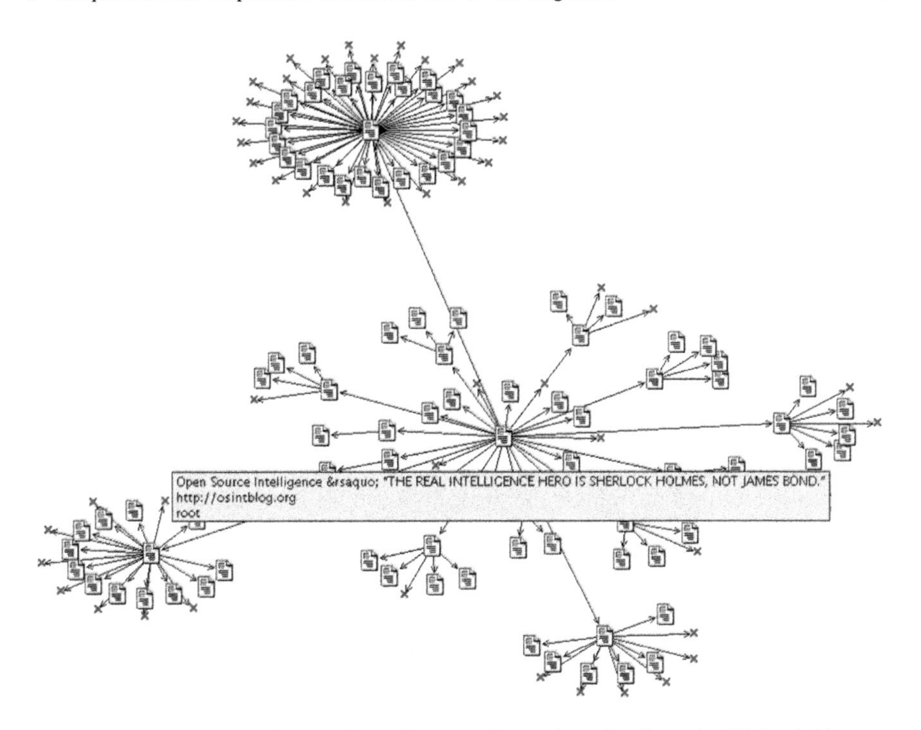

Fig. 6.3 First two layers of pages found from a web crawl starting from the URL osintblog.org

Web crawlers usually start with a number of initial seed URLs from which the crawler will begin. The crawler then scans the page, if required downloads the content, identifies new URLs on that page and then proceeds to navigate to those URLs and repeats the process. Web crawlers can be set up to follow a depth or a breadth first search pattern. Generally, following a breadth-first search, bounded by a specific number of levels is likely to achieve slightly better results as it constrains the crawler to being closer to the initial list of URLs. Further restrictions can be put on crawlers such as restricting them to a specific list of domains, ensuring that crawlers obey the robots.txt file (see Sect. 6.5.2.1), and limiting the types of links that can be followed, for example, ignoring links that navigate to JavaScript or other non-HTML files. Additionally, if a seed URL is identified with content which is likely to change over time, it may be prudent to have the crawl re-run itself at specific intervals in time. Depending on the type of content this could be as often as once an hour up to just once a month or less. If this data collection is time-stamped it will also provide a good opportunity to explore how specific data changes over time.

Web crawlers provide a good starting point for an OSINT investigation if the investigator knows that there is a significant amount of information on the web about the subject they are interested in, but they do not have the time to follow links manually or read each page to identify whether it is relevant or not. Teaming up a web crawler with a processor that identifies whether or not a page is going to be relevant is also a good way to reduce the result set and exclude links that have little content related to the initial search. Techniques such as *categorisation* and *information extraction* can help in identifying whether a page may be useful or not to an investigator.

6.3.2.3 Web Metadata

Metadata on web pages, sometimes also known as microdata, social structured data or 'rich snippets' provide information about the content of a web page in a specified format. It forms part of what is known as the semantic web. Use of the mark-up, such as that popularised by schema.org, means that within the HTML of the webpage there are certain tags which describe the content of that webpage. It can be as simple as including the title, author and description or more complicated including mark-up for organisations, books, TV series, products, locations and much more. As well as 'things' the mark-up may also contains information about the actions that can be taken on a webpage. Twitter and Facebook have created their own versions of metadata that can be included in webpages known as Card Markup[2] and Open Graph.[3] Furthermore, both HTML5[4] and WAI-ARIA[5] (a specific mark-up aimed as aiding those who use assistive technologies to browse the web) both have signposts that will assist HTML parsers in understanding (and machine reading) the content.

While these mark-ups are often used by marketers trying to promote their pages more effectively, we can also consider what does it mean for the OSINT investigator? Firstly, if during a web crawl numerous web pages are returned, each of these pages will have different ways of organising the content on the page making it more difficult to extract the main content of that page without any extraneous information. Utilising these tags within the information extraction process will allow to reliably extract, at the very least, an accurate title, description, author and more for the article you are searching. Of course, an investigator (and a crawler) may not always be searching through news sites that contain these mark-ups; however, they can also be used by forums and blogs.

6.3.2.4 APIs

APIs or application programming interfaces are one of the most common ways to retrieve data. For example, to obtain search result data from Bing,[6] their search API provides automated access to their results from a specific query. From Twitter you

---

[2]https://dev.twitter.com/cards/markup.

[3]https://developers.facebook.com/docs/reference/opengraph/.

[4]https://www.w3.org/TR/html5/.

[5]https://www.w3.org/WAI/intro/aria.

[6]http://datamarket.azure.com/dataset/bing/search.

might use their REST or streaming API or you might use Facebook's Graph API. Pipl (the person finder) also has its own API, and many other services who allow access to data also provide access via an API.

Access to an API usually requires first signing up for an API key for that particular service. Each key will then have limits restricting the amount of data you can request or receive within a specific time period. Once this time period has elapsed you will be able to run queries against the API and obtain the information that you are interested in (see Sect. 5.2.1 for a more in-depth discussion on API limits).

6.3.2.5 Open Data

During the last 10 years the move towards open data has gathered pace. Somewhat confusingly, open data is only a subset of the open source data we talk about within OSINT. Open data is usually made available, because an effort (or even a requirement) has been made to publish such data in a machine readable format to enhance transparency within organisations. Open data, therefore, can be another valuable resource in the open source investigator's arsenal. While of course much of this data is highly aggregated and anonymised, there can still be useful snippets hiding within. Data such as local government spend data details which small companies and sole traders have contracts for services within local government (and other areas) that can then be traced back to specific people. For instance, Companies House[7] now publishes basic company data for anyone to download and search: For small companies the correspondence address often corresponds with their home address, thus exposing this information to anyone who cares to search for it.

We also must not forget other types of open data that may not be directly relevant to the investigation, but assists in helping understand the environment surrounding the investigation. This includes geographic data such as that published by Geonames[8] or Ordnance Survey in the UK,[9] which can be used to turn names of places into latitude and longitude coordinates, or vice versa. Data such as weather and other environmental data, satellite, street view and other imagery may also be useful. Photo sharing sites such as Flickr and Instagram may also be able to provide useful contextual data.

6.3.2.6 Social Media

More than any other open source resource, data posted to social media can be a treasure trove of information about particular events, people and their relationships.

[7]https://beta.companieshouse.gov.uk.

[8]http://download.geonames.org/export.

[9]https://www.ordnancesurvey.co.uk/business-and-government/products/opendata-products-grid.html.

Social media even has its own distinct acronym within the intelligence family known as SOCMINT (Omand et al. 2012). The London Riots of 2011 highlighted law enforcement's inability to deal with information posted on social media and the fact that it lacked the manpower, procedures and processes to extract data from social media and turn it into actionable intelligence that would have allowed them to understand the dynamics of the riots and consequently allowed them to react faster and more pro-actively (HMIC 2011). As highlighted by this case (and other large crisis events such as terrorist attacks) we note that even the first step of collecting data from social media in an efficient and effective manner is not a trivial task. We briefly discuss some methods for obtaining data from the most common social media sites.

As discussed above, most social media sites make (some of) their data available through an API. Data resellers such as Gnip[10] and Datasift[11] provide more comprehensive access to social media data. However, this data also comes with a not necessarily insignificant price. Data can also be captured from social media 'on-the-fly', but it is very much data that is available in the moment and may not be what existed a week ago or a week into the future (Shein 2013).

Obtaining tweets from Twitter is perhaps the best example of this ephemeral nature of social media data, although, the time period in which the data remains accessible (without paying for it—thus it is not truly ephemeral in the way that a service like Snapchat[12] exists as the data is still accessible for those who really want it) varies depending on how the data is accessed. Twitter offers two main API services: its REST APIs[13] and the Streaming API.[14] The REST APIs allow users to interact with Twitter through both accessing and updating data. We are more interested in the accessing of data. The APIs provide the opportunity to download data about a particular user's friends and followers, the tweets they have posted or marked as favourites and the lists they have created. Twitter also makes available a Search API whereby users can download a significant proportion of all tweets using a specific keyword or hashtag. These tweets can then be further narrowed down by using specifications on geo-locations, sentiment, time periods and more.[15] The restriction on the Search API is that only tweets in approximately the last week are available.

In terms of straightforward access to the search API, Twitter Archiver[16] is an add-on to Google Sheets that allows to enter search queries and returns them in a spreadsheet table format with the tweet text, the user and username, date and time,

[10]https://gnip.com/.

[11]http://datasift.com/.

[12]www.snapchat.com.

[13]https://dev.twitter.com/rest/public.

[14]https://dev.twitter.com/streaming/overview.

[15]https://dev.twitter.com/rest/public/search.

[16]https://chrome.google.com/webstore/detail/twitter-archiver/pkanpfekacaojdncfgbjadedbggbbphi?hl=en.

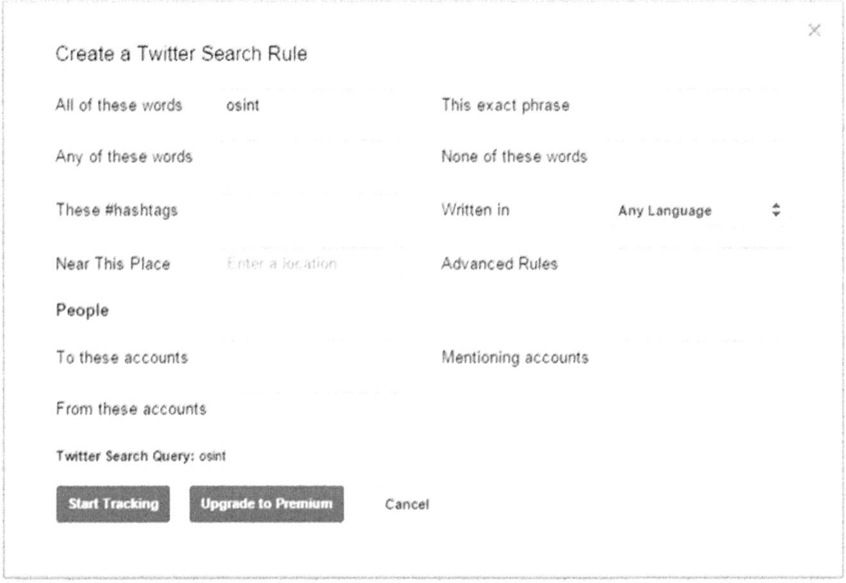

Fig. 6.4 The simple interface of the Twitter Archiver

the tweed id and some basic user information (as shown in Fig. 6.4). Similarly, NodeXL[17] is a network visualisation add-on for Microsoft Excel that provides the functionality to import data directly from the Twitter API (Hansen et al. 2010). These kinds of tools can be useful to the open source investigator as they have a low barrier to entry and provide data to a user in a familiar format, which can usually be easily imported into other tools.

Accessing Facebook[18] data through their Graph API is far more restrictive than Twitter in terms of the kinds of information that can be accessed. While information and posts made on public pages and events are readily accessible, data about the friends or from the timeline of specific people are not provided whether this information within their profile is set to public or not. Thus a wider amount of data is available to the investigator through interrogating a user's profile page manually and moving from link to link. However, capturing and managing this information is far more difficult than accessing it through an API, and there does not yet appear to exist solutions that effectively capture that data and remain within Facebook's terms of service—crawling Facebook itself is strictly forbidden (Warden 2010). Furthermore, the amount of information you can read on a person's Facebook page depends on whether you are signed into the service or not. The amount of information available is restricted when you are not signed in.

---

[17]http://nodexl.codeplex.com/.

[18]www.facebook.com.

Although, LinkedIn[19] is a professional social network, the information that is shared on the platform is usually detailed as illustrated by ICWatch,[20] which set out the personal information made available online by those in the intelligence community. Thus, if those who work in the intelligence community are potentially so lax with their own privacy then it is likely that many other citizens will be too. Therefore, in certain situations LinkedIn should also be considered a valuable source of OSINT.

As we can see, the rich amount of information made available on social media sites and its relative ease of access in obtaining or at least viewing this data makes it a goldmine in terms of investigations.

6.3.2.7 Traditional Media

Access to traditional media is easier than ever with most newspapers and media organisations having an online presence where they reproduce the articles that, for example, may be included in that day's newspaper. These news sources are often catalogued through RSS feeds and some even have their own APIs such as the BBC,[21] Guardian,[22] Associated Press[23] and the New York Times,[24] amongst others. This ease of access, which ranges from large organisation such as those listed above to much smaller local newspapers which may contain more specific information particularly relevant to investigators, dramatically improves how quickly information can be disseminated anywhere in the world.

As well as searching sites individually, news aggregators such as the European Media Monitor,[25] the Global Database of Events, Language and Tone (GDELT)[26] and to a lesser extent BBC Monitoring[27] all provide wider access to searchable databases from across the globe, which may already have been pre-translated into English.

6.3.2.8 RSS

RSS or Rally Simple Syndication is a machine-readable method, based on an XML format, of publishing information about which new articles, posts, etc. have been

[19]www.linkedin.com.

[20]https://transparencytoolkit.org/project/icwatch/.

[21]https://developer.bbc.co.uk/content/bbc-platform-api/.

[22]http://open-platform.theguardian.com/.

[23]https://developer.ap.org/ap-content-api.

[24]http://developer.nytimes.com/docs/times_newswire_api/.

[25]http://emm.newsbrief.eu/overview.html.

[26]http://gdeltproject.org/.

[27]http://www.bbc.co.uk/monitoring.

added to a website. A single RSS feed may cover one particular blog or it may cover one specific topic on a larger news site. Monitoring RSS feeds is possible either through an RSS reader such as those offered by Feedly[28] or Digg[29] or as part of a larger monitoring system.

While RSS may be helpful if you are monitoring a specific topic, it may be more useful if you want to capture the posts made to a specific blog. Blogs are often used to air personal opinions and sometimes personal grievances, especially when users feel that their opinions are not being heard through other more official channels. Thus monitoring and being alerted to these new posts and (if the functionality exists) to new comments made on these posts is potentially another vital information source.

6.3.2.9 Grey Literature

Grey literature is defined as articles, reports, white papers and other literature that does not fall into the category of normal open sources nor into the consented data, but may still contain useful information for open source investigations. These reports may often be in pdf or word documents and are not necessarily easily accessible or their existence may not be well signposted, especially as the links that they are hosted at are susceptible to change as companies and institutions update their sites and links are not maintained.

For an open-source investigator the usefulness of grey literature will depend on the context in which they are investigating. If they are following up on very specific crimes or activities perhaps information held in these grey sources may not be particularly useful. However, for the type of OSINT required for Industry data contained within grey literature may be enough to provide competitive advantage.

6.3.2.10 Paid Data and Consented Data

The word ‘open’ in open source intelligence must not be confused with the word ‘free’. Thus it is perfectly acceptable to consider sources that exist only behind a paywall as key open source. In fact, this data in particular may give investigators an advantage because people cannot necessarily control, or even be aware of, the data these private companies hold on them and consequently, they cannot take steps to remove it. This includes data such as Thomson Reuters’ World Check,[30] which compiles data on individuals deemed to be at high risk, such as suspected terrorists or members of organised crime.

---

[28]https://feedly.com.

[29]https://digg.com/reader.

[30]https://risk.thomsonreuters.com/products/world-check.

Consented data is a subset of paid data. Consented databases such as those provided by LexisNexis,[31] GBG,[32] 192.com and Experian[33] usually require payment for the access of specific records or a subscription which allows a certain number of searches per month, for example. These databases may contain information on companies, people, phone numbers, addresses, email address and other personal information that people have consented to making available. These databases can then be used by law enforcement professionals to either validate or extend the knowledge they may have about a person of interest.

6.3.2.11 Data on the Deep and Dark Web

The deep web is all content on the internet that is not indexable by Google or other search engines. This includes information on forums and other sites that are not accessible without usernames and passwords as well as pages with dynamically generated content. In fact, it is estimated that between 80 and 90 % of the internet is not available on traditional search engines (Bradbury 2011).

The dark web is a specific part of the deep web that can only be accessed through the use of specific browsers such as Tor[34] or even specific operating systems such as Tails.[35] Because content hosted on the deep and dark web is generally not indexed or easily searchable, new techniques and tools have (and still need to be further) developed to facilitate the easier access to this information.

Mirroring the path followed by the surface web over the last 10–15 years, we are now beginning to see the advent of crawlers and search engines for these sites such as those developed by DARPA known as Memex (DARPA 2014) (of which some modules were recently open sourced).[36] Researchers have also developed methods for data mining (Chen 2011) and methods to crawl forums (Fu et al. 2010) on the dark web. Google has also published the methods they have used to gather content on the deep web within their search engine, whereby they pre-computed the data required to fill in the HTML forms that would allow them to access additional content (Madhavan et al. 2008). Thus the potential for anyone to be able to access content on the dark web much more easily in the future is becoming a real possibility. (A more comprehensive discussion around OSINT and the deep and dark web can be found in Chap. 8.)

This section has given an overview of the types of data that is available to open source investigators if they know where to look for it. This compilation is definitely not exhaustive but should provide a good starting point for investigating avenues

[31]https://www.tracesmart.co.uk/datasets.

[32]http://www.gbgplc.com/uk/products.

[33]https://www.experianplc.com/.

[34]www.torproject.org.

[35]https://tails.boum.org.

[36]http://opencatalog.darpa.mil/MEMEX.html.

for acquiring open source data. The next section details the first steps in the analysis of these types of data by explaining some methods to move from unstructured to—at least—semi-structured data.

6.4 Information Extraction

While the techniques for accessing data listed above will allow you to identify the sources and extract the content, at this point we are still a distance from having an output we could call OSINT. Thus, the next stage in the process is to take the data acquired and attempt to extract pertinent information and put this into a standardised format that will enable the aggregation and visualisation processes taken further down the line to take place efficiently and effectively (see Chap. 7).

Information extraction can be defined as the process of taking data from an unstructured state into a structured state. The most common example of this process is the parsing of natural language text and the extraction of specific entities and events or the categorisation of the text. Due to the large amount of open source web-based and textual data that can be used as a starting point for amassing OSINT, it is essential that an open source investigator has a good understanding of how information can be effectively extracted from textual sources and how they can make use of existing tools as well being able to develop their own extraction solutions.

6.4.1 Natural Language Processing

A significant proportion of the information extracted from open sources is in free-text format whereby there is little restriction on what can be contained in such fields, except perhaps a character limit. However, even within natural language processing (NLP) itself there are a number of nuances to consider: namely the way that people write online on social media, on blogs, in comments or on forums varies considerably compared to the structure of textual content scraped from news articles, official press releases and other such similar documents. These differences affect the way that we extract information from the text and the ways in which we have to consider the semantics of language as well as the likelihood of spelling errors, colloquial synonyms and the use of nicknames and usernames rather than 'real names'. These idiosyncrasies are potentially admissible as evidence (see Chap. 18).

A number of libraries and APIs exist to assist in this process such as Python's NLTK (Bird and Loper Bird 2006), Gate (Cunningham et al. 2013), and the AlchemyAPI,[37] amongst many, many others. For the Open Source Investigator who

[37]http://www.alchemyapi.com/.

cannot dedicate the time and resources to learning how to take advantages of such languages and libraries resources such as parts of the Stanford NLP tools (Manning et al. 2014) exist, which provide desktop software applications to perform extraction. However, many of these tools are at their most powerful when incorporated into part of a larger toolkit.

6.4.1.1 Main Body Extraction

With the majority of open sources being online, being able to effectively extract the main content of a web page is a vital step in the process of moving towards entity extraction and other information extraction techniques to prepare the data for further analysis. Compared to even just a few years ago, today's web pages are relying more heavily on JavaScript and dynamically generated content as well as including advertising and links to popular or related articles. This results in an increasingly complex structure of HTML tags that makes it more and more difficult for computers to recognise which is the main content and which is the associated content. Thus being able to apply methods and techniques that are able to accurately and efficiently extract this information in order to have a clean document to perform NLP is the first key step in the information extraction process.

Main body extraction is the process of taking a web page's HTML structure and extracting from it only the text that makes up the article and not the surrounding images and links that you would see if you viewed the web page on a browser. Main body extraction is what tools such as Flipboard[38] and Evernote's WebClipper[39] do when you view articles using their functionality. Main body extraction is an important component in the information processing pipeline, as if we perform entity extraction without it we are likely to extract extraneous entities relevant, for example, to the top stories of the day or adverts and not to the article we are interested in; thus diluting the quality of our information extraction processes (see Fig. 6.5 for an example).

Tools and libraries to carry out such a process include Goose (available in both Scala[40] and Python[41] flavours, amongst others), the AlchemyAPI[42] and Aylien.[43]

---

[38]https://flipboard.com/.

[39]https://evernote.com/webclipper/.

[40]https://github.com/GravityLabs/goose.

[41]https://pypi.python.org/pypi/goose-extractor/.

[42]http://www.alchemyapi.com/products/alchemylanguage/text-extraction.

[43]http://aylien.com/article-extraction/.

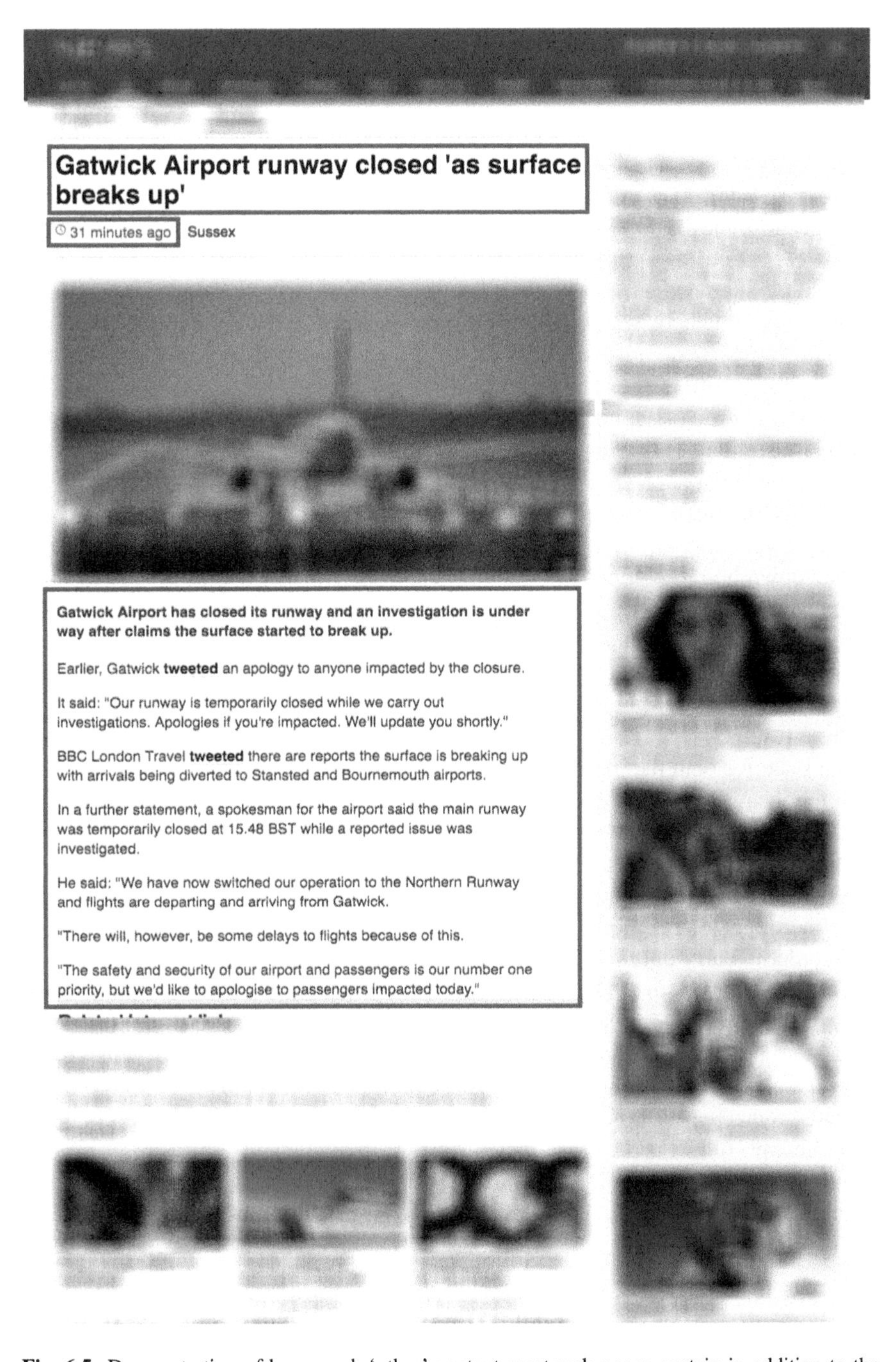

Gatwick Airport runway closed 'as surface breaks up'

31 minutes ago Sussex

Gatwick Airport has closed its runway and an investigation is under way after claims the surface started to break up.

Earlier, Gatwick **tweeted** an apology to anyone impacted by the closure.

It said: "Our runway is temporarily closed while we carry out investigations. Apologies if you're impacted. We'll update you shortly."

BBC London Travel **tweeted** there are reports the surface is breaking up with arrivals being diverted to Stansted and Bournemouth airports.

In a further statement, a spokesman for the airport said the main runway was temporarily closed at 15.48 BST while a reported issue was investigated.

He said: "We have now switched our operation to the Northern Runway and flights are departing and arriving from Gatwick.

"There will, however, be some delays to flights because of this.

"The safety and security of our airport and passengers is our number one priority, but we'd like to apologise to passengers impacted today."

Fig. 6.5 Demonstration of how much 'other' content most web pages contain in addition to the key information (*highlighted*): title, time and date, main article

6.4.1.2 Entity Extraction

Entities are real objects such as people (i.e., names), organisations and places mentioned in text. However, they can also include objects such as dates and times, telephone numbers, email addresses, URLs, products and even credit card numbers. These pieces of information are typically key parts of the text and the story that it is trying to tell and by extracting them we can utilise them in our further analysis. Entity extraction, also called named entity recognition, can be performed using linguistic, pattern based or a statistical machine learning methods. Linguistic and pattern based methods rely on accurately deconstructing the component parts of a sentence into nouns, verbs, adjectives, etc. and then using these patterns to identify which parts of the sentence may contain names or organisations. They may be assisted by the use of capital letters and other punctuation marks as well as potential word lists if the data being extracted belongs to a specific domain area with a concise vocabulary.

Machine learning techniques use existing data from the same domain as the one you wish to extract information from and uses it as a training set. Text may still be broken down into its component grammatical parts. Training data is usually labelled (thus it has a high start-up cost) as it often requires many human trainers and a significant amount of data to begin the analysis. Imran et al. (2013) showed that training data around particular crises was not necessarily effective even when the nature of the crisis was the same, meaning that some caution may have to be applied to these results.

An issue for both the machine learning and pattern-based techniques is that they both rely heavily on the type of source data they receive. For example, if a number of rules are defined based on the text found in news articles or a number of models are trained on the same basis then using this data for text posted on social media may have a much lower accuracy than if they were used against the same source. Thus the origin of the models used must be kept in mind at all times.

In order to carry out this type of entity extraction a number of software tools are available. The AlchemyAPI, Aylien, and Rosette[44] amongst many others are tools that use the 'freemium' model for their text analytics services and support entity extraction and many other types of textual analysis. However, most of these services are generic and not specific to a particular domain and thus may not give you the fine-grained entity extraction that you require. The alternatives are then to begin constructing rules and models yourself. Libraries and software such as StanfordNLP[45] Named entity recogniser (Finkel et al. 2005) which can be run as a simple desktop application, the NLTK in Python, GATE and OpeNER[46] are some

[44]http://www.rosette.com/.

[45]http://nlp.stanford.edu/software/.

[46]http://www.opener-project.eu/.

of the more popular applications. Commercial programs such as SAS's Text Analytics suite[47] also provide methods for entity extraction.

Entity extraction within a document, be it a news article or a single tweet only ever takes you so far. Typically entity extraction only extracts the entities (as would be expected). However, just as valuable as the entities themselves are the context in which they are mentioned. The next section introduces entity-relation modelling within text analytics that aims to provide more information than simple entity extraction.

6.4.2 Modelling

While the extraction of entities is no doubt useful, when we consider open sources we may want to be 'smarter' than just extracting all entities. Most textual content is built on at least some kind of structure. For example, a news article may begin with the salient and most recently discovered facts while the latter part of the article may give background information. Social media posts often do not just exist in their own context, they are a response to something happening in the world today or to a post made by someone else. Knowing this information gives context to the article, the post, the image, etc. Thus, if we know the context in which we are trying to extract data we carry it out more effectively. Therefore alongside entity extract we are also looking for indications of specific events happening, and it is even more useful if we can tie these events to the entities we have extracted.

6.4.2.1 Entity Relation Modelling

Entity relation modelling uses the idea that natural language follows a specific structure, namely that a sentence follows the pattern Subject—Predicate—Object where predicate is sometimes called a relation. The subject is the person who carries out the action, the predicate is the action itself and the object is the who/what/where the action was carried out. Entity relation modelling allows us to identify not only the entity but the action that it is associated with. This information is infinitely more valuable than simple entity extraction as it immediately gives information about the context the entity appears in and provides more options for the subsequent analysis.

6.4.3 Feedback Loops

Extracting data is rarely a one-time process, especially when carrying out intelligence-led investigations. It is more than likely that one piece of information

[47]http://www.sas.com/en_us/software/analytics.html#text-analytics.

or intelligence discovered will lead to a new line of investigation, a new enquiry or just a piece of knowledge that can enrich our existing model. Therefore it is vital that having extracted the information and passed the data on for further analysis that we do not just stop there. Either from the entity extraction itself or from the results of the analysis new information may have come to light that can improve our initial search terms, narrow down or even widen the scope of what is being looked at or allow to update the models determining what is extracted. This ensures that our OSINT investigation is following the standard intelligence cycle of direction, collection, processing, analysis and dissemination which can then led to feedback (DTIC 2007).

6.4.4 Validation Processes

Validation of your open sources needs to take place at all stages of the open source investigation process. Investigators who follow an automated information extraction process will, more than likely, suffer from the problem of having too much information rather than too little. Therefore any steps they can take to reduce this information overload will help them get to the meaning of their data more quickly. One of the primary ways they can achieve this is through the validation of the relevancy of the sources they have found which then allows them to disregard those sources and consequently narrow the pool of information they are searching in.

Each of the above steps can be utilised in the validation process. For example, if no entities or events are extracted using your model it may be this article is not relevant to your scope. This helps to determine the relevancy of the information collected. Other validation processes may include the assessment of credibility and levels of corroboration and are discussed in more depth in Chap. 7.

6.4.5 Disinformation and Malicious Intent

When accessing data from open-sources we have to realise that it is not just us that has the ability to search for and retrieve this data—those who we are investigating also have the potential to possess the same capabilities. Furthermore, if they know that open sources are being investigated, they may deliberately try to falsify information online.

The motives for incorrectly spreading information can vary in a number of different ways. First, people may share things that they have read innocently believing it to be true and not applying their own validation checks before re-sharing—a common occurrence with news stories on platforms such as Twitter and Facebook. This, whilst potentially annoying to an open source investigator, is not done by people attempting to misdirect the investigator, but rather through carelessness and lack of critical thought motivated by the desire to be the 'first' one to share that information with their friends or followers.

Secondly, people subtly edit their personal information online in order to protect themselves from a privacy point of view (Bayerl and Akhgar 2015). Then thirdly, the intentional spread of disinformation maliciously is a much more serious matter and can be seen as an attempt to deliberately subvert detection online or to point investigators in the wrong direction, so as to distract from interest in themselves or to prevent investigators from connecting the dots between them and their associates.

6.4.6 *Software Tools for Data Collection and Preparation*

Within the previous sections we have discussed data collection and preparation methods for beginning the open source intelligence cycle. However, many open source investigators cannot rely or are not able to build the tools that they require themselves. There are a number of both commercial and openly available tools to assist in this process. By using pre-built tools investigators may lose some fine-grained control over the information accessed and extracted; however, the may make up for this in the speed of carrying out this process.

Commercial tools that can carry out open source data collection and preparation include i2 Analyst Notebook,[48] Maltego and CaseFile[49] Palentir[50] and AxisPro.[51] Although Maltego has been described as not doing "much more than someone could do with technical skills and a browser" the same author also notes that "its beauty lies in its ability to scale" (Bradbury 2011). Each of these tools contain methods for data collection, both in real-time as well as being able to import offline data as well as entity extraction and relation modelling, in depth analysis and visualisation. Consequently, these tools are not just for data collection and information extraction but they really impact on every step on the intelligence cycle.

As well as commercial tools there is now a greater presence of open source software tools that can be utilised with little or no setup cost (in both financial and resource terms), although Maltego does offer a community edition of its platform. The Open Source Internet Toolkit also provides scraping functionality while browsing the web allowing the content to be downloaded easily.[52] The Firefox plug-in DataWake,[53] which forms part of the Memex catalogue, watches as you browse the web catching the sites that you browse to and some of the entities present on their page.

---

[48]http://www-03.ibm.com/software/products/en/i2-analyze.

[49]https://www.paterva.com/web6/.

[50]https://www.palantir.com/.

[51]http://www.textronsystems.com/products/advanced-information/axis-pro.

[52]http://osirtbrowser.com/.

[53]http://sotera.github.io/Datawake/.

6.5 Privacy and Ethical Issues

The collection of data from open sources raises a number of privacy, legal and ethical issues (more of which are elaborated on in Chaps. 17 and 18) related to how the data is obtained, stored and the purpose behind its collection. This is especially important within the EU as it moves from the Data Protection Directive[54] to the General Data Protection Regulation (GDPR),[55] which mandates that anyone collecting personal data in the EU or about EU citizens must have a valid reason to do and not keep hold of this data for any longer than is necessary. Unethical or unlawful methods of collection by law enforcement agencies may jeopardise any subsequent evidential use of the material.

6.5.1 Privacy by Design

A philosophy that has gained traction and is mentioned in the GDPR is Privacy by Design. *Privacy by Design* is the concept of embedding privacy protection into products and services from the initial stage through to the completed product. PbD is underpinned by seven fundamental principles (Cavoukian 2011):

- Proactive not reactive, Preventative not Remedial
- Privacy as the Default Setting
- Privacy Embedded into Design
- Full functionality—Positive-Sum not Zero-Sum
- End-to-End Security—Full Lifecycle Protection
- Visibility and Transparency—Keep it Open
- Respect for User Privacy—Keep it User-Centric

The use of privacy design principles can be extended to the collection of data for open source investigations. In addition to this, Hoepman (2014) has also identified eight privacy design strategies that users can take into account when developing their software. These are *minimise*, *hide*, *separate*, *aggregate*, *inform*, *control*, *enforce* and *demonstrate*. As well as in the development of software, many of these strategies can also be applied to the data collection process. For example, data collection still requires minimising the amount of personal data collected and processed; even when it is collected, personal data should not be available to everyone and hidden from view; and data should be aggregated at the highest possible level. Furthermore, we are not saying that these are the only privacy design

[54]2008. Data Protection Directive 95/46/EC—EUR-Lex—Europa.eu. http://eur-lex.europa.eu/LexUriServ/LexUriServ.do?uri=CELEX:31995L0046:en:HTML.

[55]2011. Protection of personal data—European Commission. http://ec.europa.eu/justice/data-protection/.

strategies that should apply to an OSINT investigators but that, perhaps, they are the most generalizable to all situations.

6.5.2 *Being Polite Online*

Despite the fact that investigators naturally want to get to the information they desire as quickly as possible, as a responsible user of the web even investigators should be only obtaining data in such a way that it is respectful to others; especially, since the creator of web content is not necessarily the same as the person(s) who run the site and pay for the bandwidth, server access, etc.

6.5.2.1 Monitor Web Crawls and Respecting robots.txt

It is not unheard of for web crawlers to get stuck in a loop of following links within the same site—and, indeed, this may even be what you want to happen. Nevertheless, in doing so, a crawler should not be accessing a particular site so many times that it begins to put a strain on the web server. Instead, reasonable limits should be set for how many times a crawler can revisit a page within a set period of time.

Another method of controlling how (and even if) a page should be visited is the robots.txt file. This file, set up by the webmaster, contains instructions for web crawlers giving permissions for where a crawler is permitted to access information on the site. In some cases it may be the whole site and in others it may just be certain directories. Be warned, that if you choose to ignore the robots.txt instructions your IP may end up being blocked by the webmaster of that site halting any further investigation down that avenue.

6.5.2.2 Keeping to API Limits

Most access to APIs comes with limits. These limits may be a cut-off point between free access and paid access (and more expensive paid access) or they may simply be the point where you are locked out of accessing data until a certain amount of time has passed. Respecting API limits has two clear advantages. Firstly, it shows that you are a responsible user of such a service, repeatedly trying to find workarounds for API limits will not make you popular with the service provider. Secondly, knowing that there is a limit to respect forces the user to consider what data they actually need, how much and how often they need it. Therefore, these limits can ensure that the user is thinking about what data they need to collect and carefully constructs the queries to only access the data they require, thus also preventing them from contravening data protection regulations.

6.6 Conclusion

In this chapter we have presented some of the methods that an investigator can use to obtain data from open sources and the reasons that they might have for doing so. We believe that this setting of the scene for open source data collection provides a basis for the rest of the chapters in this section, which will go on to show how once collected open source data can be effectively utilised for providing open source intelligence.

References

Bayerl PS, Akhgar B (2015) Surveillance and falsification implications for open source intelligence investigations. Commun ACM 58(8):62–69

Bazzell M (2016) Open source intelligence techniques: resources for searching and analyzing online information. CCI Publishing

Bird S (2006) NLTK: the natural language toolkit. In: Proceedings of the COLING/ACL on interactive presentation sessions. Association for Computational Linguistics, July 2006, pp 69–72

Bradbury D (2011) In plain view: open source intelligence. Comput Fraud Secur 2011(4):5–9

Cavoukian A (2011) 7 Foundational principles of privacy by design. https://www.ipc.on.ca/images/Resources/7foundationalprinciples.pdf

Chen H (2011) Dark Web: exploring and mining the dark side of the web. In: 2011 European intelligence and security informatics conference (EISIC). IEEE, Sept 2011, pp 1–2

College of Policing (2013) Investigation process. In: Authorised professional practice. https://www.app.college.police.uk/app-content/investigations/investigation-process/#material

College of Policing (2015) Intelligence cycle. In: Authorised professional practice. https://www.app.college.police.uk/app-content/intelligence-management/intelligence-cycle/

Cunningham H, Tablan V, Roberts A, Bontcheva K (2013) Getting more out of biomedical documents with GATE's full lifecycle open source text analytics. PLoS Comput Biol 9(2): e1002854

DARPA (2014) Memex aims to create a new paradigm for domain-specific search. In: Defense Advanced Research Projects Agency. http://www.darpa.mil/news-events/2014-02-09

Defense Technical Information Center (DTIC), Department of Defense (2007) Joint intelligence. http://www.dtic.mil/doctrine/new_pubs/jp2_0.pdf

FBI Intelligence Cycle (n.d.) In: Federal Bureau of Investigation. https://www.fbi.gov/about-us/intelligence/intelligence-cycle

Finkel JR, Grenager T, Manning C (2005) Incorporating non-local information into information extraction systems by Gibbs sampling. In: Proceedings of the 43rd annual meeting on Association for Computational Linguistics. Association for Computational Linguistics, June 2005, pp 363–370

Fu T, Abbasi A, Chen H (2010) A focused crawler for Dark Web forums. J Am Soc Inform Sci Technol 61(6):1213–1231

Gibson S (2004) Open source intelligence. RUSI J 149:16–22

Greenwald G, MacAskill E, Poitras L (2013) Edward Snowden: the whistleblower behind the NSA surveillance revelations. In: The guardian. http://www.theguardian.com/world/2013/jun/09/edward-snowden-nsa-whistleblower-surveillance

Hansen D, Shneiderman B, Smith MA (2010) Analyzing social media networks with NodeXL: insights from a connected world. Morgan Kaufmann, Los Altos

HMIC (Her Majesty's Inspectorate of Constabulary) (2011) The rules of engagement: a review of the August 2011 riots. https://www.justiceinspectorates.gov.uk/hmic/media/a-review-of-the-august-2011-disorders-20111220.pdf
Hoepman JH (2014) Privacy design strategies. In: IFIP international information security conference. Springer, Berlin, June 2014, pp 446–459
Imran M, Elbassuoni S, Castillo C, Diaz F, Meier P (2013) Practical extraction of disaster-relevant information from social media. In: Proceedings of the 22nd international conference on World Wide Web. ACM, May 2013, pp 1021–1024
Lohr S (2014) For big-data scientists, "Janitor Work" is key hurdle to insights. In: The New York Times. http://mobile.nytimes.com/2014/08/18/technology/for-big-data-scientists-hurdle-to-insights-is-janitor-work.html?_r=2
Madhavan J, Ko D, Kot Ł, Ganapathy V, Rasmussen A, Halevy A (2008) Google's deep web crawl. Proc VLDB Endowment 1(2):1241–1252
Manning CD, Surdeanu M, Bauer J, Finkel JR, Bethard S, McClosky D (2014) The Stanford CoreNLP Natural Language Processing Toolkit. In ACL (System Demonstrations), June 2014, pp 55–60
Mercado SC (2009) Sailing the sea of OSINT in the information age. Secret Intell Reader 78
NATO (2001) NATO open source intelligence handbook
Omand D, Bartlett J, Miller C (2012) Introducing social media intelligence (SOCMINT). Intell Natl Secur 27(6):801–823
Pallaris C (2008) Open source intelligence: a strategic enabler of national security. CSS Analyses Secur Policy 3(32):1–3
Rogers C, Lewis R (eds) (2013) Introduction to police work. Routledge, London
Shein E (2013) Ephemeral data. Commun ACM 56:20
Warden P (2010) How I got sued by Facebook. In: Pete Warden's blog. https://petewarden.com/2010/04/05/how-i-got-sued-by-facebook/

Chapter 7
Analysis, Interpretation and Validation of Open Source Data

Helen Gibson, Steve Ramwell and Tony Day

Abstract A key component for turning open source data and information into open source intelligence occurs during the analysis and interpretation stages. In addition, verification and validation stages can turn this OSINT into validated OSINT, which has a higher degree of credibility. Due to the wide range of data types that can be extracted from open information sources, the types of data analysis that can be performed on this data is specific to the type of data that we have. This chapter presents a set of analysis processes that can be used when encountering specific types of data regardless of what that data is concerning. These methods will assist an open source investigator in getting the most from their data as well as preparing it for further analysis using visualisation and visual analytics techniques for exploration and presentation.

7.1 Introduction

A key component for turning open source data and information into open source intelligence occurs during the analysis and interpretation stages. In fact, it has been said that "*the major difference between basic and excellent OSINT 'operations' lies in the analytical process*" (Hribar et al. 2014). Furthermore, verification and validation stages can turn this OSINT into validated OSINT, which has a higher degree of credibility. Due to the wide range of data types that can be extracted from open information sources (as was covered in Chap. 6), the types of data analysis that can be performed on this data is specific to the type of data that we have. In this chapter we will present a set of general analysis processes that can be used when

H. Gibson (✉) · S. Ramwell · T. Day
CENTRIC/Sheffield Hallam University, Sheffield, UK
e-mail: H.Gibson@shu.ac.uk

S. Ramwell
e-mail: S.Ramwell@shu.ac.uk

T. Day
e-mail: T.Day@shu.ac.uk

B. Akhgar et al. (eds.), *Open Source Intelligence Investigation*,
Advanced Sciences and Technologies for Security Applications,
DOI 10.1007/978-3-319-47671-1_7

encountering specific types of data regardless of what that data is concerning. These methods will assist an open source investigator in getting the most from their data as well as preparing it for further analysis using visualisation and visual analytics techniques for exploration and presentation.

7.2 Types of Data Analysis

A theme surrounding the analysis of open source data is the prevalence of data in an unstructured textual format. The previous chapter discussed some initial methods such as entity extraction, event detection and entity-relationship modelling as a way to begin to move from unstructured data into something more structured. This section will look at some alternative methods of text analysis as well as methods that can take this analysis alongside entity extraction and use it either as a form of intelligence itself or as a product that can be used for further analysis or visualisation.

Despite the importance of text based sources, they are not the whole of OSINT. There is also often a significant amount of other data such as times, dates, locations and other metadata that can be analysed, not mentioning the extensive amount of data that is contained in image, video and audio sources that is much harder to analyse. Perhaps the greater complexities in analysing such data may also then offer greater rewards due to fewer organisations with the capability to extract intelligence from such information.

7.2.1 Textual Analysis

As was covered in the previous chapter, we know that a high percentage of open sources contain large amounts of unstructured textual data that can be processed and analysed in a number of different ways in order to extract relevant information. Natural language processing (NLP) has already been identified as a key ingredient in the OSINT intelligence framework (Noubours et al. 2013). A number of NLP techniques and methods are discussed in this section that may help an investigator to achieve this.

7.2.1.1 Text Processing

The most simple text processing model is the *bag of words* model whereby each word in each sentence in your document (here a document may be anything from a whole word document, a news article or as short as a single tweet) is broken down into a list of words. Various analyses can then be applied to this list starting with the very simple counting of the number of occurrences of each different term to identify which are the most commonly appearing words. This can be used to get a quick overview of a document in a very simple calculation. It can also serve as a basis for keyword extraction.

This model can also be used for more complex calculations such as *concordance*, which allows the identification of the context that a particular word used throughout a document. Concordance can then be extended to identify other words, which may be used in the same or similar context and thus helps build a list of synonyms or increase corroboration by enabling the realisation of different words that may have been used to describe the same event. Concordance is also sometimes referred to as the 'keyword in context' (KWIC) (Manning and Schütze 1999). Concordance has already been used in the context of analysing crime-related documents supported by visualisations (Rauscher et al. 2013).

Another advantage of using the bag of words model is the ability to calculate collocations. *Collocations* identify words that frequently appear together recognising that sometimes this is more useful than single words.

A more advanced model than the bag of words model is the *Vector Space Model*. This model analyses documents as part of a corpus, i.e., as part of a set of documents, by identifying all the distinct terms within the corpus and for each document in the corpus giving each word a rating. The most common form of rating is to use TF-IDF (term frequency–inverse document frequency) (Salton and McGill 1986), which rather than identifying how common a word is in the whole corpus, or even that document, looks at how important a particular term is in that document compared to the rest of the corpus. Consequently, it enables us to identify what are the key terms for each document—which may not be the same as the most common terms. These terms give us another set of keywords that we may be able to use in our later aggregation and analysis.

Examples of the use of text processing within OSINT include the use of word counting models that produce word clouds for quick overviews of documents or corpora. Meanwhile the use of TF-IDF is demonstrated by Federica Fragapane in her master's thesis on organised crime networks in northern Italy in determining which types of crime occurred more frequently in different cities (Bajak 2015)

In many cases, the use of these types of textual analysis within the OSINT cycle may not produce the intelligence directly itself, but it can be used as another pointer to identify where interesting information may occur or to identify certain keywords, topics and entities that may not be easily extracted by a human user simply due to the volume of information they would have to read in order to catalogue the currently available information.

7.2.1.2 Word Sense Disambiguation

A key issue when interpreting text is that the same words can be used to express very different concepts. For example, consider the phrase "The man opened fire at a supermarket in London." versus "The man is on fire at a supermarket in London." Both contain information about two places: a supermarket and London, and about a person: a man. However, the short phrase ‘fire at’ means two completely different things in the two sentences: The first is reporting a shooting, and the second is talking about someone being on fire. A key problem in text analysis is trying to tease out the context that is obvious to a human reader, but much more difficult when simply looking at the keywords that make up the sentence. This is the issue of word sense disambiguation, i.e., the problem of identifying the true meaning of a word when it has multiple definitions. One way to solve this is through machine learning techniques, where the model eventually learns, by exposure to many correctly tagged examples, the context each word is most likely to appear in.

Being aware of this problem within OSINT is essential, as when we are relying on automated text analytics techniques we must be as aware of the pitfalls of such techniques as we are of the possible advantages they can bring. Capturing the concordance of the specific word(s) we are extracting will help to make the context more explicit when required.

A related problem is that of name disambiguation, whereby more than one name can be used to refer to the same person, for example a friend may use a nickname rather than someone’s formal name or only the full name may be referred to initially in the document and all subsequent mentions may only repeat the first or surname or even just the pronoun (known as a co-reference). The challenge is trying to match all these instances with the same person. This issue then increases vastly in complexity when the names have to be matched across multiple documents rather than within a single one.

7.2.1.3 Sentiment Analysis

Sentiment Analysis is another popular text analysis technique that aims to be able to assess the sentiment or ‘feeling expressed’ in a text. Sentiment can be calculated in a number of ways: as either positive or negative, for identifying the expression of a range of emotions such as fear, anger, happiness, etc., for expressing sentiments about a particular feature or as part of a comparison. The accuracy of sentiment analysis itself is questioned (Burn-Murdoch 2013), and the application of sentiment analysis lends itself far more readily to reviews—where sentiment is far more likely to be expressed—than it does to simply identifying what is expressed within a document. It has the potential to be useful in OSINT, as the tone of the document may indicate an underlying bias not detected when only extracting certain keywords.

Examples of current usage of sentiment within the open source intelligence and law enforcement environment include that of London’s Metropolitan Police who began utilising it following the 2011 London Riots and in the lead up to the Olympic games, although, as pointed out within the article, even they recognise there is still much room for improvement in terms of accuracy (Palmer 2013). Other up and coming areas of sentiment analysis include use in the detection of violent extremism and online radicalisation (Bermingham et al. 2009) and its application to political results including both elections and referendums. This is exemplified by

Cognovi Labs' claim that using Twitter sentiment they were able to predict that the UK would vote to leave the EU 6 h ahead of other polls and results (Donovan 2016).

7.2.2 *Aggregation*

The reason we are performing text analytics is often due to the fact that the corpus of documents we have access to is too large to manually read and extract the information from within. However, the techniques we have described thus far still occur only at the document level, i.e., what are the keywords, entities, relationships or sentiment within this specific document. The power of an automated analytical system does not come at this point though, it comes when the corpus is analysed as a whole (what are the keywords, entities, relationships that exist in multiple documents) or to allow us to identify relationships between documents (e.g., how do these keywords change over time or depend on different authorship or the sources they were extracted from). Answering these types of questions is what makes analytics powerful and begins the process of turning information into intelligence.

7.2.2.1 Document Clustering

One method of aggregation is document clustering. That is, finding similarities within documents that are strong enough to allow them to be grouped together. Depending on the method used, documents can either belong to a single cluster or they may belong to multiple clusters. Clustering, as opposed to classification, groups similar documents into clusters without the expectation that the algorithm is explicit about what is similar about those particular clusters; however, it may be implicit.

Dimension reduction methods can take advantage of a corpus of documents that have been mapped to a high dimensional data model using either the aggregated form of the simple bag of words model or through more complex calculations such as those used by the vector space model. Numerous techniques such as principle component analysis (also known as the singular value decomposition) and multi-dimensional scaling facilitate both the projection of the data onto two dimensions and the identification of discriminant features that distinguish between different document clusters. Furthermore, by reducing them to two dimensions the documents can also be easily visualised on a two-dimensional scatter plot similar to the solutions provided by IN-SPIRE,[1] and Jigsaw[2] (Görg et al. 2013; see also Gan et al. 2014 for an overview of document visualisation techniques).

---

[1]http://in-spire.pnnl.gov/.

[2]http://www.cc.gatech.edu/gvu/ii/jigsaw/index.html.

Formal Concept Analysis (FCA) (Ganter and Wille 1999) is another clustering technique that can be applied to documents with the exception that the clustering is performed using the data extracted from the document rather than the information contained in the whole document itself. This includes information such as entities, relationships, sentiment, dates and any other document deemed relevant. Albeit not in the open source arena, FCA has been applied successfully to the detection of human trafficking activities using police reports (Poelmans et al. 2011), and the same principles have been applied in principle to data obtained from open sources (or even as a method for combining OSINT and non-OSINT; Andrews et al. 2013). The results of formal concept analysis provide the user with a number of sets of terms and the documents in which they appear, i.e., a cluster, with the additional advantage of being able to extract a hierarchy such that the clusters gradually get smaller in terms of the number of documents but also more specific in their commonalities.

7.2.3 Connecting the Dots

An integral part of the intelligence gathering and creation process is figuring out who is connected to whom, who has carried out certain actions, who is located in a specific place or places and more. These relationships may be explicit in the information collected or they may be inferred by other forms of co-occurrence. Furthermore, the detection of these relationships may be simple to a human investigator, when they have access to all the information, but achieving the same results simply through automated means is more complex. Additionally, sometimes simply computing those relationships and then visualising them provides insight that is not necessarily apparent through reading the documents.

Network analysis of OSINT data (and in fact all intelligence data) came to prominence when Krebs (2002a, b) used only sources from the web in order to create the network of the 9/11 hijackers. Starting with only the names of the hijackers that were published in news media in the wake of the attacks he was able to start mapping the relationships between the hijackers and their associates including the strengths of such relationships. Through the visualisation of the network's structure and the calculation of network statistics (like those that are discussed in the section below) he was able to realise who formed the main part of the network and, more crucially, just how poorly it appeared that the members were connected to each other, although in fact these were very strong connections that were lying dormant. Furthermore, the statistical analysis of the network was able to demonstrate the importance of Mohammed Atta, who was known to be the main conspirator behind the attack.

7.2.3.1 Network Analysis

Network analysis and social network analysis provide a number of tools to the investigator to aid their understanding of how the information to which they have access is related. Networks can be *single-moded*, i.e., only contain a particular type of entity such as the relationships between people, or they may be *multivariate* and contain many different types of entities, e.g., people, locations, phone numbers as well as multiple types of relations. In a network an entity is called a *node* and a relationship between two nodes as an *edge* or a *link*. A specific type of multivariate network where there are two types of nodes, e.g., people and organisations, and links only exist between the different types is called *bipartite*. The UK police have already identified network analysis to be potentially useful for intelligence analysis, and the UK Home Office has recently issued a 'how to' guide for police forces who want to improve their social network analysis skills (Home Office 2016).

There are a number of statistics and measures associated with network analysis that provide information that help us to understand how the positions of different entities in the network affect how it works (Krebs 2013). The most simple of these metrics is *degree centrality*, which is simply the entity in the network with the most connections to other entities in the network, giving one measure of which entity is the most highly connected within the network.

Betweenness centrality is a measure of how many shortest paths within the network flow through that particular node. An entity with a higher betweenness centrality indicates that a lot of the information within the network flows through that particular node meaning that, although they may not be the most highly connected entity within the network, they may have access to a higher proportion of the key information. These people are sometimes called gatekeepers, i.e., they control who gets to know which pieces of information. A similar measure can be calculated for the edges in the network which would indicate where, if a particular communication channel was cut within the network, this may disrupt how certain people with the network receive information or even cut off a component of the network completely.

Eigenvector centrality is a measure of not only how central that particular entity is to the network but also how central the entities that it connects to are. This is based on the idea that if a person is connected to other people, who also have an important position in the network, this further increases their own importance.

Closeness centrality uses a calculation of all shortest paths within a network. A shortest path is the least number of links you have to pass through in order to reach your desired node. A node with a high closeness centrality has a shorter shortest path to other nodes in the network.

Community detection algorithms also facilitate the identification of particular communities or clusters within a network. They indicate which nodes are more heavily linked to each other compared to the rest of the nodes in the network. There are a number of different community detection algorithms that can be applied to a graph (see Fortunato 2010) and each of these algorithms may determine the community structure differently.

Network analysis has already been used in a number of case studies that signal its potential for both open and closed source investigations. A UK Home Office report (Gunnell et al. 2016) demonstrates how network analysis could be applied to understanding the patterns in the relationships between street gang members showing that a number of members of the gang could be seen as gatekeepers and enabling them to identify which members of the gang could be potential targets for interventions. Similar network analyses have also been performed on members of the mafia (Mastrobuoni and Patacchini 2012), money laundering operations (Malm and Bichler 2013), criminal hacker communities (Lu et al. 2010), and illicit drug production (Malm et al. 2008). Consequently, the potential applications for OSINT are clear.

Bipartite networks can also be converted into two regular networks using the other entity to infer relationships. Isah et al. (2015) demonstrated this approach using open source data from the medical quality database to identify pharmaceutical crime carried out by rogue manufacturers and vendor to vendor relationships on the darknet. By following this conversion approach and then applying a number of community detection algorithms they were able to identify some potential future avenues for investigation. These same techniques could easily be applied using data obtained from open sources to produce OSINT.

Openly available tools such as Gephi[3] and NodeXL facilitate both network analysis as well as network visualisation where datasets can be easily imported as csv files and calculations of different network analysis metrics made within the software and then made immediately available to map onto the visualisation.

7.2.3.2 Co-occurrence Networks

Co-occurrence networks map relationships between documents through the analysis of shared keywords, appearance of particular entities, authors, sources or any other data that we may have collected. Co-occurrence networks can include relationships between multiple entities, relationships between a single type of entity or, what is known as a bipartite network, relationship between two particular types of entities, for example, suspects and locations. Co-occurrence networks have already been used for various types of criminal analysis, although current research reports that these have usually been on closed rather than open source data; however, the main principles remain the same. Chen et al. (2004) used co-occurrence analysis of suspects from incident summaries, whereby suspects are more strongly linked if they co-occur in the incident summaries more frequently. From the network they created the police were able to confirm that they had represented they gang's structure accurately demonstrating the potential for such networks. They were also able to use the network statistics described below to identify who were the key members of the network and whether they were involved in the violent part of the gang or the part associated with drugs. A similar approach, using co-offending

[3]www.gephi.org.

networks over time in combination with network analysis was able to identify the central players in the crime network and, due to the temporal nature of the dataset, they were also able to use the information to note that the more central a node was the more likely they were to reoffend in the future (Tayebi et al. 2011). Xu and Chen (2004) tried using shortest path (the fewest number of edges to follow in order to get from one specific node to another) algorithm analysis for the identification of interesting parts of criminal networks. Similar to other methods they calculated the co-occurrence network using information found in crime reports and then applied the shortest path algorithm in order to identify certain relationships within the criminal network.

7.3 Location Resolution

When faced with data relating to locations the default approach is to wonder how this data may look on a map. A necessary first step in this process is obtaining the latitude and longitude points that describe the location. Unfortunately, it would require a minor miracle for all open (and probably closed) source data to come ready-tagged with geolocation data in the form of accurate latitude and longitude coordinates. As a result we require a process for getting from a named location to a geo-coded location and vice versa.

Structured data may come with clear location values, whether named locations or geo-coordinates; however, unstructured data often contains various numbers of named locations at various regional levels (local, administrative, national, etc.). Such information can easily be extracted from a single document through named entity extraction, but before it can be properly analysed and visualised in a geospatial manner, these named locations must be resolved, that is, finding their location as listed in the geographic coordinate system.

Such resolution calls for the use of a standard geographical coordinate system, where in the case of open source data, decimal degrees (DD) is often used to represent geographical coordinates. For instance, under degrees, minutes, seconds (DMS) the Tower of London is located at:

51° 30′ 29″N, 0° 4′ 34″W

which can easily be represented as decimal degrees:

51.508056, −0.076111

These decimal degrees coordinates provide an excellent way of working with geospatial data in an automated manner and are also accepted by the emerging GeoJSON draft standard,[4] which is seeing increased support throughout open source data and supporting technologies such as databases.

[4] http://geojson.org/.

7.3.1 Geocoding

The process of resolving a named location in a timely manner involves providing an indexed collection of named locations and their geo-coordinate counterparts. There is a popular open source dataset available for this process called Geonames that contains around 10 million features (countries, cities, towns, villages, administrative areas, places of interest, amongst others) under 645 specific classifications.[5] Furthermore, each feature has English and a local language version of its name listed where available, and the dataset also defines alternative location names that may be used.

Using such a dataset or service allows the lookup of a given named location with the result being the geographical coordinate as well as local names, regional grouping and classifications. Once resolved, the data may be geospatially represented in visualisations.

Such a process does have its flaws, however. First, it is often difficult to disambiguate between common location names. An interesting example of this problem is Springfield in the United States, which is present not only in Illinois, Massachusetts *and* Missouri, but in a total of 41 locations within the country. Wisconsin alone has five individual 'Springfields'. On top of this, it can be difficult to deal with colloquial variations and abbreviations of location names.

7.3.2 Reverse Geocoding

Where a geospatial coordinate is available, it may be necessary and useful to code the given coordinates back to a meaningful named location. In doing so, the process of geospatially clustering data by actual geographical regions becomes reasonably straightforward. Returning to the previous example, when given a decimal degrees coordinate in the region of 51.508056, −0.076111, reverse geocoding can be used to resolve this to any or all of the following (see Fig. 7.1):

Tower of London,
London Borough of Tower Hamlets,
London,
England.

Rich levels of information such as this then allows further in depth analysis of the given dataset by way of clustering and aggregation, whereby the aggregations discussed in Sect. 6.2.2 above can also be carried out by partitioning the data geographically and then at national, regional and local levels.

The issue of the difficulty in carrying out accurate geocoding within the area of crime mapping is already widely recognised. Geocoding, in particular, is affected

---

[5]http://www.geonames.org/about.html.

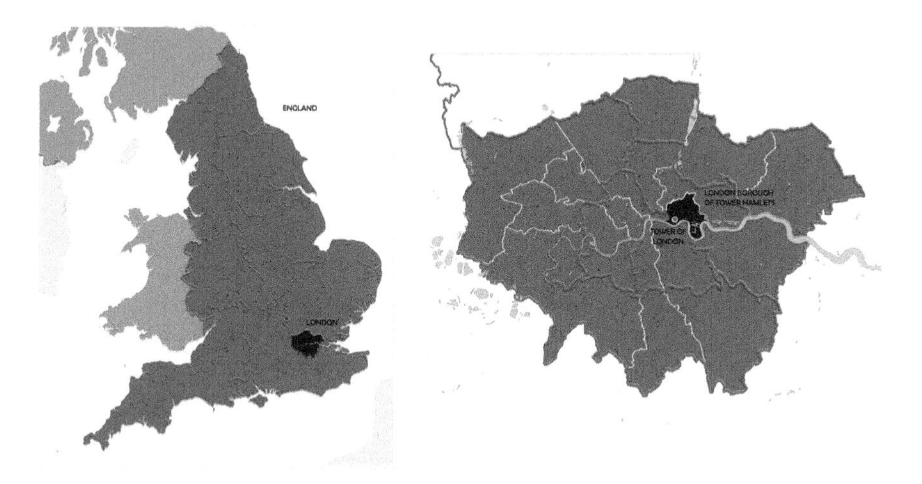

Fig. 7.1 Layers of location resolution

by issues such as spelling errors, abbreviations such as St., Rd., Av. for street, road and avenue respectively; address duplication whereby many towns will have roads with the same street names; the use of qualifiers, e.g., *north of*, *2 miles from*, etc. and missing data that all reduce the effectiveness of the geocoding. McCarthy and Radcliffe (2005) refer to this as the 'garbage in, garbage out' maxim within the context of crime mapping. However, the principles are relevant for any geocoding problem. Given the prevalence of hotspot analysis of crime data (both within the police and for informing the public), the accuracy of geocoding can have a significant impact of a person's perception of a particular area if they go by statistics only. Indeed, for the open data from data.police.uk a study has already examined at what level of spatial resolution is the data accurate identifying that at the postcode level there is "considerable spatial error in the data" (Tompson et al. 2015). This means that even if geo-coordinates are provided within OSINT data, they cannot necessarily be trusted to be completely accurate.[6]

7.4 Validating Open Source Information

As set out in NATO's Open Source Intelligence Handbook (2001) there is a convention for reviewing the status of a particular open source. They suggest that one should assess:

[6]We note that within open police data a degree of geo-masking must take place in order to protect victims, but it is then important that investigators and any other users of the data are aware of the limitations of such a dataset.

1. The *authority* of the source
2. The *accuracy* (by validating it against other sources)
3. The *objectivity* of the source (which is where sentiment analysis may be able to assist)
4. The *currency* (i.e., the provision of a timestamp for publication and the presence of an author)
5. The *coverage* (the degree of relevancy)

As mentioned in the introduction to this chapter, a key part of the analysis of open source data, once the initial extraction has taken place, is to make the move from open source intelligence to *validated* open source intelligence. By designing and utilising methods for priority, credibility and corroboration of sources, specific pieces of information and ultimately intelligence we can increase the likelihood that the final intelligence we present can be considered as having a high degree of reliability. This gives those who act on this intelligence the belief and confidence to act upon such information knowing that the analysis that has produced it has been rigorous.

In the past, and maybe still today, there was a degree of scepticism surrounding data that had come from open sources in that it may be less reliable or less likely to be accurate as well as believing that due to its open nature it does not provide security services with an answer (Pallaris 2008); and that since it is easier (and even perhaps safer) to obtain than closed source intelligence, it is somehow a less valuable form of intelligence (Schaurer and Störger 2013). Thus methods and techniques that can help to evaluate open source intelligence can only benefit OSINT and intelligence as a whole.

7.4.1 Methods for Assigning Priority

Most LEAs are receiving more information than they can possibly investigate and the ever growing constraints on resources are making tasks more difficult. When it comes to OSINT, the amount of data available is even greater and although methods for filtering reduce the signal-to-noise ratio investigators still require additional techniques to identify which pieces of intelligence they should investigate further.

Priority may be assigned based on some predefined information such as the investigator looking for information on a specific person or place, and any mentions of that entity is automatically given a higher priority. Priority may also be decided on volume: A particular entity that is appearing more frequently than other entities is important to deal with, whether or not it is necessarily relevant to the investigation. This is due to the fact that dealing with information that is incorrect as quickly as possible prevents it from polluting other information or allowing malicious information to spread.

7.4.2 Approaches for Recognising Credibility

All open sources are not created equal; there are some that we are able to assign a high degree of credibility to immediately because of the source the information originates from. For example, we may naturally assume information that has come from a large news organisation such as Reuters, the Associated Press or AFP to be immediately more credible than, for example, a single blog post. However, even these major news agencies can get things wrong, especially in the midst of a large crisis (Carter 2013). Thus we know that we can evaluate the source of the information as a starting point, but this cannot be the be all and end all of credibility assessment.

There are a number of ways that provenance (the combination of credibility and reliability) is determined manually in the creation of intelligence reports. Although not in law enforcement, the US army (DOA 2012) has published some of their methods for establishing reliability and credibility of a particular piece of information. They use a six-point reliability scale that judges data from reliable: "*no doubt of authenticity, trustworthiness, or competency; has a history of complete reliability*" to unreliable: "*lacking authenticity, trustworthiness, and competency; history of invalid information*" with the final category being a catch-all: "*cannot be judged*". A similar, but eight-point scale for credibility is used moving from confirmed ("*confirmed by other independent sources; logical in itself; consistent with other information on the subject*"), probably, possible, doubtful, improbable, misinformation (unintentionally false), deception (deliberately false) and cannot be judged. These issues can become very significant if the material is to be relied on evidentially in a subsequent prosecution (see Chap. 18).

NATO also has a similar system of reliability and credibility assessments using a six-point scale for each. Abbott (n.d.) has recommended an additional criterion of assessment to this scale—a confidence assessment whereby credibility and reliability ratings are given a confidence interval of high, medium and low. Furthermore the UK's College of Policing suggest two five-point evaluation scales, using similar definitions to above, as part of the 5 × 5 × 5 intelligence/information report as an assessment of provenance (College of Policing 2015).

As well as their rating scales, NATO (2002) has also published a set of source evaluation checklists for a number of web sources in the domain of advocacy, business and marketing, news, information and personal information. These checklists encourage open source intelligence analysts to scrutinize the authority, accuracy, objectivity, currency and coverage of the webpage by considering criteria such as who is responsible for the content, using grammatical checks as an additional measure of accuracy, identifying their goals for having such a site or article, looking at the dates the page was published and how extensive the content contained on it is.

These methods above have all discussed how credibility is achieved when an analyst can eye-ball the data or may have even obtained it themselves and so has intimate knowledge of the source. However, when data is collected through an automated process on a much larger scale, it is not cost effective to manually analyse each source individually and, while automated solutions may not be perfect, neither are people, and even simple assessments can give analysts a head start.

7.4.3 Methods for Identifying Corroboration

Another method of determining credibility is through the use of corroboration; the more sources that make the same claims or the same statements the more likely these claims are to be true. For example, the more witnesses you have to an incident agreeing on the causes, actions and results the more certain you are that you are getting a true picture of what actually happened. The same can be true of OSINT; the more open sources you can find to verify your intelligence, the stronger that intelligence is—although as we will describe later extra caution may need to be applied.

OSINT has two potential drawbacks: One is the 'echo' effect (Best and Cumming 2007), whereby a source appears highly credible because of multiple information sources proffering the same information also exist; however, these reports are all based on the same original source and so the corroboration seems higher than it actually is. The second is the tendency for misinformation (that is not necessarily intentionally false) to spread very rapidly especially on social media. For example, the reports of the London Eye being on fire and a tiger on the loose during the London Riots of 2011 (Procter et al. 2011). And even though those rumours were counteracted it still took many hours, in some cases, to do so.

Thus, in identifying methods for corroboration we also should take into account the validation of the sources in the corroboration. Sometimes corroboration can be a by-product of other techniques used to analyse the data. For example, the clustering and formal concept analysis methods mentioned above naturally group similar data and therefore the larger the group the more highly corroborated it will be.

7.5 Conclusion

This chapter has reviewed the most common concepts and techniques for analysing open source intelligence data. It has built on the results of the previous chapter by showing how methods such as entity extraction can feed into the analysis process and generate key insights for the OSINT investigator even when dealing with much larger amounts of data. The chapter has also considered the importance of not only identifying what information or intelligence is available within a source document but also strategies for determining the provenance of that information; another vital component of the intelligence cycle.

References

Abbott C (n.d.) RC(C) evaluation system. Open briefing. http://www.openbriefing.org/intelligenceunit/intelligencemethod/rccsystem/

Andrews S, Yates S, Akhgar B, Fortune D (2013) The ATHENA project: using formal concept analysis to facilitate the actions of crisis responders. In: Akghar B, Yates S (eds) Strategic intelligence management. Elsvier, Oxford, pp 167–180

Bajak A (2015) How Federica Fragapane visualized organized crime in northern Italy. In: Storybench. http://www.storybench.org/how-federica-fragapane-visualized-organized-crime-in-northern-italy/

Bermingham A, Conway M, McInerney L, O'Hare N, Smeaton AF (2009) Combining social network analysis and sentiment analysis to explore the potential for online radicalisation. In: International conference on advances in social network analysis and mining, 2009. ASONAM'09, Athens, pp 231–236

Best Jr RA, Cumming A (2007) Open source intelligence (OSINT): issues for congress, vol 5, Dec 2007

Burn-Murdoch J (2013) Social media analytics: are we nearly there yet? In: The guardian. http://www.theguardian.com/news/datablog/2013/jun/10/social-media-analytics-sentiment-analysis

Carter B (2013) The F.B.I. criticizes the news media after several mistaken reports of an arrest. In: The New York Times. http://www.nytimes.com/2013/04/18/business/media/fbi-criticizes-false-reports-of-a-bombing-arrest.html

Chen H, Chung W, Xu JJ, Wang G, Qin Y, Chau M (2004) Crime data mining: a general framework and some examples. Computer 37(4):50–56

College of Policing (2015) Intelligence report. In: Authorised professional practice. https://www.app.college.police.uk/app-content/intelligence-management/intelligence-report/

Department of the Army (2012) Open source intelligence. http://fas.org/irp/doddir/army/atp2-22-9.pdf

Donovan J (2016) The Twitris sentiment analysis tool by Cognovi Labs predicted the Brexit hours earlier than polls. In: TechCrunch. https://techcrunch.com/2016/06/29/the-twitris-sentiment-analysis-tool-by-cognovi-labs-predicted-the-brexit-hours-earlier-than-polls/

Fortunato S (2010) Community detection in graphs. Phys Rep 486(3):75–174

Gan Q, Zhu M, Li M, Liang T, Cao Y, Zhou B (2014) Document visualization: an overview of current research. Wiley Interdisc Rev Comput Stat 6(1):19–36

Ganter B, Wille R (1999) Formal concept analysis: mathematical foundations. Springer Science & Business Media

Görg C, ah Kang YA, Liu Z, Stasko JT (2013) Visual analytics support for intelligence analysis. IEEE Comput 46(7):30–38

Gunnell D, Hillier J, Blakeborough L (2016) Social network analysis of an urban street gang using police intelligence data. https://www.gov.uk/government/publications/social-network-analysis-of-an-urban-street-gang-using-police-intelligence-data

Home Office (2016) Social network analysis: "How to guide." https://www.gov.uk/government/uploads/system/uploads/attachment_data/file/491572/socnet_howto.pdf

Hribar G, Podbregar I, Ivanuša T (2014) OSINT: a "Grey Zone"? Int J Intell CounterIntell 27(3):529–549

Isah H, Neagu D, Trundle P (2015) Bipartite network model for inferring hidden ties in crime data. In: 2015 IEEE/ACM international conference on advances in social networks analysis and mining (ASONAM). IEEE, Aug 2015, pp 994–1001

Krebs V (2002a) Uncloaking terrorist networks. First Monday 7(4). http://pear.accc.uic.edu/ojs/index.php/fm/article/view/941/863

Krebs VE (2002b) Mapping networks of terrorist cells. Connections 24(3):43–52

Krebs V (2013) Social network analysis: an introduction. In: OrgNet

Lu Y, Luo X, Polgar M, Cao Y (2010) Social network analysis of a criminal hacker community. J Comput Inf Syst 51(2):31–41

Malm A, Bichler G (2013) Using friends for money: the positional importance of money-launderers in organized crime. Trends Organized Crime 16(4):365–381

Malm AE, Kinney JB, Pollard NR (2008) Social network and distance correlates of criminal associates involved in illicit drug production. Secur J 21(1):77–94

Manning CD, Schütze H (1999) Foundations of statistical natural language processing, vol 999. MIT press, Cambridge

Mastrobuoni G, Patacchini E (2012) Organized crime networks: an application of network analysis techniques to the American mafia. Rev Network Econ 11(3):1–43

McCarthy T, Ratcliffe J (2005) Garbage in, garbage out: geocoding accuracy and spatial analysis of crime. In: Geographic information systems and crime analysis, IGI Global

NATO (2001) NATO open source intelligence handbook

NATO (2002) Exploitation of intelligence on the internet. http://www.oss.net/dynamaster/file_archive/030201/1c0160cde7302e1c718edb08884ca7d7/Intelligence Exploitation of the Internet FINAL 18NOV02.pdf

Noubours S, Pritzkau A, Schade U (2013) NLP as an essential ingredient of effective OSINT frameworks. In: Military communications and information systems conference (MCC), Oct 2013. IEEE, pp 1–7

Pallaris C (2008) Open source intelligence: a strategic enabler of national security. CSS Analyses Secur Policy 3(32):1–3

Palmer C (2013) Police tap social media in wake of London attack. In: ITNews. http://www.itnews.com.au/news/police-tap-social-media-in-wake-of-london-attack-344319

Poelmans J, Elzinga P, Dedene G, Viaene S, Kuznetsov SO (2011) A concept discovery approach for fighting human trafficking and forced prostitution. In: International conference on conceptual structures. Springer, Berlin, July 2011, pp 201–214

Procter R, Vis F, Voss A (2011) Reading the riots: investigating London's summer of disorder. In: The guardian. http://www.theguardian.com/uk/interactive/2011/dec/07/london-riots-twitter

Rauscher J, Swiezinski L, Riedl M, Biemann C (2013) Exploring cities in crime: significant concordance and co-occurrence in quantitative literary analysis. In: Proceedings of the computational linguistics for literature workshop at NAACL-HLT, June 2013

Salton G, McGill MJ (1986) Introduction to modern information retrieval

Schaurer F, Störger J (2013) Intelligencer guide to the study of intelligence. The evolution of open source intelligence (OSINT). J US Intell Stud 19:53–56

Tayebi MA, Bakker L, Glasser U, Dabbaghian V (2011) Locating central actors in co-offending networks. In: 2011 International Conference on advances in social networks analysis and mining (ASONAM), July 2011. IEEE, pp 171–179

Tompson L, Johnson S, Ashby M, Perkins C, Edwards P (2015) UK open source crime data: accuracy and possibilities for research. Cartogr Geogr Inf Sci 42(2):97–111

Xu JJ, Chen H (2004) Fighting organized crimes: using shortest-path algorithms to identify associations in criminal networks. Decis Support Syst 38(3):473–487

Chapter 8
OSINT and the Dark Web

George Kalpakis, Theodora Tsikrika, Neil Cunningham, Christos Iliou, Stefanos Vrochidis, Jonathan Middleton and Ioannis Kompatsiaris

Abstract The Dark Web, a part of the Deep Web that consists of several darknets (e.g. Tor, I2P, and Freenet), provides users with the opportunity of hiding their identity when surfing or publishing information. This anonymity facilitates the communication of sensitive data for legitimate purposes, but also provides the ideal environment for transferring information, goods, and services with potentially illegal intentions. Therefore, Law Enforcement Agencies (LEAs) are very much interested in gathering OSINT on the Dark Web that would allow them to successfully prosecute individuals involved in criminal and terrorist activities. To this end, LEAs need appropriate technologies that would allow them to discover darknet sites that facilitate such activities and identify the users involved. This chapter presents current efforts in this direction by first providing an overview of the most prevalent darknets, their underlying technologies, their size, and the type of information they contain. This is followed by a discussion of the LEAs' perspective on OSINT on the Dark Web and the challenges they face towards discovering and de-anonymizing such information and by a review of the currently available techniques to this end. Finally, a case study on discovering terrorist-related information, such as home made explosive recipes, on the Dark Web is presented.

8.1 Introduction

Over the past 20 years the World Wide Web has come to constitute the most popular tool, used by billions of users, when interacting with sources for news, entertainment, communication and several other purposes. Contrary to what the average user may believe, only a small portion of the Web is readily accessible. The

G. Kalpakis (✉) · T. Tsikrika · C. Iliou · S. Vrochidis · I. Kompatsiaris
Centre for Research and Technology Hellas,
Information Technologies Institute (CERTH-ITI), Thermi-Thessaloniki, Greece
e-mail: kalpakis@iti.gr

N. Cunningham · J. Middleton
Police Service Northern Ireland, Belfast, Ireland

B. Akhgar et al. (eds.), *Open Source Intelligence Investigation*,
Advanced Sciences and Technologies for Security Applications,
DOI 10.1007/978-3-319-47671-1_8

so-called *Surface Web* constitutes the part of the Web that is gathered and indexed by conventional general-purpose search engines such as Google,[1] Yahoo![2] or Bing[3] and is subsequently made available to the general public using typical Web browsers such as Mozilla Firefox,[4] Google Chrome[5] or Internet Explorer[6] (Ricardo and Berthier 2011). However, in the same way that only the tip of an iceberg is visible above the water, while the vast majority of its mass lies underwater, a general-purpose search engine is capable of indexing just a small portion of the information available on the Web; the rest of the non-indexable content lies in the so-called Deep Web (Bergman 2001).

The Deep Web, also known as the Hidden or Invisible Web (Sherman and Price 2003), in general comprises information that cannot be retrieved when querying a conventional general purpose search engine. The content present in the Deep Web is not necessarily hidden on purpose; in many cases it is just impossible to be detected by the crawlers employed by conventional search engines for gathering the Web content due to several limitations (e.g. dynamic content returned in response to a query, private content requiring authorised access, scripted content, and unlinked content). Consider, for instance, a banking website; there is a homepage and a login page that both can be easily found by general members of the public, whereas it is only by having the correct login details that the website allows them to progress any further to the myriad of Web pages available to its customers. Because crawlers do not have the ability to enter correct login details, these Web pages are hidden from search engines and are therefore considered as part of the Deep Web. This example aims to dispel the myth that the Deep Web is a somehow undesirable place to visit consisting solely of clandestine material.

Nevertheless, a small part of the Deep Web, known as the *Dark Web*[7] (Bartlett 2014), includes content which is intentionally hidden and is inaccessible through typical Web browsers. The Dark Web consists of several *darknets* providing restricted access through the employment of specific software (i.e. special Web browsers providing access to darknets), configuration (i.e. proxy-like services permitting the communication with each darknet) or authorisation. These darknets include small peer-to-peer networks, as well as large, popular networks, such as

[1]https://www.google.com/.

[2]https://www.yahoo.com/.

[3]https://www.bing.com/.

[4]https://www.mozilla.org/en-US/firefox/products/.

[5]https://www.google.com/chrome/.

[6]http://windows.microsoft.com/en-us/internet-explorer/.

[7]The term Dark Web is often confused with Deep Web, especially in media reporting. However, it should be clear that the two terms are distinguished, and Dark Web constitutes a subset of Deep Web exhibiting specific properties. Moreover, the term Dark Web has also been used to refer to the specific content generated by international terrorist groups and made available either on the Surface Web (including on Web sites, forums, chat rooms, blogs, and social networking sites) or on the Deep Web (Chen 2011); this definition is different to the one employed in this book.

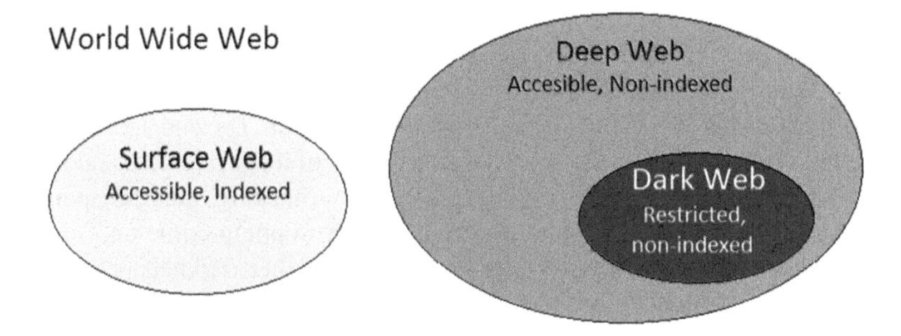

Fig. 8.1 The World Wide Web structure

Tor[8] (Dingledine et al. 2004), I2P[9] (I2P 2016) and Freenet[10] (Clarke et al. 2010), operated by organisations or individuals. Figure 8.1 illustrates the structure of the different parts of the World Wide Web.

Unlike the Surface Web and most parts of the Deep Web, the Dark Web maintains the privacy and anonymity of the network both from a user and a data perspective. This anonymous nature of the Dark Web constitutes the most important factor for attracting users since it provides them with the opportunity for hiding their identity when surfing across the Web or publishing information, i.e. it prevents someone who is observing a user from identifying the sites they are visiting and prevents the sites being visited from identifying their users. On the one hand this anonymity facilitates the communication of secret and sensitive data for legitimate purposes and, as a matter of fact, both Tor and I2P ostensibly started out as online privacy tools aimed at those who held civil libertarian views, including activists, oppressed people, journalists, and whistle-blowers. (It is also worth noting that US Naval Intelligence played a pivotal role in terms of initial expertise and funding of the Tor project, as even they realised the value of being able to transmit data anonymously and thus securely). On the other hand, this anonymity also provides the ideal environment for transferring information, goods and services with potentially illegal intentions and as a result the same technology that aims to protect civil libertarians from persecution can be exploited for protecting criminals from prosecution (also see Chap. 14).

Although such illegal intentions and actually all manners of criminality are widely observed also on the Surface Web, successful operations by Law Enforcement Agencies (LEAs) based on Open-Source Intelligence (OSINT) collected from Surface Web sources have forced some criminals and terrorists to migrate to the Dark Web as they seek more secure ways to host their criminal enterprises and share their propaganda (Biryukov et al. 2014; Owen and Savage 2015). Law Enforcement Agencies are thus investing time and effort into

[8]https://www.torproject.org/.

[9]https://geti2p.net/en/.

[10]https://freenetproject.org/.

investigating criminal activities on the Dark Web, with the best known success being the take-down of the *Tor Hidden Service* "Silk Road Underground", a marketplace that was primarily used to bring together vendors and customers for the purpose of selling illegal drugs. Such marketplaces are pervasive on Tor and LEAs strive to monitor their activities since they provide all manners of illegal services and material for purchase, including fraudulent documentation, hacking kits, weapons and explosives (Christin 2013). Despite some success on stopping crime on Tor, LEAs have had very limited success in stopping crime on other darknets, such as I2P, where several criminal activities, and in particular paedophilia, have been observed.

Therefore, LEAs are very much interested in gathering OSINT on the Dark Web that would allow them to successfully prosecute individuals involved in criminal and terrorist activities. To this end, LEAs need appropriate technologies that would allow them to discover darknet sites that facilitate such activities and identify (i.e. de-anonymize) the administrators of these sites, including the vendors in the particular case of marketplaces so as to also intercept the illegal merchandise. This is particularly challenging given the anonymity granted by darknets and requires effective and efficient discovery, traffic analysis, and de-anonymization tools. This chapter presents current efforts in this direction by first providing an overview of the most prevalent darknets on the Dark Web, the technologies underlying them, their size and popularity, and the type of information they contain (Sect. 8.2). This is followed by a discussion on the LEAs perspective on OSINT on the Dark Web and the challenges they face towards discovering and de-anonymizing such information (Sect. 8.3) and a review of the currently available techniques to this end (Sect. 8.4). A study on the particular case of discovering information on recipes for home made explosives on the Dark Web, which is a characteristic example of terrorist-related activity in darknets, is discussed next (Sect. 8.5) before concluding this work (Sect. 8.6).

8.2 Dark Web

The Dark Web contains several darknets with Tor, I2P, and Freenet being the most popular. This section describes these darknets from a high-level technical perspective so as to highlight some of the challenges that need to be addressed (Sect. 8.2.1), provides some evidence on their size and popularity (Sect. 8.2.2), and discusses the type of information they contain (Sect. 8.2.3).

8.2.1 Darknets on the Dark Web

Tor (the onion router) is currently the most popular anonymity network. It was primarily designed for enabling anonymous user access to the Surface Web. However, it has been extended so as to allow onion Web pages (.onion) and services to exist entirely within the Tor network; therefore it can also act as a darknet. The main objective of Tor is to conceal its users' identities and their online activity from surveillance and traffic analysis. The core idea is to create a hard-to-follow network pathway allowing the communication among users, so that their actual digital footprints are hidden.

Tor is based on the concept of onion routing, which dictates that instead of taking a direct route between source and destination, all messages follow a specific pathway (randomly calculated on source) consisting of several mediating network nodes acting as *relays* (selected from a centralised directory containing the currently operational relay nodes provided by volunteers around the globe), which help covering the communication tracks, so as to prevent any observer from identifying the source and the destination of the corresponding network traffic. This is realised by wrapping the messages sent in packets with encryption layers (hence the onion metaphor) as many as the number of the relay nodes existing in the pathway. This multi-layered encryption approach ensures perfect forward secrecy between relays, since each relay can only decrypt packets it is assigned with, meaning that it can only know the previous relay node the packet came from and the next relay node the packet will be sent to. No individual node (apart from the source node that created the pathway) has knowledge of the complete pathway; hence, this approach ensures user anonymity as far as network locations are concerned. Figure 8.2 illustrates how each layer of a packet sent in Tor is decrypted on each encountered relay node during its route from source to destination.

At the same time, Tor allows contributors to host their content on servers with an undisclosed location; such hidden Web nodes are referred to as *Tor Hidden Services.* Tor servers are configured to receive inbound connections only via the Tor network. Rather than revealing their IP address (as done on the Surface Web), they are accessed through their onion address. The Tor network is able to interpret these onion addresses and is thus capable of routing data to and from them, while preserving the anonymity both for the client user and the server.

The Tor Hidden Services rely on a set of steps that need to be realised so as to build a trustworthy communication path among two parties. First, a hidden service

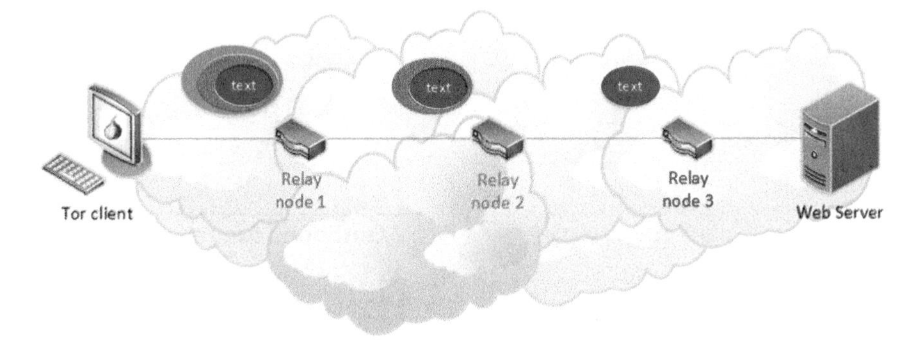

Fig. 8.2 Onion routing architecture in Tor

needs to advertise its existence in the Tor network. For that reason, it randomly picks some relays (from a list of available relays), builds circuits to them, and asks them to act as *introduction points* by disclosing to them its public key. The use of a full Tor circuit makes it difficult for anyone to associate an introduction point with the hidden server's IP address, which remains undisclosed. The hidden service assembles a hidden service descriptor, containing its public key and a summary of each introduction point, and signs this descriptor with its private key. The descriptor will be found by clients requesting a server's .onion page via a service lookup directory. The client, after getting access to the hidden service descriptor, creates a circuit to a *rendezvous point* which acts as the mediating point of communication between the client and the server. The client discloses the rendezvous point's address to the hidden service through its introduction points. At this point the establishment of the connection between the two parts is feasible and all subsequent communication will be realised through the mediating rendezvous point.

The **I2P** (Invisible Internet Project) was designed from the very beginning so as to act as an independent darknet supporting both client and server applications; nevertheless, it can also support anonymous access to the Surface Web. Its primary function is to act as a network within the Internet capable of supporting secure and anonymous communication. Similarly to Tor, I2P is based on the same core idea of building a network pathway for communicating anonymously the respective traffic based on multi-layered encryption; the approach followed, though, is significantly different. Each I2P network node (called router) builds a number of inbound and outbound tunnels consisting of a sequence of peer nodes (provided by volunteer users) for passing a message to a single direction (i.e. to and from the source node, respectively). These inbound and outbound tunnels constantly change so as to minimise the available time for attackers with de-anonymization intentions. Every node participating in the network can select the length as well as the number of these tunnels making a compromise between network throughput and anonymity.

Figure 8.3 illustrates the tunnel-oriented architecture of I2P. When a source node sends a message, it selects one of its available outbound tunnels and targets one of the available inbound tunnels of the destination node, and vice versa. This entails

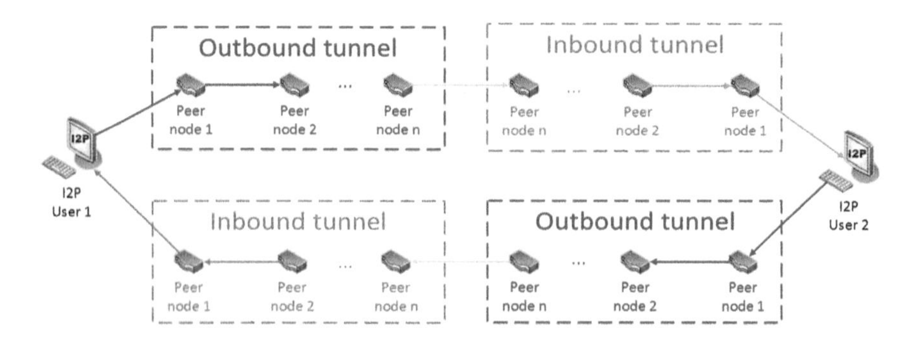

Fig. 8.3 Simple tunnel-oriented communication in I2P

that, unlike Tor, a different independent path is used for each direction of the traffic (i.e. source to destination, and vice versa), which improves anonymity in the sense that only half the traffic may be disclosed to a potential eavesdropper. One major factor differentiating I2P from Tor is that it implements garlic routing (an extension to Tor's onion routing), which encrypts multiple messages in a single packet making it harder for an eavesdropper to perform network analysis. Finally, unlike Tor's centralised approach, I2P uses a peer-to-peer distributed model for communicating the available network peer nodes, hence it helps to eliminate a single point of failure.

I2P also allows for hosting anonymously content and services within its network. The anonymous I2P Web sites, called "eepsites", require the use of I2PTunnel, a tool used for interfacing and providing services in I2P. I2PTunnel is a bridge application which allows forwarding normal TCP streams into I2P network. Along with the use of a Web server application pointing to I2PTunnel, a contributor can host content with an undisclosed location.

Freenet is a peer-to-peer network that emphasises anonymity and allows for file sharing, Web page publishing (i.e. the so-called freesites), and user communication (i.e. chat forums) without any fear of censorship. It is based on a decentralised architecture, making it less vulnerable to attacks. Unlike Tor and I2P, Freenet solely acts as a darknet where the users can only access content previously inserted in it, and does not constitute a means for anonymously entering the world of Surface Web. Freenet's primary goal is to protect the anonymity both for the users uploading content and for the users retrieving data from the network. Each user participating in Freenet constitutes a network node contributing part of their bandwidth along with an amount of their local storage. Each node can reach directly only a small number of neighbouring nodes; however no hierarchical structure exists, meaning that any node can be a neighbour of any other.

The communication between Freenet nodes is encrypted and routed through other relay peer nodes before reaching its final destination, making it very difficult for network eavesdroppers to determine the identity of the participants (i.e. the users uploading and requesting the data) or the information included. As each node passes a message to one of its neighbours, it is not aware whether this node represents the final destination or if it is a mediating node forwarding the message to its own neighbours. Latest versions of Freenet support two types of connections, known as *opennet* and *darknet* connections. The opennet connections are established automatically between all nodes supporting this connection mode, whereas the darknet connections are manually established between nodes which know and trust each other. On the one hand, opennet connections are easy to establish, however it is self-evident that the darknet connections provide a more secure environment for protecting the users' anonymity.

Unlike Tor or I2P, the uploaded content remains accessible even when the uploader goes offline. This is due to the fact that the files uploaded are broken into small chunks and are distributed anonymously among other peer network nodes. At the same time, chunk duplication is employed, so as to provide some level of redundancy. Due to the anonymous nature of Freenet, the original publisher or owner of a piece of data remains unknown. Once some content is uploaded, it will remain on Freenet indefinitely, provided that it is sufficiently popular. The least popular content is discarded so as to be replaced by new information.

In general, the core philosophy governing all darknets, including less popular networks, such as GNUnet[11] and RetroShare,[12] puts an emphasis on the security and privacy of communications and file sharing, based mainly on decentralised structures, which enable their users to maintain their anonymity while either accessing Dark Web sites or when hosting them, thus making it very challenging to de-anonymize them and monitor their traffic.

8.2.2 Dark Web Size

The anonymity provided by the Dark Web has attracted the attention of large user communities, exploiting it either for legitimate or for illegal purposes. Tor appears to be the most popular with a critical mass of users averaging around 2 million a day over the last year (May 2015–April 2016).[13] Over the same time period, Fig. 8.4 shows the number of unique .onion addresses in Tor[13], Fig. 8.5 demonstrates the number of I2P nodes,[14] and Fig. 8.6 the number of nodes in Freenet.[15] Compared to the Surface Web that contains billions of Web pages, the scale of these darknets is much smaller and in most cases well below 100,000. Tor and I2P exhibit an upward trend both starting from around 30,000 a year ago, with Tor peaking at around 100,000 unique .onion addresses for a short period of time and I2P at around 70,000 nodes, and then both decreasing at around 55,000. This indicates that many nodes (particularly Tor Hidden Services) are short-lived, as also observed in a recent study (Owen and Savage 2015). Freenet on the other hand seems to be less popular and remains stable at around 10,000 nodes over the same time period.

8.2.3 Dark Web Content

The Dark Web has proven very useful for whistle-blowers or political dissidents wishing to exchange information with journalists for uncovering significant scandals, as well as for activists or oppressed individuals for publicising information not presented by the mainstream media. On the other hand, the anonymous nature of

[11]https://gnunet.org/.

[12]http://retroshare.sourceforge.net/.

[13]As estimated by the Tor Project https://metrics.torproject.org/ on April 27, 2016.

[14]As estimated by http://stats.i2p/ on April 27, 2016.

[15]As estimated by http://asksteved.com/stats/ on April 27, 2016.

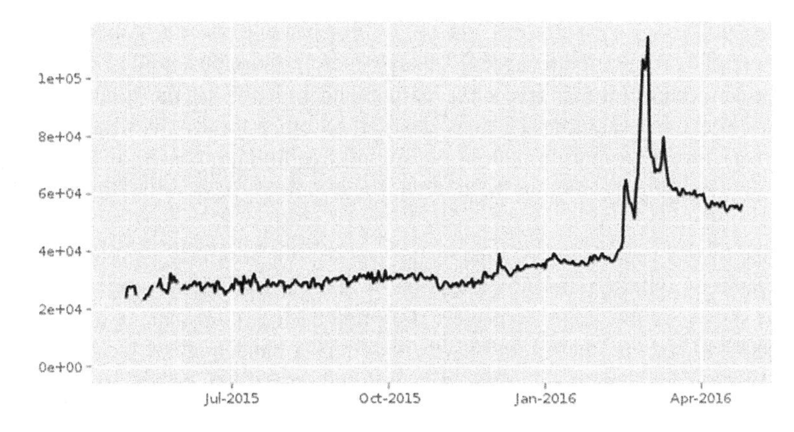

Fig. 8.4 Number of unique .onion addresses in Tor over a one-year period (May 2015–April 2016)

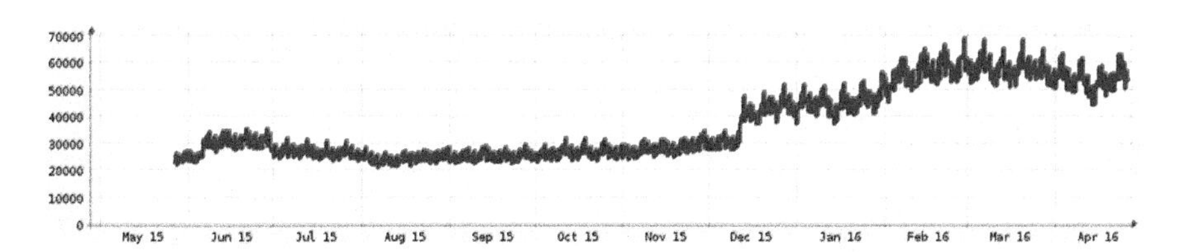

Fig. 8.5 Number I2P network nodes over a one-year period (May 2015–April 2016)

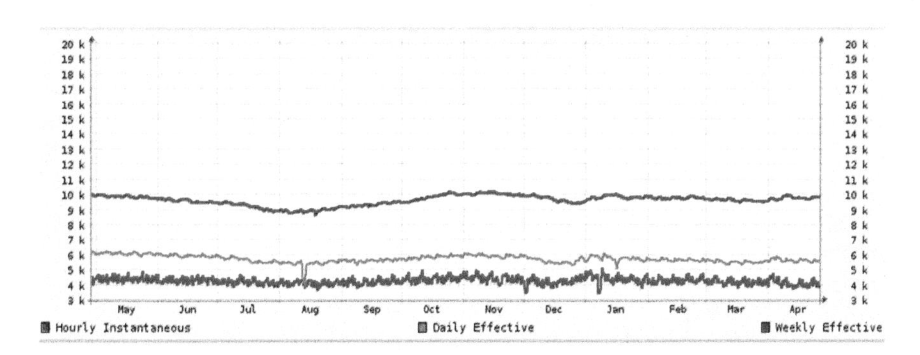

Fig. 8.6 Number of Freenet nodes over a one-year period (May 2015–April 2016)

Dark Web provides an ideal environment for transferring information, goods and services with potentially illegal intentions. In particular, Dark Web is often associated with criminal activity and is used for selling and buying drugs, promoting child pornography, providing information for synthesising home made explosives, conducting money counterfeiting and so on. A recent study that analysed the content of Tor hidden services collected within a single day has shown that Tor nodes related to illegal activities, including drugs, adult content, counterfeit (e.g., stolen credit card numbers, account hacking etc.), weapons, and services (e.g., money laundering, hitman hire etc.), correspond to 48 % of the total number of nodes, while the remaining 52 % is devoted to other topics, with politics (e.g., discussions about corruption, human rights and freedom of speech) and anonymity being the most popular among them, with 9 and 8 %, respectively (Biryukov et al. 2014).

At the same time, a comprehensive analysis of Silk Road, an anonymous online black market (currently shut-down by FBI) hosted in Tor hidden services, revealed that the most popular items sold were drug-related (i.e. marijuana, cannabis, benzodiazepines, cocaine, heroin etc.), whereas only a small part of the transactions included other products, such as books and digital products (Christin 2013). A more recent study that analysed the content of Tor hidden services crawled over an one-month period so as to take into account their volatility and short longevity (as discussed above) has shown that the content is very diverse with drugs being the dominant topic, followed by marketplaces selling items other than drugs, fraud, and bitcoin-related services (including money laundering) (Owen and Savage 2015). The aforementioned studies clearly indicate that the Dark Web provides an environment where criminality may blossom; hence it is of great interest for LEAs to monitor and investigate whenever necessary the activity on the Dark Web.

8.3 OSINT on the Dark Web

LEAs around the world, including all major UK police forces, are currently investigating darknets, primarily Tor and I2P. The existence of other darknets is known and they are also considered to be of investigative interest, but, given their popularity, Tor and I2P (and Tor in particular) are believed to be used by the majority of Dark Web criminals, and therefore offer LEAs the best opportunities for investigating Dark Web crime. This section first describes the (illegal) activities taking place on the Dark Web that are of interest to LEAs (Sect. 8.3.1) and then discusses the challenges faced by LEAs in gathering OSINT on the Dark Web (Sect. 8.3.2).

8.3.1 Landscape of Dark Web Activities of Investigative Interest

Whilst Tor and I2P were originally created for the benefit of people who required online anonymity, including activists, the oppressed, journalists and whistle-blowers, criminals have also exploited the technology for their own purposes. As illustrated by the content analysis of darknet nodes (Sect. 8.2.3), criminals freely advertise many illegal commodities including fraudulent documentation, hacking kits, weapons, illegal drugs, explosives, sexual services, and even contract killings, on what are termed *Hidden Service Marketplaces* (HSMs) that bring vendors and customers together for the exchange of goods and services.

Of these, the best known was “Silk Road Underground”, launched in the US in February 2011 by an American, Ross Ulbricht. Ulbricht, who went by the online pseudonym of “Dread Pirate Roberts”, managed to turn a fledgling criminal opportunity on the relatively unknown Tor network into a multi-million dollar world-wide enterprise in just two years. Australian and US LEAs mounted a massive investigation into Silk Road Underground and Ulbricht was ultimately arrested in October 2013 for money laundering, computer hacking, conspiracy to traffic narcotics, and attempting to have six people killed. He was convicted by a court in New York and sentenced to life imprisonment without the possibility of parole in May 2015. He was also ordered to forfeit approximately US $183 million, a testament to the extent of criminal activities taking place on HSMs. The shut-down of “Silk Road Underground” lead to the advent of new HSMs, including “Silk Road Underground 2”, and “Alphabay”, currently believed to be the largest HSM on Tor.

The reliance of Tor vendors on having a good “sellers” reputation (similar to eBay[16]) has been a surprising revelation for LEAs. It transpires that there are numerous Tor vendors and even entire HSMs that have been uncovered as scams, i.e. they simply receive the payments, while the goods or services never materialise. A Tor vendor with a good reputation and multiple positive reviews is much more likely to attract new customers. This makes easier the identification of genuine Tor vendors offering illegal commodities, and LEAs could direct their attention towards them.

At the negotiation stages between HSMs and their customers, LEAs are seeing an increase in the use of Privnote,[17] an online service to create one-time read only notes that self-destruct when read, similar to Snapchat[18] or Wickr[19] messages. There is also a much greater incidence of the use of encryption in the communications between vendors and their customers, including but not limited to the use of the PGP (Pretty Good Privacy) standard. Whilst encryption is beyond the technical capabilities of most online users, it is widely understood and implemented by the technically minded. These practices are seriously hampering investigations as these communications cannot be easily intercepted and decrypted.

Although there is a variety of cyber currencies available, Bitcoin is currently the largest and best known, and is the preferred method of payment on HSMs. During the investigation into Silk Road Underground, the US authorities seized Bitcoins with a value of approximately US $100 million. Bitcoin is a virtual currency that is

[16]http://www.ebay.com/.

[17]https://privnote.com/.

[18]https://www.snapchat.com/.

[19]https://www.wickr.com/.

unregulated and its usage can be dramatically affected by fluctuations in its value. Bitcoins are not physical in nature like normal money, but are digital "codes" that represent a complex algorithm solved by a series of linked computers relying on joint processing power; this is referred to as 'Bitcoin mining'. Bitcoin transactions are extremely hard to understand, as they are totally unlike conventional banking transactions.

Finally, terrorists, and in particular the Islamic State (IS), are also using Tor for hosting their propaganda, and to signpost how would-be jihadists can join them abroad and/or build home made explosives for domestic terrorist incidents. Some Tor sites have elicited funding for future terrorist activities and the funding has been requested as Bitcoin. This type of activity on Tor has global implications and many LEAs and intelligence agencies are actively investigating such sites and their administrators.

8.3.2 Challenges Faced by LEAs on the Dark Web

The main challenge faced by LEAs is the discovery of darknet nodes involved in illegal activities and thus of investigative interest to them. This is particularly challenging as darknet nodes are unreachable via regular search engines and existing Dark Web search engines (further discussed in Sect. 8.4.2) are still far from supporting effective searches.

Once, though, a Dark Web node of interest has been discovered, the next major challenge is to identify the individual(s) involved in the illegal activities. Unlike ordinary Web sites however, Dark Web sites do not have an easily identifiable IP address, and the resolution of a Tor Web site, for example, firstly to an ISP and then to an individual becomes exponentially more difficult because of the complicated nature of data transfer across multiple nodes on Tor. To this end, LEAs have focussed on the identification of the geographical location of such individuals. This has been successful on the Surface Web where social posts are often geo-tagged (ranging from around 1.5 % on Twitter[20] (Murdock 2011) to around 20 % on Instagram[21] (Manikonda et al. 2014) and 50 % on Flickr[22] (Thomee et al. 2016), but on the Dark Web there are no social media that use geo-tagging and the HSM vendors do not normally advertise where they are located.

Another method for geo-locating such individuals would be to examine the wording of their posts, biographies, or adverts. This has had limited success to date, as criminals on the Dark Web do not tend to give away such information, and any probing for further information by an investigator can end with them being "outed" as police. The fact that geo-locating criminals on the Dark Web is relatively

[20]https://twitter.com/.

[21]https://www.instagram.com/.

[22]https://www.flickr.com/.

problematic means that investigators have no way of initially knowing where a person is located, and invariably interact with a number of globally located criminals, before locating criminals that are domiciled within their own jurisdictions. Whilst larger LEAs may be able to sustain a significant number of "false positives" before finding a vendor who is locally located, smaller LEAs may not, due to resourcing and budgetary constraints, and management may question the viability of continuing this type of work.

To date, it is when an investigation migrates from the digital world to the physical world, that most executive actions occur, e.g. during the delivery phase of an illegal commodity. This typically means that either a Covert Internet Investigator (CII) engages with an online criminal and coerces them to meet in real life, or that a criminal attempts to purchase some form of illegal commodity advertised by an LEA, and the LEA learns of a real shipping address that affords a surveillance opportunity. However, "deconfliction" appears to be a major issue for all LEAs, particularly in regard to Dark Web investigations as there is no central control mechanism in many countries for ensuring that "blue on blue" incidents do not occur.

Nevertheless, if an LEA was to take the step of advertising illegal commodities for sale in the hope of attracting criminals to their site, the investigators would have to have expert knowledge of non-generic or official names for certain items. For example, Semtex has a chemical name that only people familiar with explosives would use, however people will search for the chemical name on HSMs. As customers generally appear to search using very specific terminology, investigators would need to carefully "frame" specific items to attract the attention of the criminally minded. Online chatter between buyers and sellers is also commonplace and there is much negotiation on the price of the goods and/or shipping costs. Therefore, good communication skills are undoubtedly required. Moreover, CIIs would need to ensure that their online presence looks realistic, e.g., by having a network of "friends" interacting with them.

The Dark Web poses major challenges for LEAs. Most OSINT investigators do not have a strong computer science/programming background and are largely self-taught when it comes to investigating darknets. This is a training issue that needs to be addressed, and LEAs may wish to start considering their recruitment policy for OSINT investigators, particularly with regard to the use of CIIs. To support them in such investigations, several technological solutions are currently being researched and developed, as discussed next.

8.4 OSINT Techniques on the Dark Web

Discovering, collecting and monitoring information are the most significant processes for OSINT (see Chaps. 6 and 7). Several different techniques (i.e., advanced search engine querying and crawling, social media mining and monitoring, restricted content access via cached results etc.) are applied to Surface Web for retrieving content of interest from the intelligence perspective. However, the distinctive nature of Dark Web, which requires special technical configuration for accessing it, while it also differentiates the way the network traffic is propagated for the sake of anonymity, dictates that the traditional OSINT techniques should be adapted to the rules, which govern the Dark Web. This section first discusses Web crawling techniques (Sect. 8.4.1) and search engine approaches for the Dark Web (Sect. 8.4.2), and then examines the major strategies for traffic analysis and user de-anonymization (Sect. 8.4.3).

8.4.1 Crawling

Web crawlers are software programs responsible for automatically traversing the World Wide Web in a methodical, automated manner for discovering the available Web resources and are usually employed by Web search engines for discovering and indexing Web resources (Olston and Najork 2010). They are typically applicable only to Surface Web; however, under special configuration they can be set to traverse the Dark Web as well. The Web crawlers operate based on a set of fundamental steps: (1) they start with a set of seed URLs (constituting the starting point of the crawling), (2) fetch and parse their content, (3) extract the hyperlinks they contain, (4) place the extracted URLs on a queue, (5) fetch each URL on the queue and repeat. This process is iteratively repeated until a termination criterion is applied (e.g. a desired number of pages are fetched). Web crawlers may follow the general approach described for retrieving Web resources regardless of their subject matter, or operate in a more focussed manner for discovering resources related to a given topic based on supervised machine learning methods that rely on (1) the hyperlinks' local context (e.g. surrounding text), and/or (2) global evidence associated with the entire parent page. Compared to general Web crawlers, focussed crawlers implement a different hyperlink selection policy by following only the links estimated to point to other Web resources relevant to the topic.

Several research efforts have developed Web crawlers for the Dark Web, mainly as a tool for collecting data and analysing the Dark Web content. In an effort to analyse the Silk Road marketplace, a Web crawler equipped with an automated authentication mechanism was developed for gathering data over a six month period in 2012, so as to characterise the items sold and determine the seller population (Christin 2013). Recent research in the context of the Artemis project, resulted in developing a multilingual Web crawler capable of traversing the Tor network and storing the information discovered in an effort to examine the core characteristics of Tor hidden services along with their evolution over time (Artemis Project 2013). The crawler was accompanied by a search engine operating as an auxiliary tool for analysing the content collected. Additionally, a Tor crawler, referred to as PunkSPIDER,[23] was built for assisting the process of uncovering vulnerabilities on onion domains hosted on Tor hidden services (Paganini 2015). It is accompanied by a vulnerability search engine where the security risk of any scanned domain can be examined. The goal of this project was to aid LEAs in their fight against illegal activities in the Dark Web. Moreover, a Web crawler for Tor hidden services has been implemented in the context of a DAPRA project, aiming at extracting and analysing hidden services content (Memex project). Furthermore, a Web crawler specifically tailored to the Tor network has been developed in order to detect the most prominent categories of content hosted on Tor hidden services (Moore and Rid 2016).

At the same time, a classifier-guided focussed crawler capable of discovering Web resources with information related to manufacturing and synthesising Home Made Explosives (HMEs) on several darknets of the Dark Web (such as Tor, I2P and Freenet), as well as on the Surface Web, has been developed in the context of the EU funded HOMER project[24] (Kalpakis et al. 2016). This focussed crawler employs a link-based classifier taking advantage of the local context surrounding each hyperlink found on a parent page, so as to estimate its relevance to the HME domain and adapt the crawler's hyperlink selection policy accordingly. It is accompanied by an interactive search engine provided to LEAs for mining, monitoring and analysing HME-related content publicly available on the Web, in an effort to further enhance and support their investigation efforts against terrorist or criminal activities which may utilise such information.

8.4.2 Search Engines

As a result of the restricted nature of the Dark Web requiring special software and/or configuration for being accessed, as well as of the volatility of the Web sites hosted in darknets (i.e., most Web sites in the Dark Web are hosted on machines that do not maintain a 24/7 uptime), conventional search engines do not index the content of the Dark Web. Nevertheless, a small number of search engines for the Dark Web exists, as well as directory listings with popular Dark Web sites. The most stable and reliable such search tools are provided for Tor. Specifically, the most popular Tor search engine is DuckDuckGo[25] (accessible both via a normal and an onion URL) emphasising user privacy by avoiding tracking and profiling its users. DuckDuckGo may return results hosted in Tor onion sites, as well as results on the Surface Web based on partnerships with other search engines, such as

[23]https://www.punkspider.org/.

[24]HOMER (Home Made Explosives and Recipes characterization—http://www.homer-project.eu/) is an EU funded project that aims to expand the knowledge of European bodies about HMEs and to improve the capacity of security and law enforcement agencies to cope with current and anticipated threats so as to reduce the probability that HMEs will be used by terrorists.

[25]https://duckduckgo.com, http://3g2upl4pq6kufc4m.onion.

Yahoo! and Bing. On the other hand, Ahmia[26] (accessible both via a normal and an onion URL) is a search engine returning only Tor-related results (after filtering out child pornography sites), and as of April 2016 it indexes more than 5000 onion Web sites.[27] Additionally, Torch[28] is also available only through its onion URL for retrieving results from Tor. Finally, several censorship-resistant directory listings, known as hidden wikis (e.g. The Hidden Wiki,[29] Tor Links[30]), containing lists of popular onion URLs are available and provide the user with an entry point to the world of Tor onion sites.

8.4.3 Traffic Analysis and de-Anonymization

The anonymity and the communication privacy provided on the Dark Web constitutes the most significant incentive for attracting users wishing to hide their identity not only for avoiding government tracking and corporate profiling, but also for executing criminal activities. It is really important for LEAs to monitor the network traffic related to criminal acts, and to identify the actual perpetrators of these cybercrimes. Therefore LEAs are greatly interested in exploiting techniques which will allow them to determine with high degree of accuracy the identity of Dark Web users participating in criminal acts. The de-anonymization of Dark Web users is accomplished either by exploiting the unique characteristics of every darknet, or based on data gathered through network traffic analysis of the communication taking place within such a darknet. In the former case, the de-anonymization process attempts to take advantage of potential weak points found in a darknet, whereas in the latter the data collected are cross-referenced so as to identify the anonymous data source.

As no existing darknet can guarantee perfect anonymity, several types of "attacks" have been proposed in the literature for de-anonymising Dark Web users (especially Tor users) after taking advantage of vulnerabilities either existing inherently within the anonymity networks and the protocols used, or being caused by the user behaviour. One of the early research studies shows that information leakage and user de-anonymization in Tor is possible to occur either due to the intrinsic design of the http protocol or due to user behaviour (i.e. users not following the Tor community directives which dictate browsing the Web via TorButton) (Huber et al. 2010). The authors of the study actually argue that https is the best countermeasure for preventing de-anonymization for http over Tor. Furthermore, another work proposes a collection of non-detectable attacks on Tor

[26]https://ahmia.fi/, http://msydqstlz2kzerdg.onion.

[27]https://ahmia.fi/documentation/indexing.

[28]http://xmh57jrzrnw6insl.onion/.

[29]http://thehiddenwiki.org/.

[30]http://torlinkbgs6aabns.onion/.

network based on the throughput of an anonymous data flow (Mittal et al. 2011). It presents attacks for identifying Tor relays participating in a connection, whereas it also shows that the relationship between two flows of data can be uncovered by simply observing their throughput.

Recent research efforts have also developed attacks for identifying the originator of a content message after taking advantage of Freenet design decisions (Tian et al. 2013). The proposed methodology requires deploying a number of monitoring nodes in Freenet for observing messages passing through the nodes. The main objective is to determine all the nodes having seen a specific message sent and identify the originating machine of the message when certain conditions are met. Finally, a survey paper (Erdin et al. 2015) that discusses well-studied potential attacks on anonymity networks which may compromise user identities presents several mechanisms against user anonymity, either *application-based*, such as plugins able to bypass proxy settings, targeted DNS lookups, URI methods, code injection, and software vulnerabilities, or *network-based*, such as intersection, timing, fingerprinting, or congestion attacks. The effectiveness of these attacks is examined, by also considering the resources they require, and an estimate is provided in each case on whether it is plausible for each attack to be successful against modern anonymous networks with limited success.

8.5 Case Study: HME-Related Information on the Dark Web

As discussed above, several techniques are currently being developed for supporting LEAs in their OSINT investigations on the Dark Web. As a case study, this section presents a crawler focussed on the HME domain developed in the context of the EU funded HOMER project. HME recipes present on the Dark Web can be exploited for subversive use by the perpetrators of terrorist attacks; hence their discovery and the monitoring of their evolution over time are of particular interest to LEAs. The focussed crawler is one of the tools incorporated in a larger framework built for discovering and retrieving Web resources including HME-related content, with particular focus on HME recipe information. It is capable of discovering and collecting HME-related information both from the Surface and the Dark Web, found on various types of Web resources, such as Web pages, forums and blog posts.

It is worth mentioning that the developed technologies are presented here in regard to the HME domain, however they can be easily applied in other domains as well (e.g. drugs, weapons, child pornography etc.) after proper training and configuration. First, this section discusses the methodology applied along with the core characteristics of the focussed crawler (Sect. 8.5.1), and then the results of the evaluation experiments are presented (Sect. 8.5.2).

8.5.1 Methodology

The developed application employs a classifier-guided focussed crawling approach for the discovery of HME Web resources. An overview of the crawling approach employed is depicted in Fig. 8.7. First, the seed pages are added to the frontier (i.e., the queue containing the URLs to-be-visited at later stages of the crawling process). In each iteration, a URL is picked from the frontier and is forwarded to the responsible fetching module based on its network type (i.e., Surface Web, Tor, I2P, Freenet). The page corresponding to this URL is fetched (i.e., downloaded) and parsed to extract its hyperlinks. Then, the focussed crawler estimates the relevance of a hyperlink to an unvisited resource based on its local context.

Our methodology is motivated by the results of an empirical study performed with the support of HME experts in the context of the HOMER project which indicated that the anchor text of hyperlinks leading to HME information often contains HME-related terms (e.g. the name of the HME), and also that the URL could be informative to some extent, since it may contain relevant information (e.g. the name of the HME). As a result, the focussed crawler follows recent research (Tsikrika et al. 2016) and represents the local context of each hyperlink using: (i) its anchor text, (ii) a text window of x characters (e.g. $x = 50$) surrounding the anchor text that does not overlap with the anchor text of adjacent links, and (iii) the terms extracted from the URL. Each sample is represented (after stopwords removal and stemming) using a *tf.df* term weighting scheme, where *tf(w, d)* is the frequency of term w in sample d, normalised by the maximum frequency of any term in that

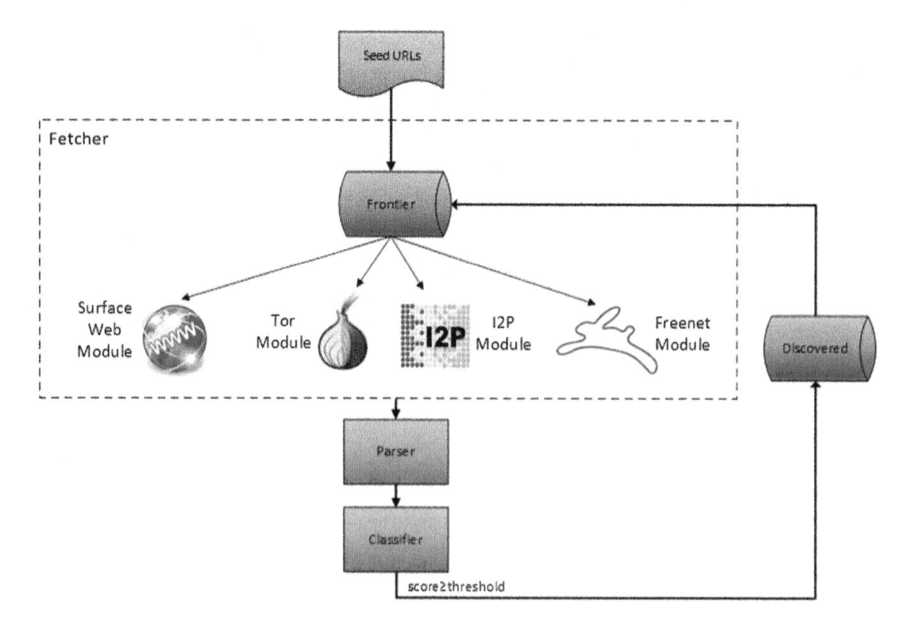

Fig. 8.7 Focussed Crawler architecture

sample, and *df(w)* is the number of samples containing that term in the collection of samples. The classification of this local context is performed using a supervised machine learning approach based on Support Vector Machines (SVMs), given their demonstrated effectiveness in such applications (Pant and Srinivasan 2005). The confidence score for each hyperlink is obtained by applying a trained classifier on its feature vector, and the page pointed by the hyperlink is fetched if its score is above a given threshold t.

The developed focussed crawler is based on a customised version of Apache Nutch (version 1.9). It has been configured so that it can be set to traverse several darknets present in the Dark Web, and specifically the most popular anonymous networks, namely Tor, I2P and Freenet. It is necessary for supporting crawling in Dark Web to enable the Tor, I2P and Freenet services respectively on the machine running the crawler.

8.5.2 Experimental Evaluation

This section provides the evaluation results of the experiments performed for assessing the effectiveness of the focussed crawler application. A set of five seed URLs was used in the experiments: one Surface Web URL, one Tor URL, two I2P URLs, and one Freenet URL. The relatively small seed set is employed so as to keep the evaluation tractable. The actual URLs are not provided here so as to avoid the inclusion of potentially sensitive information, but are available upon request. These seed URLs were obtained during an empirical study conducted with the support of law enforcement agents and domain experts after performing queries based on HME-related keywords in several search engines of the Surface and the Dark Web, such as Yahoo!, Bing, Duck-DuckGo, Ahmia, and Torch.

For evaluation purposes, a set of 543 pages fetched by the focussed crawler, when the threshold t is set to 0.5 (given that it corresponds to a superset for all other thresholds $t > 0.5$) has been assessed based on a four-point relevance scale, characterising the retrieved resources as being highly relevant (i.e. resources describing HME recipes), partially relevant (i.e. resources presenting explosive properties), weakly relevant (i.e. resources describing incidents where HMEs have been used) or non-relevant to the HME domain. These multiple relevance assessments are mapped into the two-dimensional space in three different ways: strict (only highly relevant resources are considered as relevant), lenient (highly and partially relevant resources are considered as relevant) and soft mapping (highly, weakly and partially relevant resources are considered as relevant).

The results of the experiments that evaluate the effectiveness of the focussed crawler using precision, recall, and the F-measure are presented in Table 8.1. Given that recall requires knowledge of all relevant pages on a given topic (an impossible task in the context of the Web), it is computed by manually designating a set of representative pages on the HME domain (in particular those crawled when the threshold is set to 0.5) and measuring what fraction of them are discovered by the

Table 8.1 Evaluation results

Link-based classifier		Threshold t				
		0.5	0.6	0.7	0.8	0.9
	Retrieved	452	377	270	269	87
	Rejected	7054	6792	5375	5374	4865
	All	7506	7169	5645	5643	4952
Strict	Precision	0.61	0.61	0.74	0.74	0.97
	Recall	1.00	0.92	0.88	0.84	0.43
	F-measure	0.76	0.73	0,81	0,79	0.60
Lenient	Precision	0.63	0.63	0.77	0.77	0.97
	Recall	1.00	0.91	0.87	0.84	0.42
	F-measure	0.77	0.74	0.82	0.8	0.58
Soft	Precision	0.65	0.63	0.78	0.78	0.98
	Recall	1.00	0.91	0.87	0.84	0.41
	F-measure	0.79	0.75	0.82	0.81	0.58

crawler (Olston and Najork 2010). The results have been computed for various values of threshold t at depth = 2 (i.e. the maximum distance allowed between the seed pages and the crawled pages), when applying strict, lenient or soft relevance assessments. As expected, precision increases for higher values of threshold t, whereas recall demonstrates the opposite trade-off. In particular, the difference between threshold values 0.6 and 0.7 is quite significant, with precision improving significantly, e.g., for the lenient case, it jumps from 0.63 to 0.77. The threshold value 0.8 appears to be a rather good compromise, since it achieves a high precision, while still maintaining a significant recall. Finally, as anticipated, the precision is much higher in the soft and lenient cases, compared to the strict.

8.6 Conclusions

The Dark Web has proven a very useful and reliable tool in the hands of individuals wishing to be involved in illegal, criminal or terrorist activities, setting sight on getting great economic or political benefits without being identified from government authorities and security agencies world-wide. To this end, LEAs need to become more agile when dealing with criminality on the Dark Web, and in particular on its Hidden Service Markets, and need to invest in new training and technology, if not to get ahead of the criminals, then at least to keep pace. Current technological advancements and research efforts in the fields of Information Retrieval, Network Analysis, and Digital Forensics provide LEAs with numerous opportunities to overcome the limitations and restrictions that the anonymous nature of the Dark Web imposes, so as both to prevent criminals from taking advantage of the anonymity veil existing in several darknets and to suppress any illegal acts occurring in these networks. ICT technologies have reached a substantial level of maturity so as to reliably support the LEAs, therefore it is really important that the tools provided should be exploited in day-to-day real world investigations in the upcoming years. At the same time, it is imperative that LEAs also ensure the proper use of these technologies, so as to protect freedom of speech and human rights for users who exploit the Dark Web anonymity with intentions beneficial for the society. In this context, it is clear that gathering OSINT from Dark Web is an issue of vital significance for the ongoing effort to diminish the potential threats which imperil modern societies.

References

Bartlett J (2014) The Dark Net. Random House, London

Bergman MK (2001) White paper: the deep web: surfacing hidden value. J Electron Pub 7(1)

Biryukov A, Pustogarov I, Thill F, Weinmann RP (2014) Content and popularity analysis of Tor hidden services. In: 2014 IEEE 34th International conference on distributed computing systems workshops (ICDCSW). IEEE, pp 188–193

Chen H (2011) Dark web: exploring and data mining the dark side of the web (vol 30). Springer Science and Business Media, Berlin

Christin N (2013) Traveling the silk road: a measurement analysis of a large anonymous online marketplace. In: Proceedings of the 22nd international conference on world wide web. ACM, pp 213–224

Clarke I, Sandberg O, Toseland M, Verendel V (2010) Private communication through a network of trusted connections: the dark Freenet. Available at: https://freenetproject.org/assets/papers/freenet-0.7.5-paper.pdf

Dingledine R, Mathewson N, Syverson P (2004) Tor: the second-generation onion router. Naval Research Lab Washington DC

Erdin E, Zachor C, Gunes MH (2015) How to find hidden users: a survey of attacks on anonymity networks. IEEE Commun Surv Tutorials 17(4):2296–2316

Huber M, Mulazzani M, Weippl E (2010) Tor HTTP usage and information leakage. In: IFIP international conference on communications and multimedia security. Springer, Berlin, pp 245–255

I2P (n.d.) I2P: a scalable framework for anonymous communication—I2P, from https://geti2p.net/en/docs/how/tech-intro

Kalpakis G, Tsikrika T, Iliou C, Mironidis T, Vrochidis S, Middleton J, Kompatsiaris I (2016) Interactive discovery and retrieval of web resources containing home made explosive recipes. In: International conference on human aspects of information security, privacy, and trust. Springer International Publishing, Berlin, pp 221–233

Manikonda L, Hu Y, Kambhampati S (2014) Analyzing user activities, demographics, social network structure and user-generated content on instagram. arXiv preprint arXiv:1410.8099

Memex Project (Domain-Specific Search) Open Catalog. Available at: http://opencatalog.darpa.mil/MEMEX.html

Mittal P, Khurshid A, Juen J, Caesar M, Borisov N (2011) Stealthy traffic analysis of low-latency anonymous communication using throughput fingerprinting. In: Proceedings of the 18th ACM conference on computer and communications security. ACM, pp 215–226

Moore D, Rid T (2016) Cryptopolitik and the darknet. Survival 58(1):7–38

Murdock V (2011) Your mileage may vary: on the limits of social media. SIGSPATIAL Spec 3(2):62–66

Olston C, Najork M (2010) Web crawling: foundations and trends in information retrieval

Owen G, Savage N (2015) The Tor dark net.' global commission on internet governance (No. 20)
Paganini P (2015) PunkSPIDER, the crawler that scanned the Dark Web. Retrieved 27 Jul 2016, from http://securityaffairs.co/wordpress/37632/hacking/punkspider-scanned-tor.html
Pant G, Srinivasan P (2005) Learning to crawl: comparing classification schemes. ACM Trans Inform Syst (TOIS) 23(4):430–462
Project Artemis—OSINT activities on Deep Web, infosecinstitute.com 2013, July. Available at http://resources.infosecinstitute.com/project-artemis-osint-activities-on-deep-web/
Ricardo BY, Berthier RN (2011) Modern information retrieval: the concepts and technology behind search second edition. Addision Wesley, 84, 2
Sherman C, Price G (2003) The invisible web: uncovering sources search engines can't see. Libr Trends 52(2):282–298
Thomee B, Shamma DA, Friedland G, Elizalde B, Ni K, Poland D, Li LJ (2016) YFCC100M: the new data in multimedia research. Commun ACM 59(2):64–73
Tian G, Duan Z, Baumeister T, Dong Y (2013) A traceback attack on freenet. In: INFOCOM, 2013 Proceedings IEEE. IEEE, pp 1797–1805
Tsikrika T, Moumtzidou A, Vrochidis S, Kompatsiaris I (2016) Focussed crawling of environmental web resources based on the combination of multimedia evidence. Multimedia Tools Appl 75(3):1563–1587

Chapter 9
Fusion of OSINT and Non-OSINT Data

Tony Day, Helen Gibson and Steve Ramwell

Abstract Open Source Investigations do not exist in a vacuum. Whether they are law enforcement or intelligence agency driven, private industry or business driven or the work of a private investigator, it is more than likely that the investigations began with some data that is available openly and some that is not. Thus, from the outset the investigation has some open and some closed source information attached to it. As time goes on in the investigation, the police may elicit information from both open and closed source in order to establish the details surrounding the crime and to build their case. This chapter introduces some of the available data sources for developing open source intelligence and for closed source intelligence. It then puts these data sources into context by highlighting some examples and possibilities as to how these different data types and sources may be fused together in order to enhance the intelligence picture. Lastly, it explores the extent to which these potential synergies have already been adopted by LEAs and other companies as well as future possibilities for fusion.

9.1 Introduction

Open Source Investigations do not exist in a vacuum. Whether they are law enforcement or intelligence agency driven, private industry or business driven or the work of a private investigator, it is more than likely that the investigations began with some data that is available openly and some that is not. For example, when a crime is committed the police may make available information to do with the general location of where it has been committed and what has happened. However, they may

T. Day (✉) · H. Gibson · S. Ramwell
CENTRIC/Sheffield Hallam University, Sheffield, UK
e-mail: T.Day@shu.ac.uk

H. Gibson
e-mail: H.Gibson@shu.ac.uk

S. Ramwell
e-mail: S.Ramwell@shu.ac.uk

B. Akhgar et al. (eds.), *Open Source Intelligence Investigation*,
Advanced Sciences and Technologies for Security Applications,
DOI 10.1007/978-3-319-47671-1_9

not make public who the crime has been committed against and more specific details noted at the scene of the crime. Thus, from the outset the investigation has some open and some closed source information attached to it. As time goes on in the investigation, the police may elicit information from both open and closed source in order to establish the details surrounding the crime and to build their case.

This form of investigation highlights what Hribar et al. (2014), who expand on the ideas of Johnson (2009), explain when discussing their reasoning surrounding the fusion of OSINT and non-OSINT data as part of the intelligence process:

> OSINT and information from clandestine sources can be processed separately, but without the required synergy of both kinds of data this will fail to provide adequate results/outcomes. Thus 'all-source fusion'—the combination of OSINT, HUMINT (human intelligence), and TECHINT (technical intelligence)—has become increasingly important, since it produces the required synergies.

That is the discussion we follow in this chapter, first building on discussions in Chap. 6, we introduce some of the available data sources for developing open source intelligence and for closed source intelligence. We then put these data sources into context by highlighting some examples and possibilities as to how these different data types and sources may be fused together in order to enhance the intelligence picture.

9.2 OSINT Data

An official, albeit US defence-based, definition of OSINT is: "*Open source intelligence is produced from publicly available information that is collected, exploited, and disseminated in a timely manner to an appropriate audience for the purpose of addressing a specific intelligence requirement*" (National Open Source Enterprise 2006). Therefore, any data that can be accessed without special authorisation, memberships or relationships can be considered OSINT data. Such data may be on or offline; structured or unstructured; this definition is purely by consideration of accessibility.

This section complements the discussion in Chap. 6 on the collection or acquisition of open source information by giving an overview of what may be considered open source data to give balance to the following section that will explore what are the sources of non-OSINT data. This will provide a clear view between what can be considered data that can obtained openly and what is classified.

9.2.1 Geographical Data

Throughout the last decade, more and more geographical data has become openly available through various methods. In the UK, for instance, a huge variety of such data from named locations (postcodes, street names, places) to actual mapping imagery (streets, topography, aerial) is directly available from the Ordnance Survey (Ordannce Survey 2016). There have also been efforts by the community behind the Geo Names[1] database to bring global geographical data into federated datasets providing 10 million named locations and even postcodes (or zip codes) for more than 70 countries.[2] Whilst this raw data is useful for analytics, location resolution and visualization, there are also many available online services and tools providing mapping, many based on services such as Open Street Maps, Google Maps and Bing Maps to name just a few. There are also free and proprietary data sets for resolving the rough geographical locations of IP (Internet Protocol) addresses from companies such as MaxMind.[3]

9.2.2 *Statistical Data*

As with geographical data, there has also been a huge push for more and more public data to be made available through the use of dataset downloads or APIs (application programming interface). This is in the interests of both transparency and also to provide developers and researchers with the possibility of discovering new patterns in data.[4] In the UK, the central resource for this new movement in open data is at data.gov.uk.[5] Bt there are also individual sites for localities such as Sheffield City Council Open Data.[6] In the context of law enforcement agencies, the UK police[7] have also started bringing data to the public, which reports on the geographical locations and incidence of various crime types such as burglary, possession of weapons, anti-social behaviour and vehicle crime, among others. Stepping up a notch in scale is the European statistical datasets on the Eurostat site,[8] covering a wide range of national and local data points on a great number of topics and time periods.

9.2.3 *Electoral Register*

A number of countries provide some form of electoral register, sometimes this is publicly available, perhaps even free of charge. In the UK, the electoral register

[1]http://www.geonames.org/.

[2]http://www.geonames.org/postal-codes/.

[3]https://www.maxmind.com/en/geolite2-developer-package.

[4]See, for example, detailed crime mapping: Sampson and Kinnear (2010).

[5]https://data.gov.uk/.

[6]https://data.sheffield.gov.uk/.

[7]https://data.police.uk/.

[8]http://ec.europa.eu/eurostat.

comes in two forms, an open register and a full register. Anybody may pay to access the open register, and some companies do so in order to provide this data to their customers for various reasons such as ancestry search engines. However, the full register is reserved for the operation of elections, supporting policing and also for credit reference agencies. Electoral registers may contain little or anonymised data, but should not be discounted as they may contain a whole range of information such as name, address and date of birth. Throughout the world, the existence and availability of electoral registers and at what level (local or national) they operate varies wildly.

9.2.4 Court Records

A number of countries also maintain publicly accessible court records, which may or may not be held behind pay walls. Within the UK some County Court judgements, High Court Judgements and others can be searched via the TrustOnline[9] database offered by the Registry Trust after the payment of a fee. The Registry Trust also passes information on to credit references agencies, like Experian, who may combine this data with other data they hold in order to calculate a person's credit reference. Similarly, the UK government also makes available information on bankruptcies and insolvency.

9.2.5 Social Media

Major social media platforms such as Facebook and Twitter, whilst increasing security and privacy levels all of the time and for good reason, still provide useful amounts of information and their networks. The popularity of individual social networks varies significantly from country to country and continent to continent. Whilst Facebook and Twitter see dominance in the west, services such as Qzone and Sina Weibo show significant dominance in the Far East. That being said, there are a huge number of active social media networks available with more than 20 global networks claiming to have more than 100 million registered users (Satista 2016).

9.2.6 Blogging Platforms

Similar to social media networks, blogging platforms have been widespread throughout the Web 2.0 era. WordPress is the dominant platform in the English

[9]http://www.trustonline.org.uk/.

speaking world and claims to currently handle around 55 million new posts every month.[10] One of the main distinctions between a social network and a blogging platform, however, is that the latter is not dependent on a network or community. Whilst there are a huge number of 'bloggers' using WordPress, there may be a multitude more using other platforms as well as their own custom built platform. A blogging platform may be as simple as a host HTML page updated and maintained by an individual, group or organisation.

9.2.7 Search Engines

An Internet without search engines would be akin to having a television with billions of channels, where the vast majority contain nothing but white noise. It would be extremely difficult to find anything without some form of advice or guidance. Search engines came along and filled that gap right from the early days of the internet, their main benefit being their ability to crawl efficiently and index huge amounts of the open internet data out there and to then answer queries in fractions of a second.

Today Google dominates with over 70 % of global market share in desktop search, followed not so closely by Bing with 12 % (Net Market Share 2016). Suffice it to say, Google should not be the only port of call for an open source investigator as slightly different results may be obtained depending on which search engine is used. For many search engines, their business is in big data, particularly in advertising. Therefore they will do whatever it takes to profile and in turn provide targeted results to each and every individual on the web—this includes the open source investigator.

Because of this, the use of multiple search engines (Bing, Google, Yahoo, etc.) should be considered in each investigation as well as some of the emerging ones. Examples of search engines that claim not to track users and tailor search results are DuckDuckGo[11] and Startpage.[12]

Search engines also often link to cached versions of webpages, which can be extremely useful if a website has appeared to go offline but access to the content within it is still required (see Chap. 13 for case examples in a policing context). This kind of functionality may also be provided by the internet archive.

[10]https://wordpress.com/activity/.

[11]https://duckduckgo.com/.

[12]https://www.startpage.com/.

9.2.8 Internet Archive

Founded in 1996, a non-profit organisation called Internet Archive set out on its journey to capture and archive the Internet. Provided through a tool called the Wayback Machine,[13] the service and be accessed by users to search for any given website and traverse snapshots captured at various stages of its history. Social media sites and other 'deep web' resources (i.e., those that require a user to log in) are sparsely captured, but public facing websites such as blogs (and comments on blogs) are often captured. Sometimes crucial data can be captured, which is why Internet Archive—and especially the Wayback Machine with its nearly 500 billion captured web pages—should be firmly embedded in any open source investigator's toolbox.

9.2.9 Freedom of Information

Many countries across the world have now adopted some kind of freedom of information (FOI) directive related to the public having access to information held by governments and governmental organisations (cp. Freedom Of Information Laws by Country, nd). Information obtained through FOI request almost sits on the line between OSINT and non-OSINT data. On the one hand, the public can access the data (and for free) if they ask for it; on the other hand, the public must know the right questions to ask in order to obtain that data, and access to the data depends on the restrictions put in place by that nation's government. For example, in the UK obtaining data for an FOI must cost the organisation less than £600 or they have the right to refuse the request. This means that some requests result in the production of open source data and others do not (see Chaps. 17 and 18).

9.3 Non-OSINT Data

Given the content of many of the other chapters within this book and with estimates putting open source data usage at such high-levels (80 % +) we could be forgiven for asking: after that what is left? However, that last percentage of useful, but closed source information (i.e. non-OSINT data) may be what provides us with the final pieces to the jigsaw puzzle that enable us to act on the intelligence we have already determined with more certainty or more accuracy than we had before.

In effect, non-OSINT data is that which has not been made publicly available in any way. This could be due to the data being under a proprietary or restricted license or simply held by companies and businesses, who have no need to make the

[13] http://archive.org/web/.

data available. Data which is classified or confidential can also be classed as non-OSINT data, in that it is not available to the general public by any means such as data contained in police or intelligence reports.

Such data is often of a sensitive or personal nature and is generally covered under data protection laws. That is not to say that this data cannot be exploited by the holding organisations for use in intelligence gathering. On top of this, authoritative organisations such as law enforcement agencies may have the power to obtain such proprietary data (such as bank transactions or telecommunication records) and combine it with their open data counterparts.

Information held by intelligence agencies—especially those in countries other than where the investigation is currently being conducted—usually comes with an especially high degree of secrecy and an unwillingness to share beyond the owning organisation. Thus, for example, in the UK the Security Service (MI5) may have different information to GCHQ, who again may have different information to an LEA, even though they may all be interested in the same person or the same case but for differing reasons and they all may not share this information with each other.

There are also (also see Chaps. 3 and 16) a number of other 'INTs' aside from OSINT, which form part of the intelligence domain:

- *SIGINT* (*signals intelligence*): intelligence obtained through the interception of signals be it communications or other electronic signals
- *HUMINT* (*human intelligence*): intelligence obtained through personal contact such as observation, interviews and other espionage
- *MASINT* (*measures and signature intelligence*): intelligence obtained through the analysis of signatures such as radar and CBRN sources
- *IMINT* (*imagery intelligence*): intelligence obtained through satellite imagery and aerial photography

These INTs are often more referred to within the military and defence spectres. However, they are applicable to law enforcement and other investigations as well, although they may not be spoken about using the same terms.

Generally, non-OSINT data in a law enforcement context simply covers the information and intelligence gathered as part of the existing investigation that has come from non-publicly available sources. In industry, this may be the information that they hold within the company itself related to its activities, finances or employees.

In the rest of this section we give an overview of other non-open source data that may be used in investigations.

9.3.1 Criminal Records

Many, if not all, countries have some form of criminal recording process in place, which may also be electronic and national (or at least regional or state-level). In the

UK, such data is stored in the Police National Computer (PNC), which has been in operation since 1974. As of October 2014, there were over 10 million records on the PNC containing criminal elements (Home Office 2015a, b). Needless to say, access to the PNC is tightly restricted whereby few organisations (mostly law enforcement or government agencies) have access. The PNC is held at the former National Policing Improvement Agency's data centre in Hendon, Middlesex, UK (Wissgott 2014). The PNC holds information such as convictions, vehicle information, driver's licence information, property that has been recovered and information about those who hold firearms certificates.

The PNC is complemented by the Police National Database (PND) which stores operational police information and intelligence obtained from locally held police records. The PND is specifically geared towards enabling the search for information related to people, objects, organisations, locations and events. The purpose of the PND is to enable the sharing of information between different police forces within the UK.

Both the PNC and PND link in with the two EU-wide services: the Schengen Information System (SIS) (European Commission 2015) and the European Criminal Records Information System (ECRIS) (European Commission 2016). The SIS works on the idea of alerts and the sharing of these alerts within EU Member States and the Schengen area. These alerts refer to the locating of wanted people (either because they are missing or due to criminal proceedings) or the locating of missing objects (e.g., vehicles, banknotes, documents and firearms). As of 2013 of the 45 million active alerts more than 80 % of them referred to identity documents. Rather than sharing alerts, ECRIS provides a method for LEAs across Europe to access historical criminal records of people of interest whose transgressions may have occurred outside of the current member state in which they are facing a potential conviction.

9.3.2 Financial Records

The financial sector and everything it encapsulates is tightly embedded in every part of modern day life and almost all adults in the developed world are connected to it in some way or another. Most interestingly with regard to the work of investigators is the extensive and sensitive data trail that every interaction or transaction may leave in its wake. Despite this, the financial sector is a huge and decentralised network of systems and organisations, and is not always directly accessible to LEAs in fully standardised ways.

In the UK, the financial sector is tightly regulated and there has been much progress in the way of working together with governments and LEAs for both the purpose of privacy and security. One tool that has been in use throughout the last decade has been Suspicious Activity Reports (SARs) (UK Financial Intelligence Unit 2015).

SARs not only allow organisations, but also oblige many, to report any suspicious activities with regards to any entities they deal with, whether it be customers, clients or other businesses. During the 2013–14 reporting period, the body responsible for handling the reporting of SARs, the National Crime Agency (NCA), received over 350,000 reports where the greatest proportion of reporters were accountancy firms and tax advisors (National Crime Agency 2014). Such data is considered to be highly sensitive and therefore confidential and rarely, if ever, disclosed. Direct access to SARs is provided mainly to police forces, among some other agencies, and it can often be an instigator to an investigation due to its effective 'alerting' nature (National Crime Agency 2005).

The British Bankers Association (BBA) in collaboration with the Joint Money Laundering Intelligence Taskforce (JMLIT) has more recently been developing a Financial Crime Alerts Service (FCAS), which is due to launch sometime in 2016 (Brown 2016). The JMLIT is a far reaching collaboration consisting of banks and agencies such as the BBA, NCA and Home Office. Such a system will provide advanced detection and alerting capabilities with regards to terrorist financing, money laundering, corruption, fraud and cybercrime (BBA 2014; see also Chap. 16).

9.3.3 Telecommunication Records

Similar to the financial sector, the telecommunications sector inherently and inadvertently facilitates criminal activities by nature of the server. In this context—particularly under UK law—the telecommunications sector encapsulates any service enabling electronic communication such as telephone, email, social media and video conferencing, among others. The UK, as with all other European Union Member States, has fallen under the EU Data Retention Directive 2006/24/EC,[14] which did allow the bulk gathering and storage of telecommunications for between six months and two years. However, in 2014, the directive was declared invalid by the European Court of Justice in the Digital Rights Ireland case[15] due to its failure to comply with the Charter of Fundamental Rights and it ceased to apply from April 2014 (Court of Justice of European Union 2014).

As a result of the invalidation of the EU directive, the UK government enacted the Data Retention and Investigatory Powers Act 2014 (DRIPA) covering the scope of how telecommunications data should be retained, in very much the same way as

---

[14]Directive 2006/24/EC of the European Parliament and of the Council of 15 March 2006 on the retention of data generated or processed in connection with the provision of publicly available electronic communications services or of public communications networks and amending Directive 2002/58/EC. http://eur-lex.europa.eu/legal-content/EN/NOT/?uri=CELEX:32006L0024&qid=1463735673764.

[15]Joined cases C-293/12 and C-594/12 *Digital Rights Ireland Ltd. V Minister for Communications, Marine and Natural Resources and Others and Karntner Landesregierung and Others [2014].*

the EU directive. Due to these similarities, DRIPA was declared unlawful by the UK High Court in 2015 and was referred to the European Court of Justice (EJC). A full court ruling on the legality of the legislation is due in 2016. Meanwhile, the Investigatory Powers Bill, which will replace DRIPA, progresses through Parliament and at the time of writing has reached the Committee Stage in the House of Lords. DRIPA will come to an end in any event, in December 2016 (UK Government 2014) making way for the new legislation; but there remains uncertainty for law enforcement agencies about the kinds of telecommunications data they may be able to access and retain in future.

In the meantime, the types of data specified in the 2006 EU Directive and reiterated in DRIPA mainly cover source and target entities in any given communication. Under these rules it must not contain "data revealing the content of the communication".[16] For example, a website address (or URL) will reveal the content, therefore it is unlikely to be allowed, whilst email, SMS, video conferencing content, would not be stored. The possibility of accessing source, target, time and duration data, however, can still be exceptionally powerful and revealing and would help the investigator to identify key parties in an investigation.

9.3.4 *Medical Records*

The use of medical records can be extremely sensitive and revealing, a reason why the digitisation of them has been slow around the world. It is also a topic that is unlikely to be standardised across the globe. In the US in 2001, roughly 72 % of office-based physicians had no electronic health record (EHR) system, falling to 22 % in 2013 (Hsiao 2014). Although in the UK, there may still be various approaches by different doctors' offices with regards to the entirety of a patient's medical history, every UK patient will have what is called a Summary Care Record (SCR).

Each SCR contains an overview of a given patient's care requirements. It covers, among other information: conditions, allergies and medications (NHS 2016). It is also clear that, where serious crimes are being investigated, a court order may grant law enforcement agencies access to these records (NHS 2011). Medical records are classed as 'sensitive personal data' in all data protection laws and are subject to stricter conditions in relation to accessing and processing than ordinary personal data. In the UK the Data Protection Act 1998 allows LEAs to access these records only providing certain conditions are met.[16] Alternatively, the Police and Criminal Evidence Act 1984[17] provides a procedure for officers to obtain an order to access this type of material, again only if specific conditions are met. Such data could reveal the potential movements or actions of the subjects of an investigation, for example, where they may require treatment for specific health conditions.

---

[16]Schedule 3 to the Data Protection Act 1998.

[17]Schedule 1.

For a while in the UK there has been a programme of work to centralise UK medical records, though it is unclear whether this is only in aggregated and anonymised form or in raw form. The programme, known as Care.data, has been through much criticism due to the unclear regulation on to whom and how the data will be shared (Vallance 2014). Since 2015, there has been little mention of progress with the programme, leading to uncertainty about its longevity.

9.3.5 Imagery, Sensors and Video Data

A significant amount of satellite imagery is now available commercially and thus cannot be considered non-OSINT (see also Chap. 12). However, some satellites exist purely for reconnaissance and are operated by individual nations, who may not expose the information collected using such data. Nowadays, many police forces may operate or have access to a dedicated helicopter, which can perform thermal imaging as it flies and can transmit this video directly back to command and control. In the future, forces may be able to deploy drones with similar capabilities at a significantly reduced cost, and so the extent of imagery owned by the police may increase.

Environmental sensors exist all around us. The proliferation of Internet of Things (IoT) devices will dramatically increase the data available about the status of our cities in the near future. They may track air quality, detect seismic activity, monitor river levels, check for radiation levels and report on the structural integrity of buildings, amongst others. However, the data from these sensors is often owned by their providers and not made available openly. However, if these companies begin to either open this data up, so it can be fused with other open or closed data, the potential for its use become even greater.

Law enforcement agencies have access to another powerful closed source data set. They can access many of the CCTV cameras and the associated footage from a specific area within a specific time period. These recordings, although varying in quality, can be used both to enhance situational awareness in real-time by obtaining direct feeds as well as looking back through previous footage to augment intelligence already gathered through other closed and open sources. For example, footage from a scene of a crime or of a suspect as they cross the city before or after they commit such a crime.

9.4 Fusion Opportunities

In the above we looked at some of the different open and closed source that may be available to an investigator depending on who and what they are working on. In this section we consider how the combination of open and closed source data together might provide superior intelligence compared to simply using one or the other.

9.4.1 Targeted Search

An investigation usually starts for a reason: a crime has or may have been committed, is about to be committed, an unexpected drop or increase in profit for a business has been observed or intelligence on a particular (private or criminal) target is required. It may be the case that partial information is known about a specific suspect, but not enough information is known to yet act on that information. The use of consented databases such as those provided by LexisNexis, GBG and Experian specialise in being able to narrow down the identification of targets from limited information. As Detective Constable Edwards mentions in the transcript of his talk at EuroIntel '99, (Edwards et al. 1999): "*officers are astonished when they come to us with nothing but a name and we return address lists, family names and addresses, companies and directorships, financial details and associates.*" This is a clear example of how open source information can augment and enrich with very specific details the intelligence that already exists through closed sources.

9.4.2 Validation of Other 'INTs'

One of the most critical uses of OSINT is in the validation of information obtained from other INTs (SIGINT, HUMINT, IMINT, MASINT, etc.) (Gibson 2004). Rather than the traditional idea of fusion being to bring two pieces of data or information together to make them greater than the sum of their parts, this fusion instead protects classified sources and informants. As explained by Gibson (2004) when discussing the advantages of exploiting Open Source Intelligence as a whole, OSINT can contextualise information obtained from clandestine sources, provide cover for information obtained through other INTs—that is, although information may initially obtained from closed sources, if the same information can also be obtained through open sources then this information can be shared more freely and the closed source can remain protected; OSINT also provides 'horizon scanning' for the closed source intelligence, i.e., it can give direction to what specific information should be elucidated from other, more costly, intelligence sources.

9.4.3 Filling in the Missing Links

Owing to the extensiveness of the information that is now shared online, a reasonable intelligence picture can often be created using only open sources. However, there will always be the last 5–10 % that cannot be accessed openly, and intelligence agencies and investigators must revert to covert sources in order to 'fill in the gaps'. Many of the analysis techniques discussed within the previous chapters can be applied to the fusion of OSINT and non-OSINT data, since the actual method of fusion, especially when dealing with larger datasets, may be the same whether the investigator is dealing with open or closed data. The main discrepancies are related to how, once fused, that data is able to be shared. That is, once open source data is fused with classified data it itself becomes classified and no-longer open source and shareable with other agencies.

9.4.3.1 Identity Matching

In the online world, and despite sites' terms of services,[18] people do not always identify themselves by their given first name and surname. Instead, they use other nicknames and aliases, which may or may not have a relation to their actual name. Furthermore, it is not guaranteed that a user will use the same alias across all the social media sites that they sign up to. This creates problems for investigators but also provides an opportunity for fusion.

First, investigators must try to match a person's real name to any offline aliases they use; it is not uncommon for someone to be in the records under one name, but still go by another name, especially if there have been complexities in a person's family situation including marriage, divorce, where a child switches which parent's surname they use, whether or not this is recorded officially or they simply go by their middle name or a non-obvious shortened nickname. Secondly, an investigator may try to match or link social media profiles together. This can sometimes be straightforward; for example, a user may use the same aliases or username across all social media profiles. This is becoming more common as users want to build their personal brand (Senft 2013), although not all users are necessarily aware that this comes at the expense of their privacy.

Another method investigators can use is to follow the links between social profiles. For example, people may link their Facebook and Instagram profiles given that they are both owned by the same company. In fact you can even log into your Instagram account using your Facebook credentials and so the two become inextricably linked. Otherwise a user may share posts from Facebook or Instagram on their Twitter timeline and thus provide a hook for investigators to discover links between the two. Finally, an investigator may be able to leverage the connections of the 'friends' that the user has made online in order to discover these associations. These links and connections can be particularly important if the OSINT material is to be used in evidence during the course of subsequent criminal proceedings (see Chap. 18).

The final problem brings the two themes discussed above together. That is, the fusion of offline and online identities to create one single identity. Ascertaining which people link to which social media profiles can then become a very powerful tool in leveraging more information about that person, and further fusion

[18]https://en-gb.facebook.com/help/112146705538576.

opportunities may present themselves. These opportunities may include matching photos, location tagged updates or check-ins posted on their timelines to places of interest within the investigation, identifying confirming or disproving alibis or supposed whereabouts (although some criminals are now wise to LEAs' use of social media and are attempting to subvert such investigations; Bowcott 2015) or using them as basis for the next fusion opportunity we will discuss: improved social network identification.

In the US, some LEAs already take advantage of such identity matching to provide enhanced situational awareness to officers attending an incident. Beware by West[19] provides a threat assessment to law enforcement based on the fusion of both open and closed source information such as existing criminal records, property, information in consented databases as well as data from social media profiles. The company highlights how the fusion of this information is able to give advanced warning to the potential threat posed at the scene of an incident (Jouvenal 2016).

9.4.3.2 Enhanced Social Network Creation

Social networks, and even the use of network analysis in general, is an active area for fusion between OSINT and non-OSINT data sources. The creation of a network can be efficiently complemented by the use of information obtained from both open and closed sources. Nodes can represent many different types of information in the network including people, locations, phone numbers and incidents, while edges can represent friendships, communication channels, transactions and many more. Within the network each of these entities can come from either an open source or a closed source and the network representation facilitates the understanding of how information from different sources is able to complement one another. Additionally, visualisation of the network in this way may enable the investigator to see where there are overlaps within their network and thus contribute to the identity matching discussed above.

A person's or a group's online and offline networks may have significant overlaps. A person may be a friend or a follower of someone on multiple social networks as well as them being part of their offline network. Police investigation may also identify patterns in telecommunications or other transactions. Overlaying these networks on top of each other and identifying the relationships that are present in the different networks, of which some may be the same and some may be different, may give the investigator further understanding of where there are gaps in his intelligence or leads that may need to be followed up more closely. If a person appears prominently in an offline network, but then does not appear in the investigator's picture of the online network, then this may indicate that either that person has not been resolved to a specific profile or that they are purposely keeping a lower profile and not exposing themselves on social media. Any of these conclusions that

[19]https://www.west.com/safety-services/public-safety/powerdata/beware/.

can be drawn by the investigator enhances their own intelligence picture and potentially progresses the investigation.

i2's Analyst's Notebook[20] is a commercial software solution that supports this kind of analysis allowing the investigator to construct the network themselves using the information they have available to them as well as being able to link it into their existing systems. It specifically supports network analysis, and these networks can be projected on to maps or timelines as well as supporting a freeform layout. Openly available software that provides similar functionality includes NodeXL, Gephi, and ORA-LITE (Carley 2014; which has now developed a commercial version called Netanomics[21]).

9.4.4 Environmental Scanning

In intelligence-led policing, LEAs are not reacting to one current event or a simple crime: it is about building an intelligence picture, which can give an overview of the situational awareness within a particular topic. This topic may be one of the local police's current priorities. They may be interested in organised crime (Chap. 16), child-sexual exploitation (Chap. 15), counter-terrorism (Chap. 2), radicalisation and drug trafficking as well as tracking, perhaps more minor incidents such as anti-social behaviour or community issues.

Each of these areas (and many more) has the potential to provide a rich dataset containing information derived from classified crime reports held by the police and have them augmented by data that is available openly such as that found in local news reports, social media and other sources. These may also be enhanced by the use of statistics derived from local government data or even national and international data.

The ePOOLICE[22] system (Brewster et al. 2014) was developed as a method to obtain strategic early warnings for law enforcement using environmental scanning for the detection of organised crime threats through the identification of weak signals. These weak signals may appear in social media or in news media across the world. The detection of multiple similar weak signals within the same or similar areas would serve as an early warning to LEAs of the potential for organised crime activity to be taking place. The CAPER project[23] was formed on a similar idea but also aimed to integrate private or classified data into their information streams (Aliprandi et al. 2014).

It is not hard to see how using the many data analysis techniques put forward in the previous chapters could be applied to both open and closed source intelligence,

[20]http://www-03.ibm.com/software/products/en/analysts-notebook.

[21]http://netanomics.com/.

[22]www.epoolice.eu.

[23]http://www.fp7-caper.eu/.

if the investigator has the authority to access such systems. Natural language processing can be used to extract entities from any type of data, the aggregation and analysis techniques can be applied no matter where the data comes from.

Consequently, an environmental scanning application can take the information first found in open sources and augment it with the information already held. This may be done automatically or only when a trigger point of a certain number of open sources reporting similar information reaches a specified threshold. Thus what may appear only to be a few random incidents in the news and some logs in the crime reporting system suddenly starts to indicate that a pattern may be developing, which may in turn lead to a fully-fledged investigation.

9.4.5 *Predictive Policing*

In the so-called 'big data revolution', predictive policing is another hot-topic, that, if reports are to be believed, will revolutionise the future of policing and move it into the modern era. Predictive policing focuses on the idea of identifying crime 'hot-spots'. That is, using the data that we already have to 'predict' where and when certain types of crime may occur, so that scant police resources can be deployed most efficiently and effectively.

Predictive policing is another prime area for the integration of open and closed source data. Take, for example, a simple problem such as a spate of minor burglaries. The people who are burgled may report this to the police (although they may not do much about it; BBC News 2015) and the crime will be logged into the system. This becomes closed source data. Some of the victims of the burglaries may also report their experience on social media to warn others in the area or simply to highlight their plight. This information now exists in both open and closed source formats. In addition, a number of people within the same neighbourhood may notice some sort of damage to their property (potential evidence of a break in) and post this to social media or even do so by replying to the original post of the person who was burgled. This information only exists in an open format. If the investigator is able to bring all this information together, they may only then realise the patterns surrounding the burglaries in the area and make efforts to deploy a higher police presence to that area in order to deter the burglars or even catch them.

Predictive policing is already seeing a large amount of traction within law enforcement, through the use of tools such as PredPol[24] or HunchLab's[25] software, and within academic literature. Both PredPol and HunchLab appear to only link up with existing data held by law enforcement, whereas academic literature focuses

[24]http://www.predpol.com/.

[25]https://www.hunchlab.com/.

more on what open data can do for predictive policing (of course some open data is derived from existing police data). For example, Gerber (2014) compared using data from Chicago's open data portal alone (which can be considered as a proxy for closed source police data) versus using the same data combined with all tweets obtained through the Twitter streaming API within the Chicago area. He found that by augmenting the open data with the Twitter data improved the crime prediction model in 19 of the 25 crime types he was testing. Thus the potential advantages of fusing OSINT and non-OSINT data in predictive policing are clear.

9.4.6 Situational Awareness During Major Events

While perhaps not being an open source investigation per se, another area where law enforcement are particularly interested in leveraging the information posted on social media and augmenting it with their own operational information is in their response to major events. These events may be terrorist attacks, riots and mass public disorder, major incidents such as air, road or rail crashes or as a result of a natural disaster such as an earthquake or flooding.

In this case closed data may be information collected by investigators and other first responders at the scene, it may be gathered by accessing CCTV footage, from satellite imagery (especially in the case of a major natural disaster), they may have access to additional sensor information (such as detailed river levels in the case of (potential) flooding), and they may also be able to perform their own closed source fusion and bring this data together with existing closed intelligence they may have access to. However, a major incident often generates significantly more news and social media traffic and, although much of this will be second-hand commentary, some of this will be generated by people who are actually witnesses to the incident, but may have not contacted the police formally with their observations. Furthermore, this information may be available in near real-time to those who have the power to investigate it. Thus efficient and effective methods to deal with enormous quantities of social media data through analysis and aggregation techniques and then be able to link this into existing closed source data has massive potential to improve the emergency response to crisis situations.

Such solutions are already being investigated. The Athena[26] project is investigating how open source data such as that from social media can be combined with closed-data: information provided via a specific Athena mobile application can be utilised to improve crisis communication between citizens and law enforcement (Gibson et al. 2015). LEAs then have the power to make this information open again by publishing (non-operationally sensitive) information back out to the public

[26] http://www.projectathena.eu/.

using the app and social media. A credible extension to this project would be to begin linking it into other law enforcement analysis and situational awareness systems. A number of commercially available systems also exist in this area such as Bluejay,[27] Signa,[28] Liferaft[29] and Go360.[30] However, there still appears to be a gap for a system that effectively integrates both open and closed source data.

9.4.7 Identification and Tracking of Foreign Fighters

The rise to prominence of DAESH[31] has seen an increase in people from the West heading to Iraq and Syria to go to fight or be the wives of fighters for IS. A significant proportion of these fighters as well as IS supporters (Bergar and Morgan 2015) have been tracked through their social media footprint (BBC 2014). A fusion opportunity exists for law enforcement using this data as they can match the fighters identified on social media to any potential persons of interest on existing law enforcement or intelligence databases. Furthermore, they can go back to the social media accounts and begin to create networks of who these fighters communicate with in order to assess others who may become a victim of radicalisation and potentially initiate interventions to prevent further recruitment.

9.4.8 Child Sexual Exploitation

Child sexual exploitation and child abuse images provide another opportunity for fusion between open source and closed source intelligence. For example, LEAs may obtain a database of images that have been shared openly on the surface and also the deep/dark web (see Chap. 8). When images are recovered from a suspect's computer or mobile phone these images can be matched against those already in the database to identify which are images that are already known to police and which are new images. Sophisticated facial recognition techniques may also be applied to data from both open and closed sources to enable police to recognise specific victims or abusers. The UK Home Office has already launched the Child Abuse Image Database (CAID) (Home Office 2015) which provides some of this functionality. Chapter 15 provides further discussion on the best practices and use cases in this area.

---

[27]http://brightplanet.com/bluejay/.

[28]http://www.getsignal.info/.

[29]http://liferaftinc.com/.

[30]http://www.go360.info/.

[31]Sometimes called the Islamic State (IS) or Islamic State of Syria and the Levant (ISIL).

9.5 Conclusions

This chapter has expanded on a number of different data sources for open and closed source intelligence. It has then identified numerous opportunities for the fusion of open source and closed intelligence data, mainly in a law enforcement capacity. The extent to which these potential synergies have already been adopted by LEAs and other companies as well as future possibilities for fusion that could be considered have also been explored.

References

Aliprandi C, Arraiza Irujo J, Cuadros M, et al (2014) CAPER: Collaborative information, acquisition, processing, exploitation and reporting for the prevention of organised crime. In: Stephanidis C (ed) HCI international 2014—posters' extended abstracts, Vol 434, pp 147–152

BBA (2014) Banks team up with government to combat cyber criminals and fraudsters. In: British Banking Association. https://www.bba.org.uk/news/press-releases/banks-team-up-with-government-to-combat-cyber-criminals-and-fraudsters/

BBC (2014) Syria: report shows how foreign fighters use social media. BBC News. http://www.bbc.co.uk/news/world-27023952

BBC News (2015) Sara thornton: police may no longer attend burglaries. BBC News. http://www.bbc.co.uk/news/uk-33676308

Berger J, Morgan J (2015) The ISIS twitter census: defining and describing the population of ISIS supporters on twitter. http://www.brookings.edu/research/papers/2015/03/isis-twitter-census-berger-morgan

Bowcott O (2015) Rapists use social media to cover their tracks, police warned. In: The guardian. http://www.theguardian.com/society/2015/jan/28/rapists-social-media-cover-tracks-police-cps-rape

Brewster B, Andrews S, Polovina S et al (2014) Environmental scanning and knowledge representation for the detection of organised crime threats. In: Hernandez N, Jäschke R, Croitoru M (eds) Graph-based representation and reasoning, vol 8577. Springer, Berlin/Heidelberg, pp 275–280

Brown A (2016) Chief executive's newsletter—March 2016. In: British Banking Association. https://www.bba.org.uk/news/insight/chief-executives-newsletter-march-2016

Carley KM (2014) ORA: a toolkit for dynamic network analysis and visualization. In: Alhajj R, Rokne J (eds) Encyclopedia of social network analysis and mining. Springer, New York, pp 1219–1228

Court of Justice of the European Union (2014) The court of justice declares the data retention directive to be invalid. http://curia.europa.eu/jcms/upload/docs/application/pdf/2014-04/cp140054en.pdf

Edwards S, Constable D, Scotland N (1999) SO11 open source unit presentation. In: EuroIntel '99 proceedings of the european intelligence forum "creating a virtual previous next intelligence community in the european region, pp 1–33

European Commission (2015) Schengen information system. http://ec.europa.eu/dgs/home-affairs/what-we-do/policies/borders-and-visas/schengen-information-system/index_en.htm

European Commission (2016) ECRIS (european criminal records information system). http://ec.europa.eu/justice/criminal/european-e-justice/ecris/index_en.htm

Freedom of Information Laws by Country. In: Wikipedia. https://en.wikipedia.org/wiki/Freedom_of_information_laws_by_country. Accessed 26 May 2016

Gerber MS (2014) Predicting crime using Twitter and kernel density estimation. Decis Support Syst 61:115–125

Gibson S (2004) Open source intelligence. The RUSI Journal 149:16–22

Gibson H, Akhgar B, Domdouzis K (2015) Using social media for crisis response: The ATHENA system. In: proceedings of the 2nd european conference on social media 2015, ECSM 2015

Home Office (2015a) Nominal criminal records on the police national computer. http://www.gov.uk/government/publications/nominal-criminal-records-on-the-police-national-computer/nominal-criminal-records-on-the-police-national-computer

Home Office (2015b) Child abuse image database. https://www.gov.uk/government/publications/child-abuse-image-database

Hribar G, Podbregar I, Ivanuša T (2014) OSINT: a "grey zone"? Int J Intell Counter Intell 27:529–549

Hsiao C-J (2014) Use and characteristics of electronic health record systems among office-based physician practices: United States, 2001–2013. NCHS Data Brief No. 143

Johnson L (2009) Sketches for a theory of strategic intelligence. In: Gill P, Marrin S, Phythian M (eds) Intelligence theory: key questions and debates, pp 33–53

Jouvenal J (2016) The new way police are surveilling you: calculating your threat "score." The Washington Post

National Crime Agency (2005) Money laundering: the confidentiality and sensitivity of suspicious activity reports (SARs) and the identity of those who make them. http://www.nationalcrimeagency.gov.uk/publications/suspicious-activity-reports-sars/17-home-office-circular-53-2005-confidentiality-and-sensitivity-of-sars/file

National Crime Agency (2014) Suspicious activity reports (SARs) annual report 2014. http://www.nationalcrimeagency.gov.uk/publications/suspicious-activity-reports-sars/464-2014-sars-annual-report/file

National Open Source Enterprise (2006) Intellligence community directive number 301. Office of the Director of National Intelligence. https://fas.org/irp/dni/icd/icd-301.pdf

Net Market Share (2016) Desktop search engine market share. http://www.netmarketshare.com/. http://www.netmarketshare.com/search-engine-market-share.aspx?qprid=4&qpcustomd=0

NHS (2011) The care record guarantee. Health and Social Care Information Centre. http://systems.hscic.gov.uk/scr/library/crg.pdf

NHS (2016) Your health and care records. In: NHS Choices. http://www.nhs.uk/NHSEngland/thenhs/records/healthrecords/Pages/overview.aspx

Ordannce Survey (2016) OS open data. https://www.ordnancesurvey.co.uk/opendatadownload/products.html

Sampson F, Kinnear F (2010) Plotting crimes: too true to be good? The rationale and risks behind crime mapping in the UK. Policing A J Policy Pract 4(1):15–27

Senft TM (2013) Microcelebrity and the branded self. In: Hartley J, Burgess J, Bruns A (eds) A companion to new media dynamics. Wiley-Blackwell, Oxford, UK, pp 346–354

Statista (2016) Leading social networks worldwide as of April 2016, ranked by number of active users (in millions). http://www.statista.com/statistics/272014/global-social-networks-ranked-by-number-of-users/

UK Government (2014) Data retention and investigatory powers act 2014. http://www.legislation.gov.uk/ukpga/2014/27/section/8

UK Financial Intelligence Unit (2015) Introduction to suspicious activity reports (SARs). National Crime Agency. http://www.nationalcrimeagency.gov.uk/publications/suspicious-activity-reports-sars/550-introduction-to-suspicious-activity-reports-sars-1/file

Vallance C (2014) NHS Care.data information scheme "mishandled." BBC News. http://www.bbc.co.uk/news/health-27069553

Wissgott K (2014) Karl Wissgott, NIPA—witness statement to the leverson inquiry. UK Government Web Archive. http://webarchive.nationalarchives.gov.uk/20140122145147/ http://www.levesoninquiry.org.uk/wp-content/uploads/2012/04/Witness-Statement-of-Karl-Wissgott-taken-as-read.pdf

Chapter 10
Tools for OSINT-Based Investigations

Quentin Revell, Tom Smith and Robert Stacey

Abstract This chapter looks at the essential applications, websites and services used by practitioners to form their OSINT toolkit, which may range from simple browser plug-ins to online services, reference databases and installed applications and every practitioner will have their list of favourites. The chapter does not recommend a particular piece of software or service or give detailed advice for specific system set-ups, rather it aims to equip the reader and their organisation with a framework to assess the tools at their disposal to give them some reassurance that they are suitable for their OSINT investigation and can demonstrate that they are secure, reliable, and legal.

10.1 Introduction

In this book, we have a number of examples of best practice and strategies for use in an Open Source Investigation (see particularly Part 3 in this book). In this chapter we look at the essential applications, websites and services used by the practitioner to form their toolkit. The tools used range from simple browser plug-ins to online services, reference databases and installed applications and every practitioner will have their list of favourites.

While there is an understandable desire for a definitive list of 'approved' tools (and there are numerous lists of websites, services and databases available on the internet[1]), generating and maintaining a list of tools is a never-ending task for a website and if printed in a book would quickly become outdated as websites vanish, and new applications appear.

[1]OSINT Training by Michael Bazzell': https://inteltechniques.com/links.html, accessed April 2016; www.uk-osint.net, accessed April 2016. http://www.uk-osint.net/; 'OSINTINSIGHT': http://www.osintinsight.com/index.php, accessed April 2016; Toddington Resources: https://www.toddington.com/resources/, accessed April 2016.

Q. Revell (✉) · T. Smith · R. Stacey
Centre for Applied Science and Technology, Home Office, St Albans, UK
e-mail: Quentin.Revell@homeoffice.gsi.gov.uk

B. Akhgar et al. (eds.), *Open Source Intelligence Investigation*,
Advanced Sciences and Technologies for Security Applications,
DOI 10.1007/978-3-319-47671-1_10

Crucially, because of the variety of sectors in which OSINT is now used (e.g., law enforcement/commercial/defence/security services), each sector has different aims, subjects of investigations, technical infrastructure, risk appetites and legal frameworks. Law Enforcement practitioners may have very different roles (Association of Chief Police Officers 2013).

However, in the Law Enforcement sector practitioners are commonly unable to use their 'corporate' desktops for effective investigations. This could be for a number of reasons; organisations may have blocked social media sites or 'inappropriate material'. Additionally organisations may have secured their browser from active content (e.g. videos and Java Script), or prevented the user installing their own applications. As a result practitioners will commonly run their own 'standalone' system that allows them to access the wide variety of tools available.

Once freed from the corporate desktop, the practitioner has almost limitless applications, browser plug-ins, sites and services available to use. However, this freedom to run anything has its risks: is the browser plug-in malware? Is the anti-virus used licensed for corporate use? Is the mobile app data mining contacts? Where is the cloud service hosted, and what information can it see?

Because of the ever changing nature of these tools, corporate systems are often unable to keep up, and the time and cost of a full 'FIPS or 'Common Criteria' style evaluation may not be justified.

As a result this chapter does not recommend a particular piece of software or service or give detailed advice for specific system set-ups (there is existing CESG guidance in this area covering cloud service security, browser security guidance and device security guidance; see also Chap. 15). Rather this chapter aims to equip the reader and their organisation with a framework to assess the tools at their disposal to give them some reassurance that they are suitable for their OSINT investigation and can demonstrate that they are secure, reliable, and legal.

In the UK this demonstration may be required by a number of institutions—from local management and the Senior Information Risk Owner to the Information Commissioner, the Surveillance Regulator, or the Forensic Science Regulator.

This chapter suggests a framework to help ensure that the risks of using these tools, and in particular 'free tools', have been considered. It should assist internet practitioners when deploying tools or using services; helping to balance operational needs with good practice and organisational risk.

10.1.1 Effective Cyber-Risk Management

In their principles of cyber security risk management, CESG suggests that: 'effective cyber security risk management is built on sensible decision making, following core principles of:

- Accept there will always be uncertainty
- Make everyone part of your delivery team
- Ensure the business understands the risks it is taking

- Trust competent people to make decisions
- Security is part of every technology decision
- User experience should be fantastic—security should be good enough
- Demonstrate why you made the decisions—and no more
- Understand that decisions affect each other.

Ensuring that organisations embody these principles throughout their business processes will help to support the effective management of cyber security risks' (CESG 2015a, b).

This lightweight framework outlined above is designed to support effective risk management by:

- Highlighting the risks from using online service and tools
- Demonstrating why a decision was made
- Assisting competent people to make decisions
- Allowing fantastic user experience, with good enough security.

10.2 Key Assessment Themes

Our assessment has three key themes: *security*, *reliability*, and *legality*. By looking at each theme in turn we can consider the indicators that a tool or service will meet our needs, remembering that not all indicators outlined will be applicable to all types of tool.

10.2.1 Security

While it is obvious that we wish machines to be secure in their operation, sharing information is a necessary part of an online investigation, but should ideally be limited to only that information the user perceives as necessary. Security controls help ensure that user and corporate data are not inadvertently shared with third parties including the service provider, browser vendor or plug-in provider. Some indicators of a tools security risks are summarised in Table 10.1.

10.2.1.1 Privacy

Despite the adage 'If you are not paying for it, you're not the customer; you're the product being sold' (blue_beetle 2010), we have become used to the ability to use tools and services associated with the internet at no cost, and in many instances cases there is no alternative to using 'free' services from the internet. Many of these services rely on advertising for their revenue, and the advertisers are keen to collect

Table 10.1 Criteria for assessing security of OSINT tools

Positive indicators	Negative indicators
• Accreditation of privacy or security standards (e.g., ISO 27001, UK Cyber Essentials, FIPS…) • Protects data in transit (e.g. https) • Protection of user data from host • No evidence of sending data the user perceives as unnecessary • No user registration • The source code is freely available • Offers audit trail	• Anti-Virus/malware response • Known issues or vulnerabilities • Seen sending unnecessary data • Seen sending information to other services • Claims to be selling user data • Registration requires unnecessary identification details • Unnecessary bundled software

information about an individual to better target the adverts. This has led to an infrastructure actively working to identify the individual accessing a website no matter what device they're using, through various cookies and device fingerprinting. This has the potential to link investigations, aliases, and (in combination with poor trade craft), link individual practitioners with the subject of an investigation.

10.2.1.2 Protecting Against Malware

The CESG Browser and Devices (operating system) Security Guidance Documents (n.d.) note the importance of reducing the risk from malicious software and content based attacks: "*The most effective protections will use the native security features of the underlying platform as well as protecting against specific web threats.*" Some browsers and anti-virus software have built-in security features that aim to protect against phishing websites and malicious downloads. They work by sending information about the page being visited to the cloud, and respond with a warning if the link is known to be malicious. Organisations should consider the trade-off between privacy and security when using these services.

10.2.1.3 Unnecessary Bundled Software

Some developers, and file distributors bundle their application with additional 'bloatware' (i.e. additional unwanted software or plug-ins) as part of the installation process. This can have a range of effects. At best these are an annoyance slowing your computer down and reset browser preferences, at worst they can be a front for malware and a security risk for your computer.

10.2.1.4 Cloud-Based Services

Cloud-based services can offer an investigation more than simple look-ups and information. They offer compelling services for an Open Source investigator, such as web crawling, advanced analytics and social media monitoring. The issue comes when that third party is able to access the content of an investigation. As the service gathers more of the information it has a picture of not just an individual investigation, but potentially information across investigations or organisations. There are existing CESG cloud service security principles that should be used to give some level of reassurance when using these services for large scale applications.

10.2.2 Reliability

While initially assessing the tool for its suitability, the practitioner should note the reliability of the tools. The practitioner needs the tool to function reliably and consistently, now and into the future. Support for the user should be available if needed, and the tool should ideally offer an audit trail of user actions. Some of the indicators that should be taken into account when assessing a tool for reliability are summarised in Table 10.2.

10.2.2.1 Code Quality

There is nothing inherent in the development of either proprietary or open source software that makes one more reliable than the other. The quality of any software is likely to be more reliable if many individuals have contributed to its development, it is a well-established product and there is a strict and reliable regime in place to check for errors or problems.

10.2.2.2 Open Formats and Standards

As articulated in Schofield's First Law, 'never put data into a program unless you can see exactly how to get it out' (Schofield 2002). The use of open standards and formats allow the organisation to maintain control of its information. This has a number of benefits: allowing greater interoperability and information sharing,

Table 10.2 Criteria for assessing reliability of OSINT tools

Positive indicators	Negative indicators
• Tool requires payment or offers premium features or support • Supports open standards or API • Allows user data extraction • Community associated with the tool • The source code is freely available • Open vulnerability management	• Limited functionality • The tool repeatedly fails to function • Missing or broken links • No support available

Table 10.3 Criteria for assessing legality of OSINT tools

Positive indicators	Negative indicators
• Permissive terms and conditions • Permissive licensing conditions • Offers robust audit trail of user actions • No additional authorities required to use	• Requires additional legal authorities to use • Use would breach tool's terms and conditions • Use would breach tool's licensing conditions

information can continue to be used as new software is developed and there is no dependency on single vendors, whose tools could be discontinued at any time. Maintaining access to information is critical in a law enforcement environment where data retention time can be measured in decades.

10.2.3 Legality

While assessing the tool for its suitability, the practitioner should also have an awareness of the licensing terms of the application or service. Some of the indicators that should be taken into account when assessing a tool for legality are summarised in Table 10.3.

10.2.3.1 Licensing

While each licence should be assessed on a case-by-case basis; for many 'free' tools the terms of the licence will be dependent on the application in which the tool is used. In a simple situation 'personal' use may be free of charge, and anything else would incur a charge (including use by law enforcement). This can become more complex, for example a licences might state 'free for non-commercial use', but it may not be clear whether the work of an Open Source Unit is classed as non-commercial according to the specific licence terms.' While the work of the police may not fall under the Creative Commons definition that commercial work is 'primarily intended for or directed towards commercial advantage or monetary compensation' (Creative Commons 2014), there are many different licences available. If you are unsure, contact either your legal department, or the rights holder for clarification, or search for an alternative tool that has the required functionality and that permits commercial use.

A common example of this is anti-virus licensing[2] allowing users individuals to become familiar with the service for free, but when used in a professional

[2]http://www.avg.com/gb-en/eula—"Authorized Purposes means with respect to Free Solutions and Beta Solutions, your personal, non-commercial use".

https://www.avast.com/en-gb/eula-avast-free-products—"You are not permitted to use the Software in connection with the provision of any commercial services…".

environment the same tool requires payment. This is highlighted as of particular concern as there needs to be a high level of trust in your anti-virus provider, as they commonly collect personally identifiable information such as the URLs of any visited websites, and details of any software on your machine.

10.2.3.2 Authorities

Open Source Investigations are becoming routine, and could be seen as doing what 'any member of the public can do'. It may be argued that there is 'no expectation of privacy online'; this is not the case, and this debate varies across the different countries of the world according to their individual legal standards and jurisdictions (Sampson 2014).

On this difficult point, in the UK the Surveillance Commissioner's (tentative) view is that "if there is a systematic trawl through recorded data of a particular individual with a view to establishing, for example, a lifestyle, pattern or relationships, it is processing personal data and therefore capable of being directed surveillance if it contributes to a planned operation or focused research for which the information was not originally obtained" (Parliament 2011). The commissioner has also stated that "the ease with which an activity meets the legislative threshold demands improved supervision" (Surveillance Commissioner 2012)

This is analogous to the existing public space CCTV guidance, where use of a CCTV system to establish the events leading to an incident is not usually directed surveillance, but focused monitoring of an individual or group of individuals is capable of being directed surveillance and authorisation may be considered appropriate.[3]

As part of a 'tool review' completion of the legal authorities section of the review requires an understanding of the purpose of the tool. While most sites and services will be 'dual use' and could be used to collect private information if that were the intent of the practitioner, there are services and data sources whose sole purpose is to provide personal data, or make 'hacked' data publically available.

10.3 Completing a Tool Review

In the previous section we have identified a number of positive and negative considerations, most of which can be assessed by researching the site, tool or service under scrutiny. In this section we look at how to approach and record these considerations in a review.

The review could be completed by a group, coming to a consensus view on the tool for use in their unit. The framework should always record its findings, so that

[3]For wider challenges in a law enforcement environment see Sampson (2015).

the business understands the risk it is taking. There should be a record even if the review has found an issue that means the organisation is not willing to use a particular application.

But should we try to address all concerns for all reviews? Here we can return to our risk management principles:

- Accept there will always be uncertainty
- Make everyone part of your delivery team
- Ensure the business understands the risks it is taking
- Trust competent people to make decisions
- Demonstrate why you made the decisions—and no more.

So a key point to the assessment is to stop the assessment when you feel able to make an informed decision and accept there will always be uncertainty.

We can therefore approach the review by starting with the area that will bring the most knowledge, for the least amount of effort. Begin with the claims made by the provider. If these claims are not sufficient to make a decision, we can look at information in the public domain about the tool. If that third party information is still not sufficient then the organisation or practitioner can undertake to assess the risk for themselves. It should be noted that the final stages of this assessment are technical as it begins to stray into the professional field of Security Research.

We will need a 'Document Information' section to summarise the findings, and information about the review. By applying a risk/effort based principle we can arrange our assessment in the following order:

1. *Document Information*: findings and information to aid traceability
2. *Supplier Assessment*: claims made by the supplier
3. *External Assessments*: other people's experience of the tool
4. *Practitioner's Assessment*: practitioner's assessment of the tool, it looks at issues that may not have been captured elsewhere and requires the reviewer to use the tool.

10.4 Assessment Framework

Putting the previous sections together, we have built an assessment framework listing issues for consideration and detailing why each consideration is included in the overall process.

10.4.1 Document Information

This section is to aid the traceability of reviews as they are completed and to detail activities and conclusions carried out.

Tool name and logo	Name of the tool under assessment and copy of logo to aid identification
Functionality required	This gives context and to informs the rest of the review and gives clarity about the intended use for the tool to help answer subsequent questions and helps others understand the scope of your review
Conclusions	The result of your investigations, any highlights of your research that has brought you to a conclusion and a clear indication as to the suitability of the tool or service for use in the author's context While it may seem counter-intuitive to have the conclusion at the beginning of the document, this is to aid a future busy reader. If they just want a 'quick answer', it is on the front page if they also require the supporting information they can read on.
Review date and tool version number	Version numbers can be used subsequently to check that a review is still valid for the tool. Services used by the tool may change over time affecting the validity of the review. The review date might also be the next best thing if a tool version number is not available.
Author	Who are the author(s) of the review?
Approver	Who has authorised the use of the tool on the system?
Equipment used	Tools may operate in different ways on different equipment A reference to the equipment used may help others determine the relevance of the review to their circumstances. The following items are of interest: • Operating system • Web browser • Anti-virus/malware software • Local policies e.g. clearing of cookies/cache data • Method of connection to the internet (e.g., firewall, proxy)

10.4.2 Supplier Assessment

This section looks at claims made by the supplier and use of their tool. This information should be easily available on the tools website or from a simple search.

Registration details	Registration implies that you are providing identification details which may have privacy implications and also legal implications if a false persona is used.
Vendor's claims to privacy	Some tools come with explicit statements that they distribute the data you provide. Others may claim to protect your data and you'll need to assess the evidence for this such as adherence to standards (e.g. ISO 27001). If vendors are claiming these external assessments are they externally challenged accreditations? (e.g. Cyber Essentials Plus).
Mention of third party interests	Third party interests may indicate (or explicitly confirm) data sharing. This information may identify privacy concerns.
Licence conditions of concern	There may be many reasons why a licence condition may be of concern and this will be dependent upon your particular circumstances. Examples might include 'free for non-commercial use', 'personal use only', and 'evaluation use only'.
Direct support	The support given to a 'free' tool is often limited, and can be an area in which the vendor can make a return. This can limit the amount of documentation available to the practitioner. If there is no cost for support does the business rely on selling user data to third parties, generating a privacy issue?.
Link to terms and conditions, and any conditions of concern	Whilst desirable (for traceability), it is often impractical to take a complete copy of the conditions of use. A reasonable compromise is to record a link to the conditions and quote any particular conditions of note There are websites that try to explain these terms of use.[a]

[a]https://explore.usableprivacy.org/

10.4.3 External Assessments

This section collates other people's experience of the tool; do these verify the claims made by the vendor? Again as this is intended to be relatively light-weight assessment, before the author(s) begin their own assessment, they should look for external validation.

Ownership, hosting and popularity	Note things such as where the tool is hosted, who owns it and when it was launched. This type of information might prove useful for assessing the overall risk of using the tool. Metrics such as popularity might also be of interest. This is an indicator of reliability and privacy. The sites referenced in the footer may assist with gathering this information for websites.[a]
Known issues/vulnerabilities	Searching the internet for known issues. It is often possible to identify good reasons not to use a tool without ever exposing yourself to it by installing it or using it. Known issues will mainly relate to the privacy and security indicators.
Tool maintenance	Indicators might include: • Frequency of updates • Response to issue logs • All links in web pages work This links directly to the reliability of a tool. Is the tool a 'work-in progress', with frequent major changes, or is it stable, and not needing major updates or has it been abandoned?
Support community	An active online support community can lead to a better maintained product and assist in identifying security/privacy issues.
Support for open standards	The intention of this question is to help prevent the user community from becoming locked into a given product/vendor. For instance a product might produce files that are only readable by the product itself. This could become a reliability problem if the product can no longer be run due to a service ceasing, cost implications, legal changes or for some other reason.
Source code availability	This might reduce the risk of both privacy and reliability issues or that the tool does something malicious since it is theoretically open to peer-review. However, there is no guarantee that anyone has looked at the code for issues of interest to you.

[a]http://www.who.is/; http://www.alexa.com/

10.4.4 Practitioner's Assessment

This final section is the practitioner's assessment of the tool; it looks at issues that may not have been captured by external users, outside the law enforcement environment. This section requires the reviewer to use the tool, and some of these questions—particularly the assessment of the communication channel and data content—are complex requiring technical expertise.

The technical aspects towards the end of the practitioner assessment borders on a vulnerability assessment which can be a full time profession. While it may give an indication of the potential issues with a tool, even this level of assessment would not identify if a tool or service is maliciously trying to obscure its communication. If the users have this level of suspicion of the tool, then is should not be used.

Reliability	Describe actions carried out to assess a product's reliability, repeatability etc. and if any issues have been found.
Authorities required	This will depend upon what the tool does, how you use it and your operational procedures/regulations.
Technical permission	Particularly related to mobile devices, and whether it requires access to user data to function.
Anti-virus response	When installing or running the tool, anti-virus software might identify activity that could be associated with viruses or other malicious software. This would impact the security and reliability assessment.
Unnecessary software bundles	'Free' software is often bundled with other software packages that are unnecessary for your needs. They often perform user tracking for targeted advertising which creates a revenue stream for the development of the 'free' software. In some cases the additional software might be an opportunity to deliver viruses to your computer. Both tracking and viruses pose a risk to privacy. A carefully chosen download source can often help to avoid these issues.
Communications with external servers	One way to know whether your data or identity is being exposed to third parties is to monitor where the information is being sent. This will impact on your privacy assessment: • Do you recognise the address receiving the data? • Do you trust the owners of the address receiving the data?
Data content (sent and received)	A more rigorous assessment method is to monitor the content of the data being sent and received. In this way it is possible to see exactly what is being sent to third parties, and if it is of concern to the practitioner.

10.5 Conclusion

Professional OSINT analysis toolkits need to be agile to take advantage of the array of the ever changing services, applications and plug-ins available. However these tools present an information risk to the organisation. The best people in an organisation to answer the question 'can I use this tool?' will be the practitioners themselves. The framework in this chapter guides their efforts. For some the questions may seem as too challenging, and may never complete all of the questions, for others it is not robust enough and they may wish expand the framework. This framework of questions supports good management by highlighting the risks and demonstrating why a decision was made. The result ensures the organisation is aware of its risk from the tools used in association with an investigation. A well completed assessment is of use beyond the unit and can be shared within a community of interest. It would also demonstrate the continued professional development for an individual that they understood their tools and methods. This framework is presented as a starting point for reducing the organisational uncertainty around using 'online tools'.

References

Association of Chief Police Officers (2013). Online research and investigation [Online]. Available at: http://library.college.police.uk/docs/appref/online-research-and-investigation-guidance.pdf. Accessed May 2016

blue_beetle (2010). User-driven discontent [Online]. Available at: http://www.metafilter.com/95152/Userdriven-discontent#3256046. Accessed April 2016

CESG (2015a) Principles of effective cyber security risk management [Online]. Available at: https://www.gov.uk/government/publications/principles-of-effective-cyber-security-risk-management/principles-of-effective-cyber-security-risk-management. Accessed April 2016

CESG (2015b) Principles of effective cyber security risk management—GOV.UK [Online]. Available at: https://www.gov.uk/government/publications/principles-of-effective-cyber-security-risk-management/principles-of-effective-cyber-security-risk-management. Accessed September 2015

CESG (n.d.) End user devices security and configuration guidance—GOV.UK [Online]. Available at: https://www.gov.uk/government/collections/end-user-devices-security-guidance. Accessed September 2015

Creative Commons (2014) NonCommercial interpretation [Online]. Available at: https://wiki.creativecommons.org/wiki/NonCommercial_interpretation. Accessed September

Parliament (2011) Public bill committee—protection of freedoms bill [Online]. Available at: http://www.publications.parliament.uk/pa/cm201011/cmpublic/protection/memo/pf50.htm. Accessed April 2016

Sampson F (2014) Cyberspace: the new frontier for policing? In: Akhgar B et al (eds) Cyber crime and cyber terrorism investigator's handbook. Elsevier, Oxford

Sampson F (2015) The legal challenges of big data in law enforcement. In: Akhgar B et al (eds) Application of big data for national security. Elsevier, Oxford, pp 229–237

Schofield J (2002) Beware proprietary data formats [Online]. Available at: http://www.computerweekly.com/opinion/Beware-proprietary-data-formats. Accessed June 2016

Surveillance Commissioner (2012) Annual report of the surveillance commissioner 2011–12. Office of the Surveillance Commissioner, London

Chapter 11
Fluidity and Rigour: Addressing the Design Considerations for OSINT Tools and Processes

B.L. William Wong

Abstract In comparison with intelligence analysis, OSINT requires different methods of identifying, extracting and analyzing the data. Analysts must have the tools that enable them to flexibly, tentatively and creatively generate anchors to start a line of inquiry, develop and test their ideas, and to fluidly transition between methods and thinking and reasoning strategies to construct critical and rigorous arguments as that particular line of inquiry is finalised. This chapter illustrates how analysts think from a design perspective and discusses the integration of Fluidity and Rigour as two conflicting design requirements. It further proposes that designs for OSINT tools and processes should support the fluid and rapid construction of loose stories, a free-form approach to the assembly of data, inference making and conclusion generation to enable the rapid evolution of the story rigorous enough to withstand interrogation. We also propose that the design encourages the analyst to develop a questioning mental stance to encourage self-checking to identify and remove dubious or low reliability data.

11.1 Introduction

Fluidity and Rigour—these are two conflicting design requirements that characterise the nature of the work of intelligence analysis. By *fluidity* we mean the ease by which a system can be used to express the variability of our thinking processes; and by *rigour* we mean the ability of the system processes and results to withstand interrogation that ensures the validity of our conclusions. In this chapter we discuss what these two terms—fluidity and rigour—mean for the purpose of designing criminal intelligence analysis systems in the context of Open Source Intelligence Investigations.

OSINT has been defined differently by different communities (see Chaps. 1, 2 and 3). For example, Glassman and Kang (2012) refer to OSINT as one of the outcomes of the Open Source movement. By using the tools and ethos of the Open

B.L. William Wong (✉)
Interaction Design Centre, Middlesex University, London, UK
e-mail: W.Wong@mdx.ac.uk

B. Akhgar et al. (eds.), *Open Source Intelligence Investigation*,
Advanced Sciences and Technologies for Security Applications,
DOI 10.1007/978-3-319-47671-1_11

Source movement, it is possible to create cooperative and open problem solving communities based around the internet. Such openness and sharing is needed to support human thinking in order to extend human intelligence, or what they refer to as the 'intelligence of the culture'. Initially, information artifacts may exist as isolated and un-related patterns that are fluid, diffuse and experience dependent. They refer to this stage as fluid intelligence and the more useful of these patterns eventually solidify into crystallized intelligence or "permanent habits of thought that can be used over and over again and passed down through generations" (Glassman and Kang 2012). On the other hand, others such as NATO have taken OSINT to mean something quite different: "unclassified information that has been deliberately discovered, discriminated, distilled and disseminated to a select audience in order to address a specific question" (NATO 2002) and include publicly available information (Bazzell 2016) in newspapers, journals, radio and television, and the internet (Best and Cumming 2008; Hobbs et al. 2014). For the purposes of this chapter, we will discuss interaction and visualization design in the context of the latter definition—publicly available information, collected and processed for specific purposes.

Considering OSINT in the context of internet-based information, OSINT includes a wide variety of sources of information such as social media, Twitter, electronic discussion forums and message boards, photographs, online maps, online auction sites, documents, news media, blogs and on-line services that can assist in criminal investigations such as child sexual exploitation, prostitution or sale of stolen property. The problem faced by investigators is that there is just so much data, some streaming, of different formats, that have little or no known links to on-going investigations. For example, in a series of over 40 thefts from luxury motor vehicles where the satnavs and radios were taken, an analyst discovered that items matching the stolen satnavs and radios were on sale at a well-known online auction site. Further investigations led them to the seller of the items, and coincidentally, no further thefts from motor vehicles with that *modus operandi* were subsequently reported. This required the analyst to access information from various police-only systems, e.g., stolen property database, crime reports, witness statements or statement, and to then access a free public internet site and to make comparisons with any previous cases to narrow down possible culprits. This raises a number of questions, e.g., if the analyst understands the situation and the nature of the problem well or whether it is possible to articulate a known search. However, such a strategy cannot be relied upon, if one does not know what is needed to be known to define a search. Tools and methods of interaction are needed to support the creation of that kind of understanding and to assist the analyst in making associations that lead to more directed searches of open information sources (see Chap. 10).

In interaction design, there are two complementary problems: (1) what information is to be represented at the user interface, and (2) how that information is to be represented. *What information* refers to the constructs that the system is supposed to communicate information about, and *how* refers to the visual form that the constructs should take and how one might interact with that visual form. In this chapter, we attempt to articulate what it is that is to be represented and visually supported and how the designs might allow one to interact with it in ways that enable both fluidity and rigour.

The compatibility between what we represent and how we represent it—the constructs and the visual form—is important in ensuring that the necessary information can be communicated and assimilated rapidly by the users. One principle that describes this relationship is known as the PCP or the *Proximity-Compatibility Principle* (Wickens and Carswell 1995), who explain that "displays relevant to a common task or mental operation (close task or mental proximity) shoud be rendered close together in perceptual space (close display proximity)." If the tasks are time constrained and if the data required to describe a process or system is very large and very diverse, this compatibility becomes especially important.

Much of the research on which the PCP was developed is based in process control systems. In the cognitive systems engineering community, such systems are referred to as being 'causal' or 'causal systems' (Rasmussen et al. 1994). The outcomes from such systems are predictable by the laws of nature. For example, the combined gas law tells us that if volume and temperature are known, the pressure within the containment vessel can be calculated. In most man-made processes, we need information about the performance of the engineered facility in order to manage the process. Historically represented as *Single-Sensor-Single-Indicator* displays, these SSSI presentations required the human operator to identify relevant and related indicators and combine the information mentally in order to assess the state of the process.

Recognising the difficulty and dangers of such displays, the cognitive systems engineering community developed research and design techniques that led to ways for combining the individual indicator information meaningfully. One such approach, *Cognitive Work Analysis* (e.g., Rasmussen et al. 1994; Vicente 1999) and a corresponding interface design method, *Ecological Interface Design* (Burns and Hajdukiewicz 2004; Vicente et al. 1995), has been instrumental in the drive towards displays that represent crucial information through functional relationships that represent the phenomenon being controlled. Displays based on this approach usually map the functional relationships to visual forms such as geometric shapes. Such shapes visually relate multiple performance dimensions with contextual information such as system constraints and other higher-order objectives. This enables the human operator to quickly discern, for example, if a system or process is operating too close to a safety threshold whilst attempting to maximise profit generation.

We have also seen that user performance can increase significantly by changing the way the information is presented. For example, in laboratory conditions response times in emergency ambulance dispatching improved by over 40 % simply by re-arranging the layout of the display from an alphabetically ordered list of ambulance stations to a layout that corresponds with the geo-spatial orientation of the ambulance stations in the region. (Wong et al. 1998). We also see improvements in user performance in hydro-power control management in a de-regulated energy market (Memisevic et al. 2007) as well as in conflict detection in air traffic control (Vuckovic et al. 2013). The designs that led to the improvements in performance did not come about by chance, but by a combination of deliberate investigations into what people do, how they reason about the problems facing them and the careful application of principled design approaches such as representation design (e.g., Bennett and Flach 2011; Reising and Sanderson 2002; Vicente and Rasmussen 1992; Woods 1995). In particular, we focus on how experts in those areas carried out their work, what strategies they invoked in using the information they are presented with, and the decisions that they make using that information.

An effective and well-designed user interface is a technology that enables end-users to become powerful and efficient end-users of the software. This enables them not only to exploit computationally large data sets, but also to be cognitively empowered to unravel the complexities in the data to "discover the unexpected—the surprising anomalies, changes, patterns, and relationships that are then examined and assessed to develop new insight" (Cook et al. 2007).

11.2 Intelligence Analysis

Intelligence analysis is the process of collecting, reviewing and interpreting a range of data (NPIA 2008). In reviewing and interpreting the data, analysts search, organize and intellectually differentiate (Glassman and Kang 2012) the "significant from the insignificant, assessing them severally and jointly, and arriving at a conclusion by the exercise of judgment: part induction, part deduction (Millward 2001), and part abduction" (Moore 2011, p. 3) in order to understand what is being presented to them and to make sense of that in the context of a given problem or intelligence situation (see Chaps. 2 and 3). The primary purpose of intelligence analysis is to produce information that can produce understandings that reduce uncertainty so that decisions can be made, for example, to commit military resources into combat, in diplomatic negotiations, the setting of trade and commerce policy and law enforcement (Clark 2013). As such, intelligence analysis should also provide the decision maker with the underlying significance of selected target information (Krizan 1999).

Traditionally, intelligence analysis has mainly dealt with data collected through secure or covert means. For example, as we have seen in earlier chapters, in the military and security services, special equipment is needed for collecting SIGINT (Signals Intelligence), ELINT (Electronic Intelligence), and networks of agents are needed to manage HUMINT (Human Intelligence) assets. Once the collected data is processed and verified, they become available to the intelligence analysts where they work to discover and construct meaning from the disparate pieces of information. These data are often out of sequence, from multiple sources, are of variable quality and reliability and are in different formats such as video, structured and un-structured texts, numerical data, geo-spatial data and temporal data.

The data is then subject to various forms of analyses, e.g., video surveillance data may need to be watched by a human to identify relevant situations or subject to a degree of automated facial recognition analysis to identify a person in one or across several video feeds or applying different data mining analyses to identify similarities between reports or documents and perhaps other dimension reduction techniques to make the data more suitable for interpretation or for the discovery of patterns of behaviours. Following such analyses, it then becomes possible to assemble together the data to create stories that can be used to explain the situation and how it possibly came about and is likely to evolve into, and therefore what can or should be done to counter or militate against it.

Prior to the arrival of the internet, and as far back as the years leading to the Second World War, security and intelligence agencies have analysed publicly available information to supplement their understanding of the more covertly acquired data to provide the important context in which to interpret intelligence. Sometimes they are used to compile dossiers about people of interest, technologies or evolving political situations that are often reported in open sources such as the press or published reports (e.g., Mercado 2009).

With the development of the internet, there is now much more of such information readily available. The US Government has recognized its potential value and in the 2006 National Defence Authorization Act, has referred to it as Open-source intelligence or OSINT. It is defined as 'intelligence that is produced from publicly available information and is collected, exploited, and disseminated in a timely manner to an appropriate audience for the purpose of addressing a specific intelligence requirement'. In addition, it is considered important enough that OSINT be resourced, methods be developed and improved, and given adequate over-sight, so that it can be 'fused with all other intelligence sources'.

There is a tremendous amount of publicly available information available on the internet such as public news sites, blogs, social media, imagery and geographical data. For example, a UK-based website, www.upmystreet.com used to provide not only prices of houses in a particular post code area, but also brought together a range of different data sources that helped you piece together that intricate picture of what it might be like to live in that area, including neighbourhood statistics, demographics of the area and facilities in the area. Bradbury (2011) describes how a digital forensics analyst was able to construct the profile of a person from a single Flickr photograph—who the person is, what work they currently do, what work they did previously and the names, aliases and phone numbers of that person posted in various online forums. OSINT software such as Maltego makes it possible to very easily trace and track where email addresses appear online, how and in relation to what devices. There are many other tools that analyse meta-data and trawl through social media (see Chaps. 6, 7 and 8). While such tools are very powerful, they still very much represent and support the functions of collection, analysis and to a certain extent collation and organization, and much less so the cognitive function of differentiation, making sense of the data and the formulation of hypotheses (rather than the test of hypotheses).

OSINT can be incorporated into the general process of intelligence analysis, if we consider OSINT as another source of intelligence data with specialist methods and tools. For the purposes of this chapter, it is suggested that just like any other source of intelligence, the data needs to be collated, organized, differentiated and made sense of by the assembly of data in ways that enable analysts to construct explanations or in ways that engender the formulation of hypotheses. Such hypotheses would initiate a variety of analytical reasoning approaches, inference making strategies and further lines of inquiry.

While there is much we can learn from the cognitive engineering community and much we can apply, the problem in intelligence analysis is different in a subtle but significant way. Take for example, criminal intelligence analysis, which addresses the "identification of and provision of insight into the relationship between crime data and other potentially relevant data" (www.interpol.int). Intelligence analysis is not merely reorganizing data and information into a new format (Krizan 1999); intelligence analysis is not about controlling the performance of a process such as hydro-electricity generation, where the objects and components of the process produce information that needs to be monitored and controlled so that they meet performance criteria. Rather, it is about analysing data, understanding what each piece means and then assembling that understanding together with supporting information in ways that provide strong explanations about a crime or situation.

In causal systems such as the hydro-electricity generation plant mentioned above, the functional relationships needed to manage the process are known a priori. However, in intelligence analysis, there are no such functional relationships known or knowable in advance. An intelligence analysis system needs to support the analyst in assembling and constructing the relationships between various findings and observations, where meanings and insights are ad hoc and constructed during the course of the analysis and is heavily context dependent. Such systems are not governed by laws of nature. They are instead, governed by logic and argumentation: Does the assembly of data in a particular way make sense to the case at hand?

The intelligence analyst seeks to understand the separate pieces of information that may come from multiple sources, be it in different formats, may be structured or un-unstructured, quantitative, qualitative and subjective, from which to assemble an explanation that could be used to direct inquiries or lines of investigation. This suggests that a user interface and visualization design must support a different type of process; not one of monitoring and controlling a process, but one to support logical construction and composition.

11.3 What Do We Design?

So, in supporting intelligence analysis, what is it that needs to be designed? There are many good systems that support the search and retrieval, and statistical and semantic analysis. However, there are much fewer for supporting problem formulation and the creation of hypotheses. This is part of the ecology of cognitive work that I refer to as the *Reasoning Workspace*—a conceptual framework that comprises three sub-spaces: the Data Space, the Analysis Space, and the Hypothesis Space (Fig. 11.1). In the *Data Space*, we need to know what exists in the datasets, so that one may know what to look for in it or across several datasets. The *Analysis Space* is where various statistical and semantic analyses take place and produce results that help the analyst understand and identify patterns, behaviours, implications and so

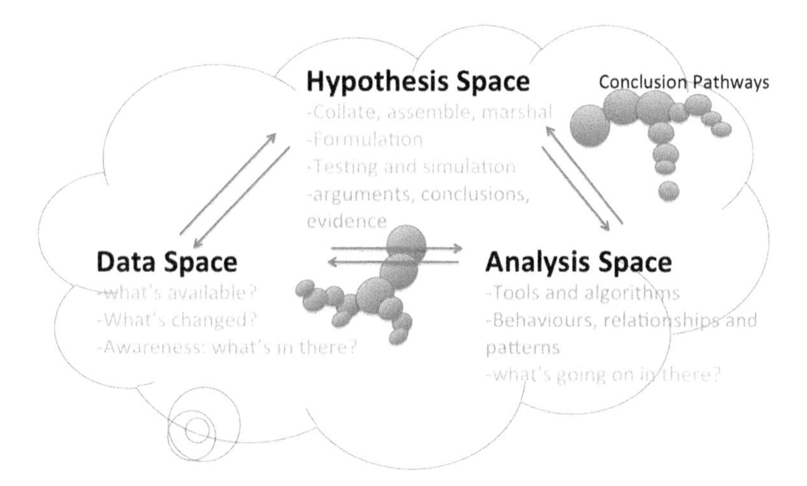

Fig. 11.1 The reasoning workspace

forth. The *Hypothesis Space* is where the analysts assemble the data, create tentative possible explanations and attempt to justify the various arguments.

In this chapter, we refer to designing to support the Hypothesis Space. We should design to support human thinking and reasoning that occurs during the collation, assembly and argument formation. Specifically, the kinds of thinking that occur during sense-making, the generation of insight and the formulation of initial ideas that can be the basis on which strong arguments can be built upon.

Sense-making is the process by which we come to understand the meaning offered by the combination of cues of a situation (Klein et al. 2006, 2007). This understanding then allows us to reason about and project further implications, causality and attribution. Klein and colleagues explain that people make sense of a situation by interpreting the data they are presented with in relation to their basic or general understanding. They refer to this understanding as a frame. This frame can be any sort of prior knowledge based on one's experiences such as training, socio-cultural background and so forth, that helps them interpret what the cues of a situation mean. In this process, people learn what the situation means, which in turn contributes to developing their frames, which in turn guide the person in determining what other cues can or should be considered. This process is known as *connect*, i.e., when a connection is made between the data that one sees and one's frame.

As the person understands the situation better, connecting with more data that informs him of the situation, the person embarks on the process of *elaborate*—searching for more relevant information that can add to his understanding of the situation, learning that there may be other dimensions to the problem than originally thought, therefore driving the demand for more information. As they understand the situation better, they then realize that perhaps some aspects of their understanding are incorrect, leading them to *question* their conclusions and the assumptions they made in order to arrive at those conclusions. If their understanding is flawed, they may engage in the process of *reframing* their understanding of the problem or situation. Although described in a linear fashion, the Data-Frame Model does not describe a linear process—depending upon one's knowledge and understanding of the situation, the sense-making process can start at any point in the Data-Frame Model.

We have found that this approach to explaining sense-making affords clearer indicators about the nature of the process that systems designed for intelligence analysis need to support. In our study with a number of intelligence analysts,[1] we have observations that suggest that Klein's et al. Data-Frame model of sense-making is a good approximation of the sense-making that goes on in the intelligence analysis process. We also observed that in addition to this, analysts also invoke other strategies when attempting to establish causation and correlation. Analysts also invoke counter-factual reasoning (Klein and Hoffman 2009) when attempting to determine how significant elements of the situation are to the outcomes and conclusions. We also observe that analysts make use of the various inference making strategies—induction, deduction and abduction—depending upon what data they have, the rules for interpreting the data and premises they are starting with and the conclusions they would make or would like to make (presented as the I, A and D in Fig. 11.2; see also Wong and Kodagoda 2016, and Gerber et al. 2016). Furthermore, very often they would test the validity of the propositions they arrive at by practicing critical thinking—where they attempt to assess the quality and validity of their thinking and the data they use, the criteria they use for forming judgments, and so forth. In fact, critical thinking is so important that many intelligence analysis training schools have introduced it into their training.

One other thing we observed that happens alongside all of this is somewhat more subtle: analysts are constantly trying to explain the situation, sometimes re-constructing the situation from pieces of data and from inferential claims and then carrying out searches or further analysis to find necessary data to back the claims. This process of explanation is crucial to making sense and how it is used to link data, context and inferences. It often starts off as a highly tentative explanation that is based on very weak data or hunches. As the analyst explores this possibility, making conjectures, suppositions and inferential claims, from which they then connect with additional data and then testing their relevance and significance. In this process they elaborate, question, and often reframe and discard, their ideas, and eventually building and re-building the story so that it becomes robust enough to withstand interrogation.

We suggest that there is a progression—though, not necessarily in a linear manner—where explanations begin as tentative, creative, playful, and generative thinking. Then as the explanations mature, they transition towards thinking strategies that are more critical, evaluative, deliberate and final. If we assume a continuum where at one end we have a tentative explanation we call a 'loose story'

---

[1]Focus group studies with 20 intelligence analysts (Wong et al. 2012), Think-aloud studies with 6 analysts performing a simulated intelligence task (Rooney et al. 2014), Think-aloud studies with 6 librarians carrying out a surrogate task of creating explanations from a literature review task (Kodagoda et al. 2013b).

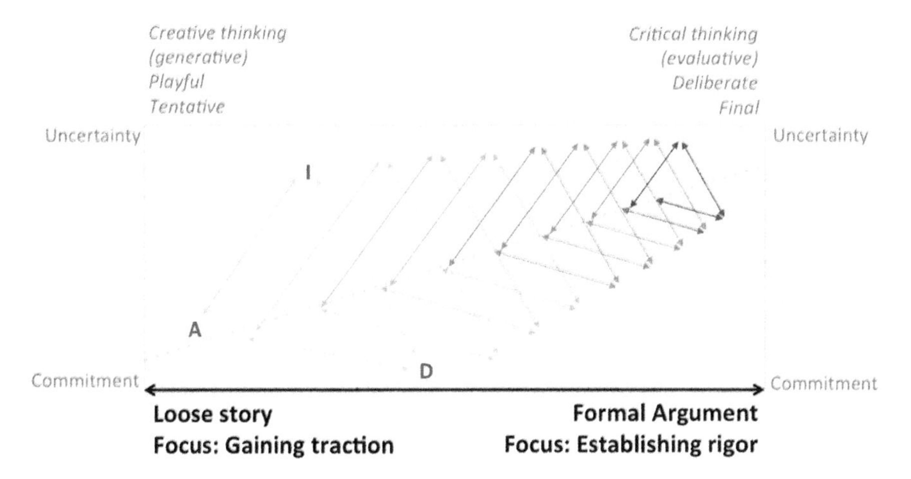

Fig. 11.2 How analysts think: characteristics of the thinking terrain

that accounts for the data, and the other end where the loose story has evolved into a strong and more formal argument such that it is rigorous and able to withstand interrogation, say, in a court of law.

At the 'loose story' end of the continuum, the emphasis is on gaining traction so that possibilities can be formulated and then investigated further (Kodagoda et al. 2013a). At this end there is high uncertainty about the data and the premises and conclusions; there is also very low commitment to any possibility. This mental stance enables the analyst to consider alternative possibilities, work on them, and without forcing themselves down a particular pathway at too early a stage.

At the 'formal argument' end of the continuum, there is much lower uncertainty. The analyst is more definite about what the data and their relationships mean, and very likely has become more committed to a particular path of investigation. At this end, the emphasis is on verifying that the data used to construct the conclusions, the claims being made based on the data, and the conclusions themselves, are valid. Towards this end of the spectrum, one would also be much more committed to the propositions, using techniques such as ACH (Analysis of Competing Hypotheses, Heuer 1999) and other techniques such as Wigmore charts (Wigmore 1913; Goodwin 2000) to critically and rigorously test the claims and supporting arguments. This relationship is illustrated in Fig. 11.2, the terrain of thinking.

11.4 Designing for Fluidity and Rigour

The tools supporting the generative, creative, playful and tentative exploration at the 'loose story' end of the spectrum will be different from the tools needed to support the more evaluative, critical inquiry that leads to a more deliberate, final and rigorous explanation. The tools should fluidly link these different ends of the task spectrum. At the 'loose story' end of the spectrum, the tools should be fluid enough to express our thinking and reasoning and the playful experimentation needed for one to gain cognitive traction with which to start an idea or to generate enough momentum to pursue a line of investigation. The tools should fluidly and dynamically switch modes between creative and formal, enabling the analysts to create tentative possibilities, to then assemble the data and from which to construct and to ground the explanations so that explanatory narratives become well supported and show strong arguments.

Early ideas of such a concept has been implemented in INVISQUE, Interactive Visual Search and Query Environment (Wong et al. 2011; see Fig. 11.3). Originally designed to support searching and collation with electronic library resource systems, INVISQUE represents publications (journal and conference articles) on individual index cards—a common metaphor in the library world. A search for the word 'technology' will return all publications with the word 'technology' and present it on the display as a cluster arranged in an x-y coordinate system, where the y-axis has been assigned citation numbers and the x-axis assigned the year of publication. A quick glance will show which article has the highest number of citations for a given period of time. Index cards representing articles of interest can be dragged out of the cluster and set aside (e.g., bottom left of screen in Fig. 11.2), which can activate a change in the search weighting to indicate items of greater interests. Articles which are of no interest can also be dragged off to another designated space to indicate it is to be discarded or of less interest. The system can also perform Boolean operations such as an "AND" operation by dragging one cluster onto another to create a new cluster representing the intersection or what is common between the two clusters. For example, by dragging the cluster 'science'

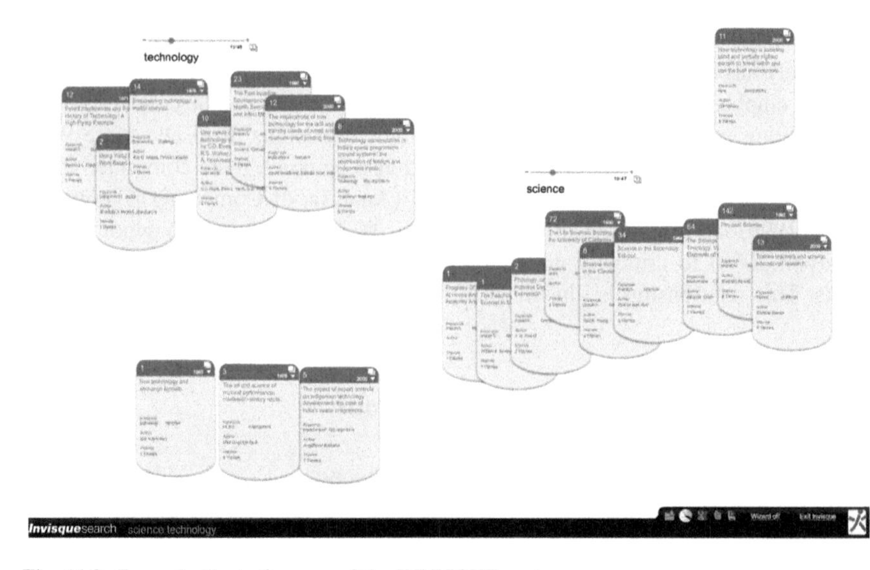

Fig. 11.3 Example illustrating use of the INVISQUE system

on to the cluster 'technology' we create a new cluster representing the intersection of the two clusters—technology AND science.

INVISQUE has many other features, such as a clicking on an index card drills down to retrieve the article it represents, a scatter plot showing all the papers contained within the cluster distributed as dots over time that can also be used as the slider set immediately above each cluster to move the viewing window to another part of the data set. To facilitate the identification of where else an author might be mentioned in the data set, simply click on the name of the author, and other occurrences of the author will be highlighted in the immediate cluster as well as in any other cluster where the author might be mentioned. To carry out a subsequent search, the user only needs to drag and drop the keyword onto the canvas. A query will be initiated and the system will retrieve all records with this author name or keyword, showing up as a new cluster.

The clusters can be fluidly moved around the canvas, and individual cards can also be freely assembled in ways that are meaningful to the user to support the creation of narratives that help the user explain what a group of cards means. These groups can in turn be electronically associated into new clusters. While clusters can be minimised, all the cards generally are persistently displayed on the canvas, allowing the user to associate meaning with the spatial placement of the cards or clusters on the canvas and to also help them in remembering the sequence of searches they have been engaged in, or that a particular cluster is temporary and is serving initially as an anchor for generating or even directing further investigations.

INVISQUE has provided a test-bed with which we attempted to study how an analyst thinks, reasons and interacts with such a free-form system (e.g., Kodagoda et al. 2013a; Wong et al. 2013). In the following sections, we reflect on our work in the hope that we articulate the needs of interactive visualisation for intelligence analysis.

11.4.1 Fluidity as a Design Concept for OSINT Investigations

In the earlier sections we have hinted a little at what fluidity means. In this section we will attempt to explain the notion of fluidity as a design concept. In Baber et al. (2013), we observed a number of teams engaged in a realistic but fictional police investigation about drug smuggling to determine when an arrest should be made and who should be arrested. At the initial stages the participants spent some time learning what the case is about by setting out the timeline of events, the key characters and their relationships and activities and where these characters were located. They used a number of paper-based or white-board representations, including tables showing telephone conversations, sketch maps showing rough geographies of the area indicating where the different people were based, dates observed, movements and travel routes and so forth. We observed that the participants often transition between different views to re-confirm data or to update their charts or diagrams, or to evaluate and test their possible explanations by talking through the re-constructed event, exploring various combinations and making inferences or suppositions where there were missing or unclear data. The participants did this individually as well as with members of their team. During this process of trying to establish a baseline story, they were also observed to draw or mark their sketches and data. They corrected and refined their thinking. They make new connections or setting aside or discarding those elements now understood to be incorrect. Uncertainty is common at this stage: not only uncertainty with the data, its origins and its quality, but more so with the way the data relates to each other and how they are constructed into stories. As such, the explanations proposed are often very loose and the analyst was seldom committed to a single explanation or story at this stage. Before they could even construct and propose their final conclusions, there was a lot of 'maybe … what if … is it possible that …?' type of discussions.

Toulmin (1958) made similar observations when he discussed phases of an argument. He explains that arguments are said to have modality. At the "first stage of dealing with any sort of problem, there will be an initial stage at which we have to admit that a number of different suggestions are entitled to be considered" (p. 17). He explains that given the trustworthiness of available information and how they are developed into candidate arguments, those possibilities that lack credibility are eventually dismissed. The arguments evolve and develop. System and user interface designs need to cater for such evolutionary progression in arguments.

But what form should the system and user interface designs take? In a study by Takken and Wong (2013), they investigated how the ability to move pieces of information around freely, creating groups of sensibly related cards and tentative groupings that can be easily re-grouped, encourages experimentation and sense-making. The study required participants to construct food chains, given a set of cards with a variety of animals, plants and habitats. Participants were placed in two groups—a first group where participants were allowed to use their hands to move the cards around and a second group where participants were not allowed to use their hands. Participants in both groups were asked to think aloud, i.e., talk about what they were thinking about as they performed their tasks. Their utterances and actions and were audio and video recorded.

The study found that in the group that was allowed to use their hands, there were a total of 99 instances of utterances and actions that represented a sense-making action, whereas there were only 50 observations of utterances and actions that represented sense-making in the group that was not permitted to use their hands. It was proposed that this difference is perhaps because the ability to freely pick up, move and re-arrange the cards into and out of groups is an epistemic action—an externalization of cognition—that helps the participant to quickly construct relationships and to test if the relationships are sensible, especially when there are more cards and information that can easily be handled by one's working memory.

In another study, the INVISQUE system virtual (electronic) representation of cards, was used to identify the most influential authors in a particular subject area.

Kodagoda et al. (2013a) discovered that during the early stages of analysis, once the participants discovered what terms existed in the data set, they were able to determine what questions they could ask or carry out searches. They would then identify anchors or key terms that they believe would spawn useful paths of investigation. These anchors are often tentative. They are experimented with, combined and reframed, helping the analyst gain cognitive traction, a footing with which to assemble and develop the loose story or the loose collection of pieces of data and weave that into a more rigorous and coherent argument.

11.4.2 Rigour as a Design Concept for OSINT Investigations

Rigour should not be considered the antithesis of fluidity. Instead, it should be considered as being complementary. Analysts need fluidity to assemble creatively and quickly, and test the validity of the arguments produced. Fluidity, in terms of design, is primarily concerned with the way the user interactions are designed to support the creative and traction-establishing nature of early stage reasoning.

By rigour we refer to the rules that govern the logic for the assembly of data in constructing the story and for improving the credibility and trustworthiness of the assembled data, the inferences and conclusions as the story evolves into a formal argument. The rules of logic when implemented in support of rigour as a design concept should be careful not to hinder the early experimentation, tentativeness and creative exploration needed for an analyst to gain traction and to produce possible candidate pathways for investigation.

Rigour is often associated with critical thinking (e.g., Moore 2007). In critical thinking the emphasis is on ensuring that the data used, the sources of the data and the assumptions made are sound. In intelligence analysis, by definition, we will have data that is incomplete, possibly deceptive and un-reliable, and messy. Strictly applying the rules of critical thinking could lead to much data being discarded due to their low quality or reliability, especially at the early stages where the data is un-assembled and viewed largely in isolation. This could make the generation of plausible scenarios that rely on expertise to bridge the reliability gap in the data difficult or impossible—because the data or inferential claims at this early stage cannot yet be backed up or adequately grounded. This hinders the gaining of traction that could lead the investigation into other avenues generating leads or uncovering other sources of data not previously considered. We are not arguing for the removal of critical thinking, but instead to apply the laws of logic and critical analysis in ways that do not hinder the creative exploration and generative cognitive strategies needed at the early stages.

The intelligence analyst is often confronted by missing data, deceptive data and low quality data. However, the problem is not the missing data, deceptive data or the low quality data—but rather in being un-aware that there are missing data or that it is from dubious sources. Being aware that data could potentially be missing enables the analyst to ask questions about what information could be missing, which could lead on to questions of where to get such information and how to verify it or how experts might make a 'leap of faith' initially by filling in the gaps by using storytelling techniques. This helps the analyst gain traction. The early assembly of data into a story is used as a starting place to explore the possible profitability of this line of investigation. What is bad is that if the analyst treats all the data used in his or her story as equal, and where another analyst is not able to determine the validity of the conclusions from the story.

One method to address the missing data problem was proposed in Wong and Varga (2012). It is called 'black holes'. The human visual system is very good at picking out breaks in regularity. So, instead of simply showing a list of data available, we can present that set of data in a way that regularizes it. Data can be presented by time, and where there are gaps or black holes in the sequence, it becomes readily detectable by visual inspection. This enables the analyst to choose to ask questions about the black holes (Fig. 11.4).

For example, data from a surveillance operation may indicate a particular regular pattern of behaviour where Person X is regularly observed to have lunch at a particular restaurant. Drawing on the concept of emergent features (Bennett et al. 1993), black holes emerge if the observations are regularized and presented along a timeline, rather than as a list where timing has to be read and interpreted, rather than simply viewed. The black holes create an entity that encourages the analyst to ask 'What happened? What are the reasons for the black holes? Did the surveillance operation miss making or reporting a lunch meeting, i.e. a potential failure of the system, or perhaps Person X was actually away doing something else at another location?' This knowledge that the data is missing now becomes data that can generate other possibilities or be incorporated into other lines of investigation.

The next step is to use this method of representing black holes in combination with the methods for representing the logic of arguments, and to make visible the reasoning process. Toulmin (1958) provides a way to layout the logic of an argument. Toulmin's layout shows how a conclusion is supported by grounding data, and the warrant that connects and grounds the conclusion in that data, and how further supporting data can be associated with the grounding data. It is possible to conceive that in complex arguments, there may be several layers connected together. Wigmore (1913) provides a similar graph-style representation of evidence

x x x x x x x x x x x x x x x x x x x x x x x x x x x x x x x x x x x

Sequential but not temporal (aggregation)

x x x x x x x   x x x x x x x x x x x x x x x x x   x x x x x x   x x x x   x

Sequential and temporal <= showing the ***blackholes***

Fig. 11.4 Black holes (Wong and Varga 2012)

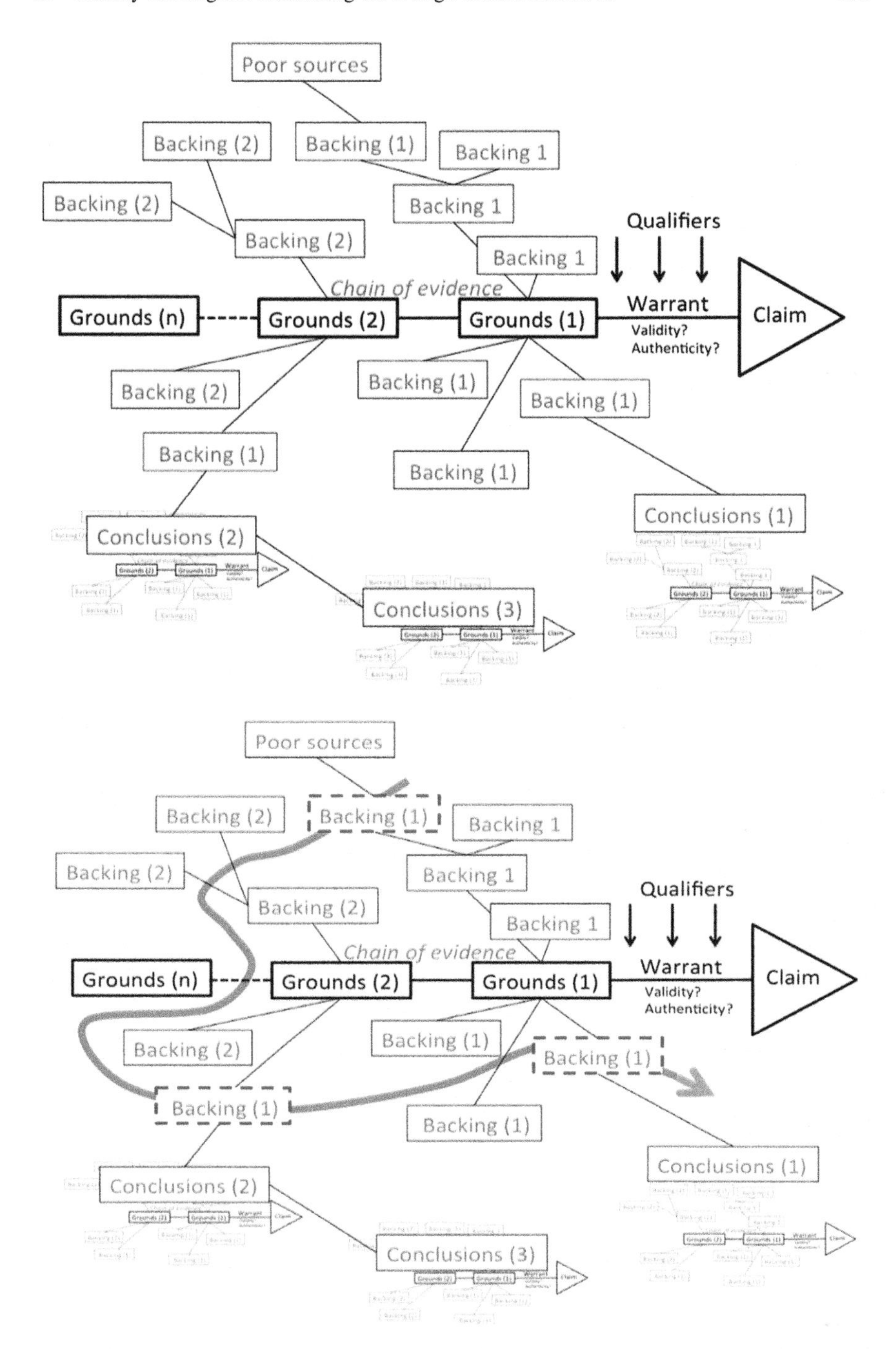

Fig. 11.5 *Top* A generic layout showing parts of an argument. *Bottom* shows a 'brown worm' linking weak data or data with dubious provenance (Wong and Varga 2012)

used in courts, called Wigmore Charts. It can represent the evidence used by the defence and the prosecution and the arguments for and against, given various pieces of evidence. The visual form makes visible the arguments and the supporting evidence and the claims made that link the arguments. This enables the lawyers to study the cases and to identify weaknesses or gaps in their arguments. CMaps (IHMC) and Compendium (KMi) are software that support some of these functionalities.

To address the issue of not knowing what data is of poor quality or the way by which a conclusion came about, Wong and Varga (2012) proposed the concept of 'brown worms'. Figure 11.5-top, lays out the claims made, the grounds for the claim, the backing data and the warrants that guide the interpretation of the relationships. The figure also shows how some backing data are based on conclusions drawn from someone else's analysis or from other inferential claims. The diagram makes it clear, where such weaknesses are located in the conclusion pathway. It also indicates that some conclusions are based on other conclusions, just as it indicates poor data in the link. This introduces weaknesses as it may not be clear what evidence was used in basing that claim on. This might suggest that these sets of data are of questionable quality. If so, the analyst can draw a line through them linking all the weak and questionable data to create what we have referred to as a 'brown worm'. We can then collectively and temporarily remove these weak and questionable data by pulling out the brown worm—effectively deleting the weak and questionable data from the set temporarily. This then allows the analyst to engage in analysis strategies such as counterfactual reasoning—to consider the effect of the removal of the data on the outcome: Has the removal of the weak data made a material difference to the outcome? If not, then it is probably safe to continue the analysis in the absence of such data. Brown worms make visible the weaknesses in the argument, the inferences made and the data used.

These interface design concepts are based on human factors' principles of perception and attention, some of which have been encapsulated in the design approaches known as Ecological Interface Design (Burns and Hajdukiewicz 2004; Vicente et al. 1995) and Representation Design (Bennett and Flach 2011). In this chapter, we have attempted to apply these principles in a way that reflects the demands of the intelligence analysis problem.

11.5 Conclusions: Guidance for Designing Analysts' Tools

Instead of mainly supporting the final stage of reasoning, where we formalise our explanations as arguments, we propose that the system design should also facilitate and support the easy and rapid construction and assembly of data collection in ways that enable the creation of early loose stories. In this way we encourage the exploration of alternatives, appreciation of the context and the avoidance of pre-mature commitment.

To support fluidity and rigor, the formal argument structures should be retained for the final formal stages. The model for the early stage should be aligned with the free-form approach to the assembly of data, inference making and conclusion generation that allows easy and rapid evolution of the story by interaction and visualisation design techniques. These techniques should facilitate modification of the relationships, groupings, themes and even spatial-temporal arrangements; and the questioning of the data and the story to support the understanding, conceptualization of new ideas, that support the elaboration, questioning and reframing processes of the Data-Frame Model.

One useful realization from the use of the Data-Frame Model of sense-making is that it makes one aware of the constant need to have a questioning mental stance; and therefore, from a design perspective, how this might translate into user interface and interactive visualization designs that fluidly support such questioning of the data, conclusions and assumptions made during the intelligence analysis process.

User interface and visualisation designs need to support and encourage the self-checking—not simply reminding the person whether they are biased, but for example, facilitating the identification and easy removal of dubious or low reliability pieces of information in the data set to test for mutability and propensity ('brown worm' effect; Wong and Varga 2012).

We anticipate that, as we address the conflicting yet complementary demands of fluidity and rigour, we will better understand and address the core needs that a user interface and interactive visualization would be required to support in the intelligence analysis task. We also envisage that when the user interface integrates the necessary technologies, it will help to make the job of those entrusted with the defence of our security and freedom faster and less prone to mistakes.

Acknowledgments The research leading to the results reported here has received funding from the European Union Seventh Framework Programme (FP7/2007-2013) through Project VALCRI, European Commission Grant Agreement N° FP7- IP-608142, awarded to B.L. William Wong, Middlesex University and partners.

References

Baber C, Attfield S, Wong W, Rooney C (2013) Exploring sensemaking through an intelligence analysis exercise. In: Paper presented at the 11th international conference on naturalistic decision making NDM 2013, Marseille, France, 22–24 May 2013

Bazzell M (2016) Open source intelligence techniques: resources for searching and analyzing online information, 5th edn

Bennett KB, Flach JM (2011) Display and interface design: subtle science, exact art. CRC Press, Taylor and Francis Group, Boca Raton

Bennett KB, Toms ML, Woods DD (1993) Emergent features and graphical elements: designing more effective configural displays. Hum Factors 35(1):71–97

Best RAJ, Cumming A (2008) Open source intelligence (OSINT): issues for congress. In: Paulson TM (ed) Intelligence issues and developments. Nova Science Publishers, Inc

Bradbury D (2011) In plain view: open source intelligence. Comput Fraud Secur 2011:5–9

Burns CM, Hajdukiewicz JR (2004) Ecological interface design. CRC Press, Boca Raton, FL
Clark, R. M. (2013). Intelligence analysis: A target-centric approach. Thousand Oaks, CA: CQ Press, SAGE Publications.
Cook K, Earnshaw R, Stasko J (2007) Discovering the unexpected. IEEE Comput Graph Appl 27:15–19
Gerber, M., Wong, B. L. W., & Kodagoda, N. (2016). How analysts think: Intuition, Leap of Faith and Insight, Proceedings of the Human Factors and Ergonomics Society 60th Annual Meeting, 19-23 September 2016, Washington, D.C., USA (pp. 173-177): SAGE Publications.
Glassman M, Kang MJ (2012) Intelligence in the internet age: The emergence and evolution of open source intelligence (OSINT). Comput Hum Behav 28(2):673–682
Goodwin J (2000) Wigmore's chart method. Inf Logic 20(3):223–243
Heuer RJ (1999) The psychology of intelligence analysis: center for the study of intelligence. Central Intelligence Agency
Hobbs C, Moran M, Salisbury D (eds) (2014) Open source intelligence in the twenty-first century: new approaches and opportunities. Palgrave Macmillan, London, UK
Klein G, Hoffman R (2009) Causal reasoning: Initial report of a naturalistic study of causal inferences. In: William WBL, Stanton NA (eds) Naturalistic decision making and computers, Proceedings of the 9th bi-annual international conference on Naturalistic Decision Making NDM9, 23-26 June 2009, BCS, London. BCS, London, pp 83–90
Klein G, Moon B, Hoffman RR (2006) Making sense of sensemaking 2: a macrocognitive model. IEEE Intell Syst 21(5):88–92
Klein G, Philips JK, Rall EL, Peluso DA (2007) A data-frame theory of sense-making. In: Hoffman RR (ed) Expertise out of context: proceedings of the sixth international conference on naturalistic decision making. Lawrence Erlbaum Associates, New York, pp 113–155
Kodagoda N, Attfield S, Wong BLW, Rooney C, Choudhury T (2013a) Using interactive visual reasoning to support sense-making: implications for design. IEEE Trans Visual Comput Graphics 19(12):2217–2226
Kodagoda, N., et al., Using Interactive Visual Reasoning to Support Sense-making: Implications for Design. IEEE Transactions on Visualization and Computer Graphics, 2013b. 19(12): p. 2217–2226.
Krizan L (1999) Intelligence essentials for everyone. Joint Military Intelligence College, Washington, DC
Memisevic R, Sanderson P, Wong W, Choudhury S, Li X (2007) Investigating human-system interaction with an integrated hydropower and market system simulator. IEEE Trans Power Syst 22(2):762–769
Mercado SC (2009) Sailing the sea of OSINT in the information age. Stud Intell 48(3):45–55
Millward W (2001) Life in and out of Hut 3. In: Hinsley FH, Stripp A (eds) Codebreakers: the inside story of Bletchley park. Oxford University Press, Oxford, pp 17–29
Moore DT (2007) Critical thinking and intelligence analysis, 2nd edn. National Defense Intelligence College, Washington, DC
Moore DT (2011) Critical thinking and intelligence analysis, 2nd edn. National Defense Intelligence College, Washington, DC
National Defense Authorization Act for Fiscal Year 2006, 119 STAT. 3136 (2006)
National Policing Improvement Agency (2008) Practice advice on analysis. National Policing Improvement Agency, on Behalf of the Association of Chief Police Officers, UK
NATO (2002) NATO open source intelligence handbook v1.2 January 2002
Rasmussen J, Pejtersen AM, Goodstein LP (1994) Cognitive systems engineering. Wiley, New York
Reising DVC, Sanderson PM (2002) Work domain analysis and sensors II: pasteurizer II case study. Int J Hum Comput Stud 56(6):597–637
Rooney C et al. (2014) INVISQUE as a Tool for Intelligence Analysis: the Construction of Explanatory Narratives. Int J Hum Comput Interact 30(9):703–717

Takken S, Wong BLW (2013) Tactile reasoning: hands-on vs. hands-off—what's the difference? In: Paper presented at the 11th international conference on naturalistic decision making NDM 2013, Marseille, France, 22–24 May 2013

Toulmin S (1958) The uses of argument, 8th reprint 2008 edn. Cambridge University Press, Cambridge, England

Vicente KJ (1999) Cognitive work analysis: toward safe, productive, and healthy computer-based work. Lawrence Erlbaum Associates, Mahwah, NJ

Vicente KJ, Rasmussen J (1992) Ecological interface design: theoretical foundations. IEEE Trans Syst Man Cybern 22(4):589–605

Vicente KJ, Christoffersen K, Pereklita A (1995) Supporting operator problem solving through ecological interface design. IEEE Trans Syst Man Cyber 25(4):529–545

Vuckovic A, Sanderson P, Neal A, Gaukrodger S, Wong BLW (2013) Relative position vectors: an alternative approach to conflict detection in air traffic control. Hum Factors 55(5):946–964

Wickens CD, Carswell CM (1995) The proximity compatibility principle: its psychological foundation and relevance to display design. Hum Factors 37(3):473–479

Wigmore JH (1913) The principles of judicial proof as given by logic, psychology, and general experience and illustrated in judicial trials, 1st edn. Little, Brown & Co., Boston, MA

Wong BLW, Varga M (2012) Blackholes, keyholes and brown worms: challenges in sense making. In: Proceedings of HFES 2012, the 56th annual meeting of the human factors and ergonomics society, Boston, MA, 22–26 Oct 2012). HFES Press, Santa Monica, CA, pp 287–291

Wong WBL, O'Hare D, Sallis PJ (1998) The effect of layout on dispatch planning and decision making. In: Johnson H, Nigay L, Roast C (eds) People and computers XIII, HCI '98 conference. Springer, in Collaboration with the British Computer Society, Sheffield, UK, pp 221–238

Wong BLW, Choudhury ST, Rooney C, Chen R, Xu K (2011) INVISQUE: technology and methodologies for interactive information visualization and analytics in large library collections. In: Gradmann S, Borri F, Meghini C, Schuldt H (eds) Research and advanced technology for digital libraries: international conference on theory and practice of digital library TPDL 2011, Berlin, 26–28 Sep 2011. Lecture notes in computer science, vol 6966. Springer, Berlin, pp 227–235

Wong BLW, Kodagoda N, Attfield S (2013) Trialling the SMART approach: identifying and assessing sense-making. In: Proceedings of the human factors and ergonomics society 57th annual meeting, 30 Sep–4 Oct 2005. HFES Press, San Diego, pp 215–219

Wong, B. L. W., & Kodagoda, N. (2016). How analysts think: Anchoring, Laddering and Associations, Proceedings of the Human Factors and Ergonomics Society 60th Annual Meeting, 19-23 September 2016, Washington, D.C., USA (pp. 178-182): SAGE Publications.

Woods DD (1995) Toward a theoretical base for representation design in the computer medium: ecological perception and aiding human cognition. In: Flach J, Hancock P, Caird J, Vicente K (eds) Global perspectives on the ecology of human-machine systems, vol 1. Lawrence Erlbaum Associates, Inc., Publishers, Hillsdale, NJ, pp 157–188

Part III
Pratical Application and Cases

Chapter 12
A New Age of Open Source Investigation: International Examples

Eliot Higgins

Abstract Whether individuals and organisations are ready for it or not, new opportunities and challenges presented by open source information from a variety of sources are something we must face now and in the future. It should be understood that this vast volume of online open source information allows anyone, for better or for worse, to become an investigator. What they choose to investigate could range from war crimes in a far-away country, to SWAT teams on their doorstep, and awareness of this behaviour is crucial in so many different ways. There are also many new tools being developed for those individuals and organisations in the public and private sphere that are aimed at aiding the process of open source investigation, be it for discovering information in the first place, or organising the information that's discovered in a more accessible fashion. This chapter describes current cases of OSINT use by non-LEA (citizen) actors including their tools and techniques.

12.1 Introduction

Over the last several years the field of open source investigation has undergone something of a renaissance. This has been fuelled by the events of the Arab Spring, where social media was first used at protests, then later by media activists and armed groups participating in the conflicts that arose from those protests. During that period individuals and organisations began to explore ways of examining this vast volume of information coming from those countries, and new tools and platforms have been used to examine this information.

In the wake of the Arab Spring the same skills and techniques learned during that period began to be applied to new conflicts and situations. Russia's suspected involvement in incidents in Ukraine came under close scrutiny after the downing of Malaysia Airlines flight 17 (MH17) on 17 July 2014, and open source information

E. Higgins (✉)
Bellingcat, Leicester, UK
e-mail: eliothiggins@bellingcat.com

B. Akhgar et al. (eds.), *Open Source Intelligence Investigation*,
Advanced Sciences and Technologies for Security Applications,
DOI 10.1007/978-3-319-47671-1_12

—in particular that coming from social media—became a key source of information in the public debate about what had happened to MH17. Meanwhile, social media posts from local citizens in Eastern Ukraine and the Russian border region with Ukraine were used to expose Russia's suspected military involvement in the conflict in Ukraine.

What has been particularly notable in the last few years is how access to a new range of online tools has now made open source investigation something that anyone can do from their own home. Thanks to Google and other technical service companies it is now possible to access satellite imagery of the entire planet, geo-tagged photographs from across the world, street level imagery from thousands of locations in dozens of countries, and access a huge volume of video content from every corner of the Earth.

With these tools alone anyone has the potential to be an open source investigator and new (and often free) tools and platforms are being created and rapidly adopted by a growing, public, online community of open source investigators. Whether or not professionals working in the field adapt to these new changes, it is clear that there is a growing public movement already using this information.

Already open source information is being used by a wide variety of news organisations and human rights organisations in their own work. In 2013 for example the New York Times used footage posted online by Syrian armed opposition groups to identify weapons smuggled to the Syrian opposition by foreign governments, exposing what was meant to be a highly secretive operation thanks to YouTube videos.[1] Human Rights Watch and Amnesty International now use footage gathered from social media in combination with satellite imagery and in person interviews to establish the facts surrounding a wide range of events. In one example, in June 2014, Human Rights Watch used satellite imagery combined with videos posted online by the Islamic State to find the exact location of mass executions in Tikrit, Iraq.[2]

More recently, attempts to organise the vast amount of information coming through open sources in conflict zones have resulted in the creation of platforms such as PATTRN from Goldsmith's University's Forensic Architecture department,[3] used to host Amnesty International's Gaza Platform,[4] and projects such as the Syrian Archive.[5] These projects collect the vast number of open source information coming from a variety of new sources, and attempts to organise them to make the information more accessible. This allows users to gain a deeper understanding of conflict without having to do the length research required to bring order to the chaos of these new sources of open source information.

[1]http://www.nytimes.com/2013/02/26/world/middleeast/in-shift-saudis-are-said-to-arm-rebels-in-syria.html.

[2]https://www.hrw.org/news/2014/06/26/iraq-isis-execution-site-located.

[3]http://www.gold.ac.uk/news/pattrn/.

[4]http://gazaplatform.amnesty.org/.

[5]https://syrianarchive.org/.

What makes this so powerful is that it is no longer the case of one individual's or group's word against that of another, but rather the ability categorically to demonstrate a case using publically accessible open source information. While the *evidential* use of this information in legal proceedings brings some specific challenges (see Chap. 18) tit is now being used to challenge directly claims and denials by governments, creating a far more difficult environment for those governments who attempt to present their own counter-factual version of events.

When Syria was first accused of using cluster munitions in the conflict in Syria in 2012 Human Rights Watch was able to gather video footage of the cluster bombs used,[6] and when the Syrian government denied cluster bombs were being used hundreds of videos of unexploded cluster bombs posted on YouTube by groups across Syria showed the reality of the situation.[7]

In the case of the downing of MH 17 the Russian Ministry of Defence presented their own evidence (on 21 July 2014), all of which was shown to be in contradiction to available open source material.[8]

In the Atlantic Council's report Hiding in Plain Sight[9] Russia's involvement in the conflict in Ukraine was clearly demonstrated using a variety of open source material, including identifying vehicles filmed in Russia later filmed in Ukraine, identifying evidence of cross border artillery attacks from Russia into Ukraine, and tracking the movements of individual Russian soldiers in and out of Ukraine. One case study from the last example was then investigated in the multi-award winning VICE News documentary Selfie Soldiers, where the locations identified in the Atlantic Council report were visited by VICE News reporter Simon Ostrovsky, recreating the photographs posted on social media by a serving Russian soldier inside Russia and Ukraine.[10] This provided a clear demonstration that the investigation techniques not only worked, but could be built upon by people working in another field.

But where is the starting point with these investigations when there are so many new sources of information? In the case of the conflict in Syria for example social media use had been severely limited in many, if not all, opposition areas as the conflict began to escalate. This resulted in a situation where each town or armed group would have its own media centre, with a Facebook, Twitter, and YouTube page where you could find all the information being shared publically from an area. This was useful in one sense in that all the information one could expect to find coming from an area would always come from those channels.

---

[6]https://www.hrw.org/news/2012/07/12/syria-evidence-cluster-munitions-use-syrian-forces.

[7]https://www.hrw.org/news/2012/10/23/syria-despite-denials-more-cluster-bomb-attacks.

[8]https://www.bellingcat.com/news/uk-and-europe/2015/07/16/russias-colin-powell-moment-how-the-russian-governments-mh17-lies-were-exposed/.

[9]http://www.atlanticcouncil.org/publications/reports/hiding-in-plain-sight-putin-s-war-in-ukraine-and-boris-nemtsov-s-putin-war.

[10]https://news.vice.com/video/selfie-soldiers-russia-checks-in-to-ukraine.

This was particularly apparent in the wake of the Sarin attacks on 21 August 2013 in Damascus where targeted areas produced hundreds of videos, photographs, and text posts shared on the Twitter, Facebook, and YouTube accounts of local groups. No access by foreign journalists was possible to the areas attacked, so these social media posts were the only source of first-hand accounts of the attacks.

It was possible to gather a great deal of information about the attacks from these videos. Videos of the victims of the attacks showed the symptoms caused by the chemical agent used, narrowing down the range of possible chemical agents used. In addition to photographs and videos of the munitions used, it was possible to establish the likely chemical agent was Sarin, launched at the attacked sites in two different types of rockets.

Based on the videos of the impact sites it was possible to establish the exact location of several impact sites, and build a much clearer view of which areas were effected. This was particularly useful as maps of the areas effected published in the media and by governments were highly inaccurate, based on confused and unchecked information.

Following the attacks there were a great many theories proposed in attempts to blame different sides in the conflict for the attacks. With open source material it was possible to evidence certain claims, and counter narratives pushed by certain governments and individuals whose own claims clearly contradicted the open source evidence available.

In the conflict in Ukraine, with comparatively unlimited internet access for the local population, there was masses of information being shared online, much of which was totally irrelevant to the conflict, requiring a different approach to discovering the content. Location- based search tools were particularly useful for discovering content, but in some cases it will be necessary to explore social media accounts manually along with other open source information.

There are many other ways to explore open source information, in particular information from social media, but the next major question is, once that information has been obtained, how can one be sure that the information found can be trusted? Verification of social media content in particular has become a core part of open source investigation, and by combining different types of open source information it has been repeatedly demonstrated that it is possible to verify claims through the use of open source investigation.

In the case of photographs and videos geolocation has become a key methodology in the verification process. Geolocation in the context of open source investigation is using clues in the images to establish the location an image was recorded. At its most basic that might be as simple as comparing large and distinct structures visible in the image to satellite imagery of the same area. In other situations, it might mean piecing together clues from an image, such as the area codes from telephone numbers on billboards, or looking for a specific design of street furniture to narrow down the possible locations, then searching for more and more details to narrow down the location to the precise spot the image was recorded. Even when images contain geographical co-ordinates in their metadata it is expected that they would also be geolocated using these methods, and the geolocation of images is considered essential in the field of open source investigation.

This presents both opportunities to verify content, but also a threat to those who might be operating in the field. In late 2015 an incident in San Diego, where police responding to a domestic violence call came under fire, was widely reported on social media.[11] This included photographs of police taking up positions around the building, including the position of police sniper teams, all of which could be rapidly geolocated. Despite requests by the San Diego police locals and news organisations continued to share images of the police participating in the siege. Although in this incident it caused no significant problems for the police, it did demonstrate that thanks to the widespread use of smart phones and social media it's now extremely difficult to control what information is being shared online, and methodologies like geolocation have potential risks to forces operating in public environments in crisis situations.

Geolocation also presents challenges to those who are attempting to present information in a public space. As the awareness of verification techniques increases presenting information to the public now faces a further level of scrutiny. The Russian government has been a particular focus of this scrutiny thanks to the events of the conflicts in Ukraine and Syria in which they participated.

In Ukraine Russia was publicly accused by some of sending troops and equipment into Ukraine, and hundreds of images were shared on social media sites such as Facebook, YouTube, Twitter, and VKontakte of military vehicles in Ukraine and Russia. Some vehicles had identical markings and damage that showed that it was highly likely that they were the same vehicles, but had been photographed in both Russia and Ukraine. Using geolocation it was also possible to verify the exact location the images of these vehicles had been taken, demonstrating that Russian military vehicles had in all probability been sent to Ukraine.

After the downing of MH17 photographs and videos of a Buk missile launcher travelling through separatist territory on 17 and 18 July were geolocated, and the presence of the Buk missile launcher in those locations were further verified by social media posts by locals as the missile launcher travelled through the locations shown in the images. On 30 March 2015 the Joint Investigation Team investigating the incident published a video[12] using this same geolocated, open source, imagery to call for witnesses who had seen the Buk missile launcher being transported on 17 July to come forward. Without revealing any closed source information used by the criminal investigation, it was possible for the Joint Investigation Team to present clear imagery of the route it believed the missile launcher had travelled, aiding in its call for witnesses.

In addition to long form investigations, crowdsourcing can also be useful in breaking news type situations. In May 2016 for example, ISIS social media

[11]https://www.bellingcat.com/resources/2015/11/06/overly-social-media-and-risks-to-law-enforcement-san-diego-shooting/.

[12]https://www.youtube.com/watch?v=olQNpTxSnTo&sns=.

accounts encouraged supporters of ISIS to post pictures on social media from major European cities, displaying messages of support for the group. This social media campaign was in support of an imminent speech from ISIS spokesperson, Abu Mohammed al-Adnani, and was widely assumed to be about the recent crash of an EgyptAir jet on 19 May.

As noted by conflict analyst J.M. Berger, this was the "first time in months that the ISIS social media team has come out in force to push a release," and that the ISIS "fanboys" felt accomplished in getting their hashtag to trend.[13]

The images shared online all included handwritten notes expressing support for ISIS, but in a number of cases also included a view of the background behind the note. It was immediately noted that in some cases these backgrounds could be geolocated, revealing the position from which the photographs had been taken.[14]

Bellingcat presented each image to their Twitter followers, numbering over 40,000 at the time, and asked them to attempt to geolocate each image. Quickly, a group of Twitter users began debating the images, and those users where able to identify the locations the photographs were taken from in Germany, France, the UK, and the Netherlands.

The end result was an ISIS social media campaign aimed at intimidating European populations being reported as an embarrassing failure in the main stream press in multiple countries. The speech the campaign was meant to be promoting was virtually forgotten in the reporting of the incident, and local police forces confirmed they were investigating the locations provided.

Crowdsourcing can also be used to cover a social media footprint. For example, in November 2015, during an anti-terror operation in Brussels, local citizens were asked not to share images of the ongoing operations. Many of the images taken by members of the public were being shared on social media with the hashtag #BrusselsLockdown and, in response to the request not to share images, social media users began to flood the hashtag with photographs of cats. This resulted in earlier photographs of police operations shared on the hashtag being drowned out by the sheer volume of cat photographs, as well as generating positive media coverage of the event.[15] In response to the public's impromptu social media campaign the Belgian federal police responded with a thank you on Twitter—A bowl of cat food with the caption "For all those cats who helped us yesterday… please!"[16]

While not all open source investigation and research is suited to crowdsourcing or public involvement, this demonstrates that not only this technology is possible, but that there is also a growing community of people interested in open source

[13]https://twitter.com/intelwire/status/734045412886228992.

[14]https://www.bellingcat.com/news/mena/2016/05/22/isis-had-a-social-media-campaign-so-we-tracked-them-all-down/.

[15]https://www.buzzfeed.com/stephaniemcneal/brussels-lockdown-cat-pictures?utm_term=.nk7qY56nK#.qybXjr0Ld.

[16]https://twitter.com/FedPol_pers/status/668749104655302656.

investigation who are willing to give their free time to participate in crowdsourced projects, and that there are an increasing number of free and low cost tools available that empower those people to do so.

The open source investigation tool box is ever expanding, drawing on information from a growing number of social media and sharing platforms, which are themselves constantly filling with an ever increasing volume of content from the expanding global smart phone market. Sharing photographs rapidly expanded to sharing videos, and livestreaming video apps have become increasingly popular. Platforms such as Yomapic, WarWire, EchoSec, and others now offer low cost options for anyone to search for geo-tagged images and videos posted on social media sites allowing anyone to discover near real time images from the ground, an increasing feature on reporting of breaking news stories (see Chap. 10).

Not only is the situation on the ground changing, but so too is the situation above. More satellites are being launched each year increasing the amount of free satellite imagery on a variety of platforms, driving down prices of commercial satellite imagery. High resolution video from satellites will also become increasingly common, and the use of satellite imagery by investigators has become increasingly common.

With the new tools and resources available open source investigation has become an invaluable source of information to anyone investigating a huge range of topics and for a huge range of reasons. Recent years have seen human rights organisations, activists, and journalists adopting these new tools and resources, and it is already clear that open source investigation will become a key part of many investigators' work whatever their background.

12.2 Conclusion

Whether individuals and organisations are ready for it or not, new opportunities and challenges presented by open source information from a variety of sources are something we must face now and in the future. The rapid spread of smart phone technology has put a camera in everyone's pocket, allowing individuals to immediately share almost every waking moment of their existence. With an internet connection we can all access this information from anywhere in the world, so the challenge we face is understanding how this effects our work, and how we can use this information in our own work.

It should be understood that this vast volume of online open source information allows anyone, for better or for worse, to become an investigator. What they choose to investigate could range from war crimes in a far-away country, to SWAT teams on their doorstep, and awareness of this behaviour is crucial in so many different ways.

There are also many new tools being developed for those individuals and organisations in the public and private sphere that are aimed at aiding the process of open source investigation, be it for discovering information in the first place, or organising the information that's discovered in a more accessible fashion. This chapter has covered some of those tools, but this continues to be an ever-changing area, so engagement with these new developments can be very beneficial to anyone work in the area.

References

Chivers CJ, Schmitt E (2013) In shift, Saudis are said to arm rebels in Syria. The New York Times. Retrieved from http://www.nytimes.com/2013/02/26/world/middleeast/in-shift-saudis-are-said-to-arm-rebels-in-syria.html

Czuperski M, Herbst J, Higgins E, Polyokova A, Wilson D (2015) Hiding in plain sight: Putin's war in Ukraine. Retrieved 25 July 2016, from http://www.atlanticcouncil.org/publications/reports/hiding-in-plain-sight-putin-s-war-in-ukraine-and-boris-nemtsov-s-putin-war

Czuperski M, Higgins E, Hof F, Herbst J (2016) Distract, deceive, destroy. Putin at war in Syria. Retrieved 26 July 2016, from http://publications.atlanticcouncil.org/distract-deceive-destroy/

Higgins E (2015) Russia's colin Powell moment—how the Russian government's MH17 lies were exposed. Retrieved 25 July 2016, from https://www.bellingcat.com/news/uk-and-europe/2015/07/16/russias-colin-powell-moment-how-the-russian-governments-mh17-lies-were-exposed/

Higgins E (2016) ISIS Had a social media campaign, so we tracked them down. Retrieved 26 July 2016, from https://www.bellingcat.com/news/mena/2016/05/22/isis-had-a-social-media-campaign-so-we-tracked-them-all-down/

Human Rights Watch (2012a) Syria: evidence of cluster munitions use by Syrian forces. Retrieved 25 July 2016, from https://www.hrw.org/news/2012/07/12/syria-evidence-cluster-munitions-use-syrian-forces

Human Rights Watch (2012b) Syria: despite denials, more cluster bomb attacks. Retrieved 25 July 2016, from https://www.hrw.org/news/2012/10/23/syria-despite-denials-more-cluster-bomb-attacks

Human Rights Watch (2014) Iraq: ISIS execution site located. Retrieved 25 July 2016, from https://www.hrw.org/news/2014/06/26/iraq-isis-execution-site-located

McNeal S (2015) After police asked them not to tweet about the lockdown, Belgians began tweeting cute cat pictures—BuzzFeed News. Retrieved 26 July 2016, from https://www.buzzfeed.com/stephaniemcneal/brussels-lockdown-cat-pictures?utm_term=.gnjQoWeDQ#.hyxbBVjDb

Chapter 13
Use Cases and Best Practices for LEAs

Steve Ramwell, Tony Day and Helen Gibson

Abstract The dramatic increase in the use and proliferation of the internet over the last 15–20 years has seen increasingly large amounts of personal information made, not necessarily intentionally, available online. Consequently, law enforcement agencies have recognised they must open their eyes to this information and begin to use it to their advantage, especially since one of the key benefits of utilising open source information is that it is significantly less expensive to collect than other intelligence. This chapter illustrates how OSINT has become increasingly important to LEAs. It discusses how those carrying out open source intelligence investigation work online might best go about such a practice through the use of specific techniques and how an officer may protect themselves while carrying out such an investigation. It further presents exemplar case studies in how these best practices may, or already have been, exploited in order to bring about tangible results in real investigations.

13.1 Introduction

Open Source Intelligence is not a new phenomenon. Since the 1940s those working in intelligence have leveraged open sources to their advantage (Bradbury 2011). More than law enforcement agencies (LEAs), the military have led the field in realising the value of open source intelligence, the methods and techniques for obtaining it and exacting the benefits from it. BBC Monitoring (a worldwide news aggregator with translated content),[1] in one form or another, has existed since the start of the Second World War translating and examining foreign broadcasts, whilst during the Cold War both sides created vast repositories of open source information from print magazines, books and newspapers (Schaurer and Storger 2013).

[1]http://www.bbc.co.uk/monitoring.

S. Ramwell (✉) · T. Day · H. Gibson
CENTRIC/Sheffield Hallam University, Sheffield, UK
e-mail: S.Ramwell@shu.ac.uk

B. Akhgar et al. (eds.), *Open Source Intelligence Investigation*,
Advanced Sciences and Technologies for Security Applications,
DOI 10.1007/978-3-319-47671-1_13

The dramatic increase in the use and proliferation of the internet over the last 15–20 years has seen increasingly large amounts of personal information made, not necessarily intentionally, available online. Consequently, law enforcement agencies have recognised they must open their eyes to this information and begin to use it to their advantage, especially since one of the key benefits of utilising open source information is that it is significantly less expensive to collect than other intelligence.

The Metropolitan Police have been a proponent of OSINT for a number of years. In 1999, Steve Edwards and colleagues (Edwards et al. 1999) presented an overview of their open source unit and its role within the police. They noted how officers openly and quickly accepted the idea of gathering data from open sources and its advantages of speed, efficiency, availability and cost. Furthermore, they also acknowledged that open source information does not just have one use case within the police: It can support both strategic and tactical responses.

The use of open source information within law enforcement is now becoming more widespread, and subsequently there is a real need for those carrying out such work to be exposed to some initial best-practice principles that they can exploit within their own work. This chapter firstly sets out how OSINT has become increasingly important to LEAs, it then follows with a further discussion on some of these best practice principles for OSINT investigators and then goes on to illustrate their worth within the context of some real-world use cases.

13.2 OSINT in an Increasingly Digital World

Are you a tourist or native? For those of you who were born before the internet was a readily available resource then you are a tourist. Your digital footprint was conceived at a time when the internet was invented and you personally placed it there or someone else did with or without your knowledge. The rest of you are native, born and entering a world where this means of technology is without surprise and possibly any significance; you have never known any other way of life. To compound this, a quick search of open source social media sites reveals proud parents to be, displaying images of a foetus still in the womb, a human with a digital presence prior to birth!

Basic OSINT investigation, within the context of the internet, seeks to identify the online and social footprint of these users and extract data. This is an inevitable by-product of any online visit made, including but not restricted to social media use, chat rooms, etc. Information can also be obtained from resources that seek payment for their services and whilst of value, caution should be exercised as this may cause the investigator to expose their action and or intent. The investigator should also consider that a person has no awareness of their online presence. This can be demonstrated by a photograph of them appearing on a social media site and their name being allocated to the image, all done without their knowledge.

Once an individual has participated online, consensually or otherwise, the ability to remove or erase their digital footprint becomes an extremely difficult task and given the correct training, continued practice and up to date tools an investigator can usually find a user's online footprint.

The bedrock of any investigation is the initial information provided to conduct such a search. This should be as comprehensive as possible. It is recommended that a document be designed by the investigator and provided to those seeking to use their skills. This ensures that all information is obtained where possible to the satisfaction of the investigator. It cannot be stressed enough how comprehensive this first information should be. Often nicknames, street names, schools, clubs, historic mobile numbers, associates and old email addresses are the key to locating a social profile or online presence as opposed to a simple name and date of birth. Indirect links by association can often prove to be the best method of finding persons.

Other factors to consider are the ethnicity and if known the geography that can be associated to any person being sought. Ethnicity can direct investigation towards social sites used by particular nations. The site VK[2] is used by Russian nationals. To compound this 50–80 % of global OSINT is not English. Furthermore we should have awareness of what the different social media sites are being used for and what can be accessed.

From experience it should be noted that as LEAs in the UK are now extensively working in partnership roles with local authorities, and therefore LEAs ask that the author of the request seeks any additional information from these partners. One final comment is to ensure that the request has a clear and defined objective. What is it that those who conduct the search and the investigation want to achieve?

The information being sought can best be described by use of the analogy "drilling for oil". The oil field being the raw data on the internet being drilled for and extracted, then going through a series of refining processes until a pure clean product is produced. The vast majority of open source online research follows this process (see Chaps. 6 and 7).

An exception to this rule would be to create a plausible social profile and embed it within a geographic or specific crime group to monitor and extract information. This should not be entered into lightly as a considerable amount of backstop or history needs to be created. Often bank accounts are needed and mobile numbers that are active but untraceable are required, as is an address. This persona would not engage beyond making "friends". It would only post generic comments or thoughts designed and constructed to be open questions or thoughts and not directed or focused on any person or group. The longer this account remains functional the better it will become at finding the product, as it becomes immersed into the social platform(s) with age. This style of profile requires significant dedication to maintain and remain plausible with constant checks on security and data leakage. Often the pitfalls surrounding these accounts are not the other users of the social platforms, but the software vendors themselves that monitor for unusual activity. Also legal and ethical considerations need to be taken into account (see Chaps. 17 and 18).

[2]https://vk.com.

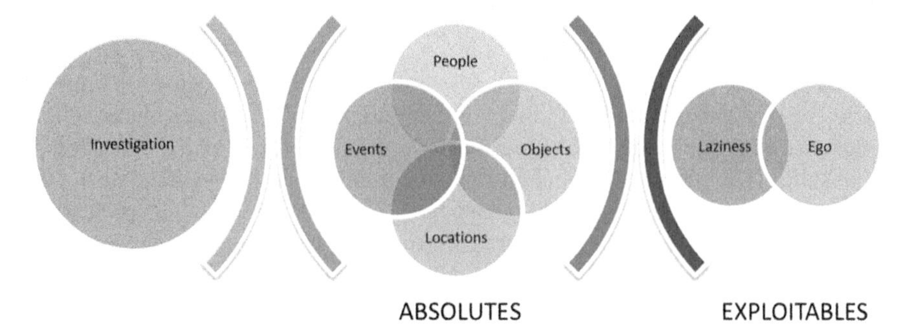

Fig. 13.1 Investigative best practices

13.3 OSINT Best Practices for LEAs

On commencing an investigation there are four points of research to be examined. At least one of these MUST exist to have an enquiry. These are *person*, *object*, *location* and *event* (abbreviated to POLE) and can be referred to as the *absolutes*. From years of experience two other areas have been established to investigate: These are *laziness* and *ego* or LEgo, referred to as the *exploitables* (Fig. 13.1).

13.3.1 Absolutes

Intelligence obtained through any technique may be attributed in terms of the people, object, location and event (POLE) data model. This is of particular importance for law enforcement agencies to remain in-line with the guide for authorised professional practice from the UK College of Policing (College of Policing 2013). Furthermore, the usage of POLE has already started a discussion (TechUK 2014) around best practices and common standards for LEAs in the sharing of information between themselves and their partners identifying that it would be instructive if the UK Home Office mandated more best practices around data within LEAs. The emergence of the use of POLE has exemplified how this approach could be beneficial.

13.3.2 Exploitables

Individuals, especially if they know they are doing something wrong, will often make an effort to cover their tracks, which can be achieved in a number of ways which includes measures such as tightening their privacy settings or using an alternative name. That being said, human beings are fallible, even when trying to be careful they make mistakes, and they often leave a trail of their activity that they may have not considered as particularly important or even be aware of at all. These weaknesses result in a number of areas that can be exploited for the purposes of open source intelligence.

The most important of these are laziness and ego (or 'LEgo' for short), which have particular relevance to microblogging and social media. *Laziness* is not just attributable to any weakness in the security of the user's online profile, but can often be more associated to insecurities in associates, friends or family of the party being sought. This openness or insecurity allows a 'backdoor' to information on the main person of enquiry, hence the need for all available information in the initial request.

Ego is the common pitfall of social media: People consciously or subconsciously use social platforms to share their emotions, thoughts and significantly images of themselves. They often display this by posing in images suggesting their mood with a comment to match. The very purpose of social media sites is to be social. Therefore ample opportunity can be found to develop this (see Chap. 6).

Tracks are often left uncovered due to individual laziness, even when social media sites such as Facebook have security and privacy settings, which would close many of these gaps. Though laziness is only one facet, egotistical behaviour on the other side is exploitable as it often results in individuals bragging about their (illegal) activities. There are many entertaining examples such as the Michael Baker case in Kentucky, U.S. who decided to publish an image of himself siphoning fuel from a police car (Murray 2012) or the case of Maxi Sopo who, after committing bank fraud, evaded police by leaving for Mexico, only to brag about his idyllic whereabouts online and to find that one of his connections was in fact a justice department affiliation (Topping 2009). Another practice often observed is where individuals publish images of themselves brandishing illegal paraphernalia such as weapons, drugs, and copious amounts of cash.

On the other hand, some individuals simply do not understand how to cover their tracks online or choose not to further understand how to. This is often seen in traditional organised crime, where the lowest ranking members of the group are likely to be less educated and have greater potential to make mistakes or leak information.

Online social networks in particular provide rich information about offline social networks, which is often highly accessible in the open source arena. In the UK, population coverage of Twitter is around a fifth (Wang 2013) and around half for Facebook (Kiss 2013), and although many users may have more friends online than offline (often twice as many; Quinn 2011), this data can provide a deep insight into actual social connections. Exploiting these connections often makes it possible to discover individual user activity, even when that user has effectively locked down their account to prying eyes. Their public communications and posts between themselves and connections are often subject to the privacy settings of the other party, providing opportunities for the open source intelligence investigator. Other instances have been observed where online connections, being friends, family members or mere acquaintances, have published compromising content for the target individual—such as a photo.

13.3.3 Information Auditing

An important but often time-consuming and cumbersome process when performing open source intelligence, is the need to audit each step of an investigation. Reasons for doing so may include legal obligation to process, identification of research gaps, and maintaining oversight into the depth of research, among others. Common to all of these cases is the need to actively choose and justify following any path in the intelligence gathering process, as without justification of potential relevance, there may be privacy and ethical complications (see Chaps. 17 and 18).

Auditing can be carried out in a number of ways. The most basic and obvious, though time consuming approach is to manually audit every step in a document or spreadsheet. In other words, to manually record each URL and media item that is accessed whilst gathering intelligence for a particular case—an approach which is error prone. Capturing such data through an automated process is straightforward and common practice. Generally, computer systems capture all or most user, application and system events in event logs or log files, but in more highly scaled environments, databases can be commonplace for this role.

Automatic screen recording on the other hand is less common, and is often a manual process carried out by the individual user. However, it is possible to provide screen recording capabilities through an automated process outside the control of the investigator. An approach that could perhaps become a standard model in law enforcement agencies or anywhere needing accountability.

Finally, the automatic capturing of specific text and media whilst the investigator carries out their investigation is also an important possibility. Such an approach could be a powerful way to combine the processing and analytical capabilities of an automated system with the oversight and direction of an investigative mind-set.

13.3.4 Strategic Data Acquisition

It is feasible in open source intelligence to gather wider, more strategic data, but restraint should also be used to avoid gathering too much or—more importantly—to ensure due process in justification carried out for legal and ethical reasons. Commonly, when the topic of web crawling is discussed, it is often assumed that the optimal approach is that utilised by search engine providers—to access links on a web page recursively. In doing so, it takes little time to become over-burdened with data, and especially data that may be completely irrelevant to the needs of the investigation. Not only does this approach produce too much noise, but it may also be considered as mass surveillance.

Storage, bandwidth and processing can be very expensive when big data starts to become involved, but these are not the only costs. Accessing data in search engines or social media sites from automated systems can often be impossible without circumventing their policies or technologies, an act in itself which can introduce legal complications. Where it is possible, there are often charges incurred for the privilege. For instance, using the Bing search engine API (application programming interface) for web searches only, a package of up to 500,000 requests currently costs around $2000 per month.[3] While other APIs such as Twitter's REST API are limited to a 15 requests per 15 min window for request pages of individual user posts[4] (see also Part 2 on methods and tools).

Where the investigation is interested in more general data, it makes sense to crawl specific start points related to the target information to a particular depth. Depth in crawling considers how many links the crawler will access recursively. Monitoring sources can be useful in a more strategic context. This consists of regularly revisiting particular web pages looking for changes or links to new unseen web pages.

13.3.5 OSINT Pitfalls

13.3.5.1 Leakage

The use of online tools that aid the investigator with finding, decoding or enriching data could be potential sources of leakage or social engineering. It is not inconceivable that there are seemingly secure and useful online tools to aid the investigator that in turn could be assisting the investigated by way of alerting them. Also, it is possible that once data has been obtained by a third party service, it could then be exploited in a manner that could compromise any investigation. As a result of these possibilities, the investigator must consider information leakage by way of ensuring that, where possible, confidential or critical data is not unintentionally provided to an unverified third party.

A run of the mill example of this is the situation in which an investigator provides an image to a third party forum or web application in order to extract exchangeable image file format (EXIF) data such as geolocation, when there are many offline tools that will do a similar job. Once the image is uploaded, it is unclear to the investigator how this image may then be used.

13.3.5.2 Anonymization

Whilst it is a popular issue that criminals often exploit the availability of online anonymization technologies to cover their tracks, the same should be true for the open source intelligence investigator.

[3]https://datamarket.azure.com/dataset/bing/search.

[4]https://dev.twitter.com/rest/public/rate-limits.

The internet protocol (IP) address, which identifies the distinct identity of the source and target of an internet request, can be and almost always is tracked by web applications such as social networks. Combining this identity with multiple instances of unusual social network behaviour—where carrying out investigative activities is unusual when compared with the behaviour patterns of average users—may lead to the compromising of the investigation, or possible refused access to the server through account or IP address blacklisting.

Owing to this apparent vulnerability on the part of the investigator, it may be necessary to take extra steps in order to protect their online identity and to do so in a manner that is easily configurable. For instance, allowing them to switch IP addresses on a regular basis in order to keep the traces for individual investigations isolated from one another. Hiding the source IP address can be achieved through the use of web proxies which simply mask the address, VPNs which route and encrypt requests via an intermediary destination and anonymity or onion networks such as TOR, which not only encrypts the request but also scrambles its route from source to target through various randomly allocated intermediary locations.

13.3.5.3 Crowd-Sourcing and Vigilantism

The growing popularity of the crowd-sourcing movement also has roots in OSINT with multiple people coming together using open source data to attempt to investigate or solve a problem or crime (see Chap. 12). However, in the past crowd-sourcing has also spilled over into vigilantism with a number of catastrophic effects. The large crowd-sourcing effort surrounding the Boston Marathon exemplified a number of these problems. First, a number of completely innocent people were incorrectly identified as potential suspects causing emotional pain and suffering to themselves and their families (Lee 2013). Furthermore, the number of incorrectly identified suspects actually caused the FBI to go public with the names and images of the Tsarnaev brothers earlier than they would normally, simply to stem the flow of people being wrongly targeted (Montgomery et al. 2013). While this is less of an issue for an individual investigator, LEAs should be aware of the consequences of how these crowd-sourced campaigns can spring up and impact on their own investigations.

13.3.5.4 Corrupting the Chain of Evidence

All data collected and the means of its collection, even open source data, must not contravene the European Convention on Human Rights (ECHR)[5] and, in particular, Article 8 which protects a person's right to privacy. Moreover, in the UK, this is

[5]http://www.coe.int/en/web/conventions/full-list/-/conventions/treaty/005.

further emphasised in section 6 of the Human Rights Act[6] which states that a public authority cannot act in a way which is incompatible with this ECHR. Thus investigators have to apply significant care to understand what personal data they are allowed access within the context of that investigation to ensure that it can be used in future. In addition to this, investigators must also abide by data protection laws (such as the Data Protection Act[7] and the General Data Protection Regulation[8]). Investigators should also ensure that they use anonymous laptop when conducting such investigations and make sure that they take at least screen captures of any digital evidence they obtain that contain both a time and date.

13.3.5.5 Source Validation

An OSINT investigator must not get caught out by the fact that all open sources are not made equal and that some may be more reputable than others. It may be that the site itself is presenting a particularly subjective opinion or that only a small component of it is not reputable (e.g., a specific tweet or Facebook post). Some methods for exploring and validating the credibility of open sources are explored in Chap. 7.

13.4 LEA Usage of OSINT in Investigations: Case Examples

The use of open source information leading to open source intelligence is having a real impact on modern day policing. Information gleaned from open sources is leading to arrests for serious criminal activities. This section describes some examples of how the practice of open source intelligence has assisted in police investigations. The activities described in this section were all conducted using freely available software and social platforms. No interaction of any kind has taken place with any of the parties involved.

13.4.1 Exploiting Friendships in an Armed Robbery Case

FL was wanted for an armed robbery. Owing to the gravity of this offence considerable resources and assets had been deployed to locate the suspect, however his current whereabouts were unknown. Specialist teams spent two days conducting

[6]http://www.legislation.gov.uk/ukpga/1998/42/section/6.

[7]http://www.legislation.gov.uk/ukpga/1998/29/contents.

[8]http://eur-lex.europa.eu/legal-content/EN/TXT/?uri=uriserv:OJ.L_.2016.119.01.0001.01.ENG&toc=OJ:L:2016:119:TOC.

observations on an associate's addresses at a significant cost, which did not result in the obtaining of the required information. In order to move the investigation forwards, help was then sought from an internet investigation unit known as the "Technology Intelligence Unit" (TIU).

Within two hours the TIU located FL on the Facebook social media site. Owing to the personal security setting used by FL his Facebook site revealed little public information. Furthermore, FL exposed little personal information meaning the team could have reached a dead end. However, postings made onto the account by another person disclosed a previously unknown girlfriend, RB who was an active member on social media at the time of viewing. The social media account of RB had little or no privacy settings enabled and displayed images of both her and FL on a beach. Through supplementary research on the account it was established that the location of the beach was in Cornwall, UK. It was further revealed that she had an infant. Additionally, images of her accommodation in the East Midlands were provided.

The consequence of these discoveries was the following positive outcomes. First, the fact that FL was not in the area for which the force was responsible allowed the specialist teams to be stood down saving significant costs. Secondly, research of RB on publicly available data sites offered several potential residential addresses where she may have been residing. These were all compared using Google Street View against an image posted on her social media site. From this image it was possible to identify the exact address of RB. All this information was provided in a readable chronological format to the investigating team. FL was subsequently arrested at RB's address in a safe and controlled manner.

This success illustrates the use of the absolute POLE points: both the person and the location were identified. Then further research using the exploitables LEgo showed the images of them posing for a selfie image on a beach in Cornwall and the image of her property.

The fact that the TIU information provided a known location also allowed the investigating team the opportunity for other tactics to be deployed in detaining FL.

13.4.2 Locating Wanted People Through Social Media

An individual, known as KS, was wanted on a Crown Court bench warrant, i.e., an order, within UK law, from the court directing that the person be arrested.

He had been sought for over three months by two different police forces, but the investigation had continued without much success. More than eight police officers had conducted research doing both house to house and council visits in order to try to locate the whereabouts of KS. These actions had amounted to a significant use of police time and costs that could be better used in matters of a greater priority. The TIU was tasked with seeing if they could help in identifying his location, and consequently he was arrested within one hour!

The TIU conducted their investigation as follows. A social media profile was located for KS. He had posted recent comments stating that he was working with chemicals for the Environment Agency. Additionally he put up a picture of a canal lock with a comment that this was his location when he wanted to take a cigarette break. Research on the internet identified the exact address of the Environment Agency and its proximity to the canal lock. Police officers were notified who travelled to the location and arrested him. The absolute used was the name of KS and the exploitables being the images and comments concerning where a place of work was and an image showing where the cigarette break was.

A second example of using OSINT in this manner was carried out by inspecting the Facebook account of a wanted person named ABC. He had breached his bail conditions and had been recalled to prison. Immediate inspection of his Facebook account provided little personal information. However, an image of him had been posted onto the account wearing a new looking, extremely bright red shiny puffa jacket. This had drawn several comments.

Owing to the uniqueness of the jacket he was wearing, a capture was made of the image and placed onto the electronic briefing system used by the police force. The following day a police officer on patrol in a local shopping centre spotted a male wearing the jacket and recognise it as being the same as the male was wearing from the briefing. Other officers were called in and a rapid and safe controlled arrest was made. Significantly the officer who spotted the male and made this arrest had never seen the person ABC before.

These examples again clearly illustrate how the LEgo effect can be easily exploited by OSINT.

13.4.3 Locating a Sex Offender

Police officers who were seeking to arrest AJ for a rape offence were unable to locate him in their home force area. Using the name and details of other family members, a Facebook account was located for AJ.

An image posted on the account showed a house surrounded by scaffolding, a metal storage container and a vehicle parked on the road. The vehicle clearly showed its registration number. The police initially attempted to check the details of the vehicle registration plate, but it failed to provide a current keeper for the vehicle. The image was inspected again, and it was noted that on the metal container was a fixed sign showing a phone number and the name of the company who had supplied it. By following up this line of enquiry with the company the police were able to establish the location of the container. This provided the investigating officers with an address that led to the arrest of AJ.

Once again LEgo plays a key role in developing the enquiry. As mentioned in Sect. 13.4.1 on the arrest of FL, paying attention to images posted may result in additional information that the individual may have provided without awareness.

13.4.4 Proactive Investigation Following a Terrorist Attack

On the afternoon of 22nd May 2013 the British soldier fusilier Lee Rigby was run down and hacked to death by Michael Adebolajo and Michael Adebowale in broad daylight outside barracks in London. Both killers were identified as being British of Nigerian descent, raised as Christians and converts to Islam. The killers made no attempt to leave the scene of the attack, and it was rapidly broadcast on media platforms globally.

This crime was being monitored by the TIU in case of any reaction that required police intervention locally. During the course of this monitoring, online activity was noted on the social media platform Tweetdeck,[9] which is a dashboard application that facilitates the monitoring of Twitter in a more accessible way than through the Twitter web interface directly. This showed a posting by an online Australian news group who claimed to have had communication with a friend of one of the people responsible for the murder: Michael Adebolajo. The webpage was inspected and a full page article was online which named this friend as Abu Nusaybah. His real name is Ibrahim Hassan. Following this claim, Twitter was searched by a member of the TIU and an account was found for Abu Nusaybah. Details from the account and other information was captured.

The TIU also tried to search Facebook, but found no account. However, using the profile image posted to the Twitter account of Nusaybah, a Google image search was conducted. This revealed a cached web page of the Facebook account for Nusaybah. Through the further inspection of this page the TIU were able to uncover videos, images, postings and friends of Nusaybah, which were all captured and utilised in further investigations. In particular, the TIU identified that the videos contained material which contravened the terrorism laws of the United Kingdom. All this information was rapidly provided to partner agencies. Nusaybah was arrested two days after the Woolwich murder and moments after giving an interview to BBC Newsnight about his friend. He was subsequently jailed after admitting two terror charges for posting lectures by fanatical Islamists online and encouraging terrorism.

The data collected was all obtained using open source tactics and the case provides a good example of the LEgo exploitables as it was an individual's ego that allowed rapid identification and collection of intelligence. In addition, in contrast to the previous examples, it shows how OSINT can be employed not only when there is a specific initial target, as in the previous four cases, but also when there is a suspicion or an expectation that criminal activity may take place. In this case, the TIU began with a proactive investigation by monitoring information posted to social media in the aftermath of the attack rather than participating in an already ongoing investigation into a specific individual.

---

[9]https://tweetdeck.twitter.com/.

13.5 Going Undercover on Social Media

The information obtained in the above use cases was only gathered through the monitoring of profiles without actually interfering or interacting with them on social media. The usage and befriending of those suspected of criminal activity on social media sites by those in law enforcement can be seen to walk the line between ethical and unethical practice and, at least in the United Kingdom, is governed strictly with the use of non-attributable computers and logging of how and when the profile is in use (Association of Chief Police Officers 2013; HMIC 2014; Tickle 2012; see also Chaps. 17 and 18). However, there are numerous examples of police, especially in the U.S., going undercover on Facebook to get closer to the criminals they are trying to catch.

For example, Officer Michael Rodrigues (Yaniv 2012) made friends with numerous members of a gang associated with burglaries in Brooklyn. He was then able to know when they were planning their next 'job' as they talked of it openly on Facebook as well as seeing the images they posted afterwards of the items that had been stolen. Again, this highlights the LEgo principle: The members were too lazy and also perhaps too egotistical to vet those requesting to be their friend on Facebook, and they wanted show off about the items they had managed to steal.

Similar undercover work has existed previously with it being common for officers to enter chat rooms (Martellozzo 2015) pretending to be young children in order to get the attention of sex offenders (Tickle 2012) or by infiltrating forums that facilitate the exchange of images (CEOP 2008). This work has also now extended to social networks, which due to their popularity amongst young people, provide an opportunity for child sexual grooming (Hope 2013), but also for officers to go undercover on such sites and catch potential offenders themselves (Silk 2015). While these tactics are not unlawful per se, LEAs need to very careful that they do not overstep the line between legitimately creating an opportunity for others to commit crime and unlawful entrapment/incitement (see Chaps. 17 and 18).

13.6 Conclusions

This chapter has explored how Open Source Intelligence is changing the way that law enforcement conduct their investigations. We discussed how those carrying out open source intelligence investigation work online might best go about such a practice through the use of specific techniques and how an officer may protect themselves while carrying out such an investigation. The second half of the chapter then went on to present some exemplar case studies in how these best practices may, or already have been, exploited in order to bring about tangible results in real investigations.

References

Association of Chief Police Officers (2013) Online research and investigation. Available online: http://library.college.police.uk/docs/appref/online-research-and-investigation-guidance.pdf

Bradbury D (2011) In plain view: open source intelligence. Comput Fraud Secur 4:5–9

CEOP (2008) UK police uncover global online paedophile network. Available online: https://www.ceop.police.uk/Media-Centre/Press-releases/2008/UK-police-uncover-global-online-paedophile-network/

College of Policing (2013) Information management: collection and recording. In: Authorised professional practice; Available online: https://www.app.college.police.uk/app-content/information-management/management-of-police-information/collection-and-recording/#categorising-police-information

Edwards S, Constable D, Scotland, N (1999) SO11 open source unit presentation. In: EuroIntel' 99 PROCEEDINGS E1-European intelligence forum "creating a virtual previous next intelligence community in the European region", pp. 1–33

HMIC (Her Majesty's Inspectorate of Constabulary) (2014) An Inspection of undercover policing in England and Wales. Available online: https://www.justiceinspectorates.gov.uk/hmic/publications/an-inspection-of-undercover-policing-in-england-and-wales/

Hope C (2013) Facebook is a "major location for online child sexual grooming", head of child protection agency says. In: The telegraph. http://www.telegraph.co.uk/technology/facebook/10380631/Facebook-is-a-major-location-for-online-child-sexual-grooming-head-of-child-protection-agency-says.html

Kiss J (2013) Facebook UK loses 600,000 users in December. In: The guardian. https://www.theguardian.com/technology/2013/jan/14/facebook-loses-uk-users-december

Lee D (2013) Boston bombing: how internet detectives got it very wrong. In: BBC News. http://www.bbc.co.uk/news/technology-22214511

Martellozzo E (2015) Policing online child sexual abuse-the British experience. Eur J Policing Stud 3(1):32–52

Montgomery D, Horwitz S, Fisher M (2013) Police, citizens and technology factor into Boston bombing probe. In: The Washington post. https://www.washingtonpost.com/world/national-security/inside-the-investigation-of-the-boston-marathon-bombing/2013/04/20/19d8c322-a8ff-11e2-b029-8fb7e977ef71_print.html

Murray R (2012) Man steals gas from cop car, gets caught after he posts pic of theft on Facebook. In: NY Daily News. http://www.nydailynews.com/news/crime/man-steals-gas-car-caught-posts-pic-theft-facebook-article-1.1063916

Quinn B (2011) Social network users have twice as many friends online as in real life. In: The guardian. http://www.theguardian.com/media/2011/may/09/social-network-users-friends-online

Schaurer F, Storger J (2013) The evolution of Open Source Intelligence (OSINT). Comput Hum Behav 19:53–56

Silk H (2015). Paedophile snared after arranging to meet 12-year-old victim who was actually an undercover police officer. In: The mirror. http://www.mirror.co.uk/news/uk-news/paedophile-snared-after-arranging-meet-6414747

TechUK (2014) TechUK launches "Breaking down barriers" report. In: TechUK. https://www.techuk.org/insights/reports/item/2302-techuk-launches-breaking-down-barriers-report

Tickle L (2012) How police investigators are catching paedophiles online. In: The guardian. http://www.theguardian.com/social-care-network/2012/aug/22/police-investigators-catching-paedophiles-online

Topping A (2009) Fugitive caught after updating his status on Facebook. In: The guardian. https://www.theguardian.com/technology/2009/oct/14/mexico-fugitive-facebook-arrest

Wang T (2013) 2+ years, 15MM users in UK, teams in 6 EU countries, excited to return to HQ and home. Proud to hand off to incredible European leadership! In: Twitter. https://twitter.com/TonyW/status/375889809153462272

Yaniv O (2012) Cop helps take down Brooklyn crew accused of burglary spree by friending them on Facebook. In: NY Daily News. http://www.nydailynews.com/new-york/helps-brooklyn-crew-accused-burglary-spree-friending-facebook-article-1.1086892

Chapter 14
OSINT in the Context of Cyber-Security

Fahimeh Tabatabaei and Douglas Wells

Abstract The impact of cyber-crime has necessitated intelligence and law enforcement agencies across the world to tackle cyber threats. All sectors are now facing similar dilemmas of how to best mitigate against cyber-crime and how to promote security effectively to people and organizations. Extracting unique and high value intelligence by harvesting public records to create a comprehensive profile of certain targets is emerging rapidly as an important means for the intelligence community. As the amount of available open sources rapidly increases, countering cyber-crime increasingly depends upon advanced software tools and techniques to collect and process the information in an effective and efficient manner. This chapter reviews current efforts of employing open source data for cyber-criminal investigations developing an integrative OSINT Cybercrime Investigation Framework.

14.1 Introduction

During the 21st century, the digital world has acted as a 'double-edged sword' (Gregory and Glance 2013; Yuan and Chen 2012). Through the revolution of publicly accessible sources (i.e., open sources), the digital world has provided modern society with enormous advantages, whilst at the same time, issues of information insecurity have brought to light vulnerabilities and weaknesses (Hobbs et al. 2014; Yuan and Chen 2012). The shared infrastructure of the internet creates the potential for interwoven vulnerabilities across all users (Appel 2011): "The viruses, hackers, leakage of secure and private information, system failures, and interruption of services" appeared in an abysmal stream (Yuan and Chen 2012).

F. Tabatabaei (✉)
Mehr Alborz University, Tehran, Iran
e-mail: ftt.tabatabaei@gmail.com

D. Wells
CENTRIC/Sheffield Hallam University, Sheffield, UK

B. Akhgar et al. (eds.), *Open Source Intelligence Investigation*,
Advanced Sciences and Technologies for Security Applications,
DOI 10.1007/978-3-319-47671-1_14

(Wall 2007; 2005) and Nykodym et al. (2005) discussed that cyberspace possess four unique features called 'transformative keys' for criminals to commit crimes:

1. Globalization, which provides offenders with new opportunities to exceed conventional boundaries
2. Distributed networks, which create new opportunities for victimization
3. Synopticism and Panopticism, which enable surveillance capability on victims remotely
4. Data trails, which may allow new opportunities for criminals to commit identity theft

In addition to the above, Hobbs et al. (2014) claim that one of the main trends of the recent years' internet development is that "connection to the Internet may be a very risky endeavour."

As well as the epidemic use and advancement of mobile communication technology, the use of open sources propagates the fields of intelligence, politics and business (Hobbs et al. 2014). Whilst traditional sources and information channels (news outlets, databases, encyclopedias, etc.) have been forced to adapt to the new virtual space to maintain their presence, many 'new' media sources (especially from social media) disseminate large amounts of user-generated content that has subsequently reshaped the information landscape. Examples of the scale of user generated information include the 500 million Tweets per day on Twitter and the 98 million daily blog posts on Tumblr (Hobbs et al. 2014) as well as millions of individual personal Facebook pages. With the evolution of the information landscape, it has been essential that law enforcement agencies now harvest relevant content through investigations and regulated surveillance, to prevent and detect terrorist activities (Koops et al. 2013).

As has been considered in earlier chapters the term Open Source Intelligence (OSINT) emanates from national security services and law enforcement agencies (Kapow Software 2013). OSINT for our purposes here is predominantly defined as, "*the scanning, finding, collecting, extracting, utilizing, validation, analysis, and sharing intelligence with intelligence-seeking consumers of open sources and publicly available data from unclassified, non-secret sources*" (Fleisher 2008; Koops et al. 2013). OSINT encompasses various public sources such as academic publications (research papers, conference publications, etc.), media sources (newspaper, radio channels, television, etc.), web content (websites, social media, etc.), and public data (open government documents, public companies announcements, etc.) (Chauhan and Panda 2015a, b).

OSINT was traditionally described by searching publicly available published sources (Burwell 2004) such as books, journals, magazines, pamphlets, reports and the like. This is often referred to *literature intelligence* or *LITINT* (Clark 2004). However, the rapid growth of digital media sources throughout the web and public communication airwaves have enlarged the scope of Open Source activities (Boncella 2003). Since there are diverse public online sources from which we can collect intelligence, this type of OSINT is described as *WEBINT* by many authors.

Indeed, the terms *WEBINT* and *OSINT* are often used interchangeably (Chauhan and Panda 2015a, b). Social media such as social networks, media sharing communities and collaborative projects are areas where the majority of user generated content is produced. Social Media Intelligence or *SOCMINT* refers to 'the intelligence that is collected from social media sites'. Some of their information may be openly accessible without any kind of authentication required prior to investigation (Omand et al. 2014; pp. 36; Chauhan and Panda 2015a, b).

Many law enforcement and security agencies are turning towards OSINT for the additional breadth and depth of information to reinforce and help validate contextual knowledge (see for instance Chap. 13). Unlike typical IT systems, which can adopt only a limited range of input, OSINT data sources are as varied as the internet itself and will continue to evolve as technology standards expand (Kapow Software 2013): "OSINT can provide a background, fill epistemic gaps and create links between seemingly unrelated sources, resulting in an altogether more complete intelligence picture" (Hobbs et al. 2014, p. 2).

OSINT increasingly depends on the assimilation of all-source collection and analysis. Such intelligence is an essential part of "national security, competitive intelligence, benchmarking, and even data mining within the enterprise" (Appel 2011, p. xvii). The process of OSINT is shown in Fig. 14.1. OSINT has been used for a long time by the government, military and in the corporate world to keep an eye on the competition and to have a competitive advantage (Chauhan and Panda 2015a, b). Also a great number of internet users' enjoy legal activities "from communications and commerce to games, dating, and blogging" (Appel 2011, p. 6), and OSINT plays a critical role in this context.

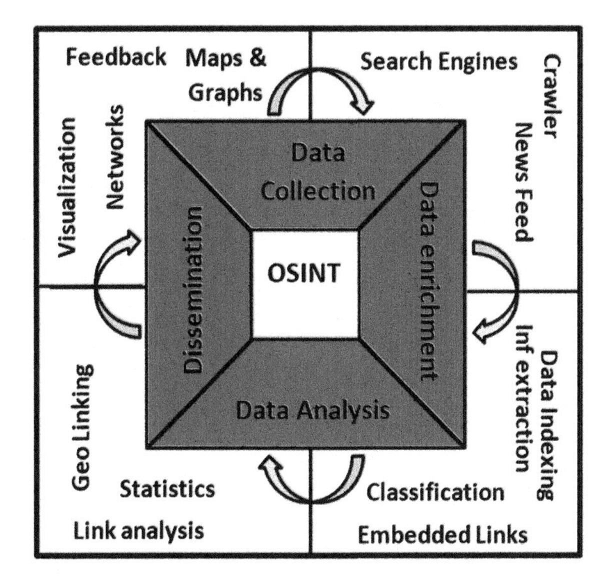

Fig. 14.1 The OSINT process

The current chapter aims to present an in-depth review of the role of OSINT in cyber security context. Cybercrime and its related applications are explored such as the concepts of the Deep and Dark Web, anonymity and cyber-attacks. Further, it will review OSINT collection and analysis tools and techniques with a glance at related works as main parts of its contribution. Finally, these related works are articulated alongside the cyber threat domain and its open sources to establish a 'big picture' of this topic.

14.2 The Importance of OSINT with a View on Cyber Security

Increases in the quantity and type of challenges for contemporary, national security, intelligence, law enforcement and security practitioners have sped up the use of open sources in the internet to help draw out a more cohesive picture of people, entities and activities (Appel 2011; also Chaps. 2, 3, 12 and 13). A recent PWC[1] American Survey (2015) entitled "Key findings from the 2015 US State of Cybercrime Survey" from more than 500 executives of US businesses, law enforcement services and government agencies articulates that "cybercrime continues to make headlines and cause headaches among business executives." 76 % of cyber-security leaders said they are more concerned about cyber threats this year: "Cybersecurity incidents are not only increasing in number, they are also becoming progressively destructive and target a broadening array of information and attack vectors" (PWC 2015).

In a report of the U.S. Office of Homeland Security, critical mission areas, wherein the adoption of OSINT is vital, include general-intelligence, advanced warnings, domestic counter-terrorism, protecting critical infrastructure (including cyberspace), defending against catastrophic terrorism and emergency preparedness and response (Chen et al. 2012). Therefore, intelligence, security and public safety agencies are gathering large volumes of data from multiple sources, including the criminal records of terrorism incidents and from cyber security threats (Chen et al. 2012).

Glassman and Kang (2012) discussed OSINT as the output of changing human–information relationships resulting from the emergence and growing dominance of the World Wide Web in everyday life. Socially inappropriate behaviour has been detected in Web sites, blogs and online-communities of all kinds "from child exploitation to fraud, extremism, radicalisation, harassment, identity theft, and private-information leaks." Identity theft and the distribution of illegally "copied films, TV shows, music, software, and hardware designs are good examples of how the Internet has magnified the impact of crime" (Hobbs et al. 2014).

---

[1]PricewaterhouseCoopers.

The globalization, speed of dissemination, anonymity, cross-border nature of the internet, and the lack of appropriate legislation or international agreements have made some of them very wide-spread, and very difficult to litigate (Kim et al. 2011). There exist different types of dark sides of the internet, but also applications to shed on the dark sides, comprising both technology-centric and non-technology-centric ones. *Technology-centric dark sides* include spam, malware, hacking, Denial of Service (DoS) attacks, phishing, click fraud and violation of digital property rights. *Non-technology-centric dark sides* include online scams and frauds, physical harm, cyber-bullying, spreading false or private information and illegal online gambling. *Non-technology responses* include legislation, law enforcement, litigation, international collaboration, civic actions, education and awareness and caution by people (Kim et al. 2011).

Computer crime and digital evidence are growing by orders that are as yet unmeasured except by occasional surveys (Hobbs et al. 2014). To an intelligence analyst, the internet is pivotal owing to the capabilities of browsers, search engines, web sites, databases, indexing, searching and analytical applications (Appel 2011). However, there are key issues which can distract from the right direction of OSINT projects such as harvesting data from big open records on the internet and the integration of data to add the capability of OSINT project parameters (Kapow Software 2013).

14.3 Cyber Threats: Terminology and Classification

Cyber-crime[2] is any illegal activity arising from one or more internet components such as Web sites, chat rooms or e-mail (Govil and Govil 2007) and commonly defined as "criminal offenses committed using the internet or another computer network as a component of the crime" (Agrawal et al. 2014). In 2007, the European Commission (EC) identified three different types of cyber-crime: traditional forms of crime using cyber relating to, for example, forgery, web shops and e-market types of fraud, illegal content such as child pornography and 'crimes unique to electronic networks' (e.g., hacking and Denial of Service attacks). Burden and Palmer (2003) distinguished 'true' cybercrime (i.e., dishonest or malicious acts, which would not exist outside of an online environment) from crimes which are simply 'e-enabled'. They presented 'true' cyber-crimes as hacking, dissemination of viruses, cyber-vandalism, domain name hijacking, Denial of Service Attacks (DoS/DDoS), in contrast to 'e-enabled' crimes such as misuse of credit cards, information theft, defamation, black mailing, cyber-pornography, hate sites, money laundering, copyright infringements, cyber-terrorism and encryption. Evidently, crime has infiltrated the Web 2.0 "along with all other types of human activities" (Hobbs et al. 2014).

[2]In this chapter, the terms computer crime, internet crime, online crimes, hi-tech crimes, information technology crime and cyber-crimes are being used interchangeably.

Cyber-attacks are increasingly being considered to be of the utmost severity for national security. Such attacks disrupt legitimate network operations and include deliberate detrimental effects towards network devices, overloading a network and denying services to a network to legitimate users. An attacker may also exploit loop holes, bugs, and misconfigurations in software services to disrupt normal network activities (Hoque et al. 2014).

The attacker's goal is to perform reconnaissance by restraining the power of freely available information extracted using different intelligence gathering ways before executing a targeted attack (Enbody and Sood 2014). Meanwhile, "secrecy" is a key part of any organized cyber-attack. Actions can be hidden behind a mask of anonymity varying from the use of ubiquitous cyber-cafes to sophisticated efforts to covert internet routing (Govil and Govil 2007). Cyber-criminals exploit opportunities for anonymity and disguise over web-based communication to navigate malicious activities such as phishing, spamming, blackmail, identity theft and drug trafficking (Gottschalk et al. 2011; Igbal et al. 2012). Network security tools facilitate network attackers in addition to network defenders in recognizing network vulnerabilities and colleting site statistics. Network attackers attempt to identify security breaches based on common services open on a host gathering relevant information for launching a successful attack.

Kshetri (2005) classified cyber-attacks into two types: *targeted* and *opportunistic* attacks. In *targeted attacks* specific tools are applied against specific cyber targets, which makes this type more dangerous than the other one. *Opportunistic attacks* entail the disseminating of worms and viruses deploying indiscriminately across the internet (Hoqu et al. 2014). Figure 14.2 provides a taxonomy of cyber-crime types (what) with their motives (why) and the tools to commit them (how).

To counter the ability of organized cyber-crime to operate remotely through untraceable accounts and compromised computers and fighting against online crime gangs it is therefore essential to supply tools to LEAs and actors in national security for the detection, classification and defence from various types of attacks (Simmons et al. 2014).

14.4 Cyber-Crime Investigations

14.4.1 Approaches, Methods and Techniques

Current information professionals draw from a variety of methods for organizing open sources including but not limited to web-link analysis, metrics, scanning methods, source mapping, text mining, ontology creation, blog analysis and pattern recognition methods. Algorithms are developed using computational topology, hyper-graphs, social network analysis (SNA), Knowledge Discovery and Data Mining (KDD), agent based simulations, dynamic information systems analysis, amongst others (Brantingham 2011).

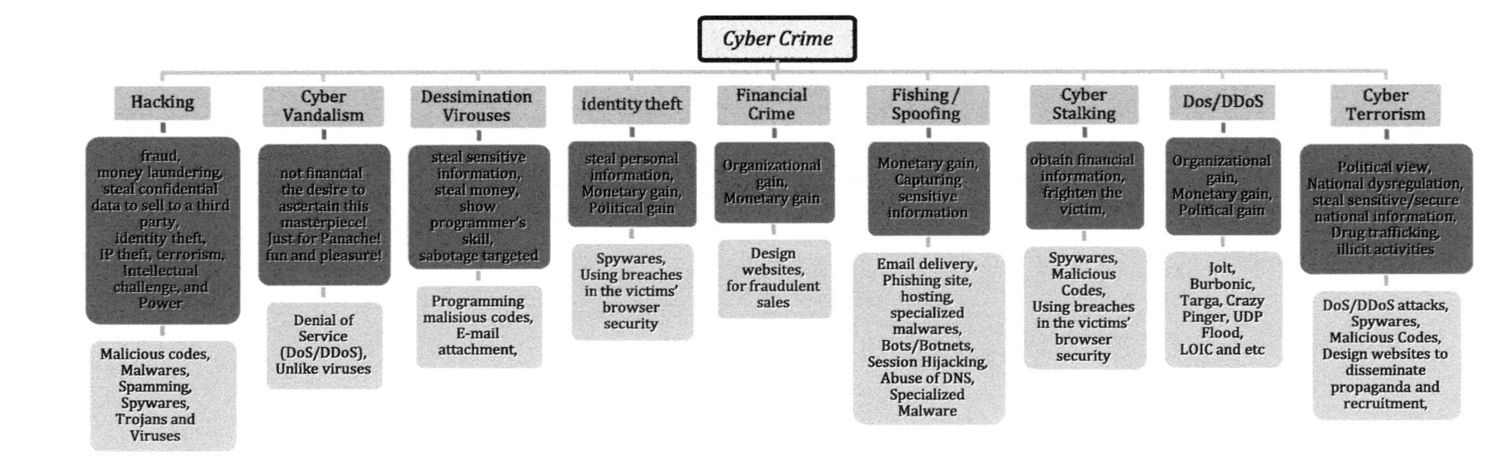

Fig. 14.2 Cyber Crime types: Which-Why-How (Type, Motives, Committing Tools and techs)

Table 14.1 Tools for the collection, storage and classification of open source data

Tools purpose	Application/description of tool(s)
Data encoding	The term *encoding* refers to the process of putting a sequence of characters into a special format for transmission or storage purposes. In a web environment, relevant datasets are recovered from data services available either locally or globally on the internet. Depending on the service and the type of information, data can be presented in different formats. Modelling platforms are required to interact with a mixture of data formats including plain text, markup languages and binary files (Vitolo et al. 2015; Webopedia.com n.d.). *Examples*: The Geoinformatics for Geochemistry System (database web services adopting plain text format), base 64online Encoder, XML encoder
Data acquisition	The automatic collection of data from various sources (e.g., sensors and readers in a factory, laboratory, medical or scientific environment). Data acquisition has usually been conducted via data access points and web links such as http or ftp pages, but required periodical updates. Using a catalogue allows a screening of available data sources before their acquisition (Ames et al. 2012; Vitolo et al. 2015). *Examples*: Meta-data catalogues
Data provenance	This term is used to refer to the process of tracing and recording the origins of data and its movement between databases. Behind the concept of provenance is the dynamic nature of data. Instead of creating different copies of the same dataset, it is important to keep track of changes and store a record of the process that led to the current state. Data provenance can, in this way, guarantee reliability of data and reproducibility of results. Provenance is now an increasingly important issue in scientific databases, where it is central to the validation of data for inspecting and verifying quality, usability and reliability of data (particularly in Semantic Web Services) (Buneman et al. 2000; Szomszor and Moreau 2003; Tilmes et al. 2010; Vitolo et al. 2015). *Examples*: Distributed version Control Systems such as Git, Mercurial[a]
Data storage	This term refers to the practice of storing electronic data with a third party service accessed via the internet. It is an alternative to traditional local storage (e.g., disk or tape drives) and portable storage (e.g., optical media or flash drives). It can also be called 'hosted storage', 'internet storage' or 'cloud storage'. Relational databases (DB) are currently the best choice in storing and sharing data (Vitolo et al. 2015; Webopedi.com n.d.). *Examples*: Postgre SQL, MySQL, Oracle, NoSQL

(continued)

Table 14.1 (continued)

Tools purpose	Application/description of tool(s)
Data curation	Data curation is aimed at data discovery and retrieval, data quality assurance, value addition, reuse and preservation over time. It involves selection and appraisal by creators and archivists; evolving provision of intellectual access; redundant storage; data transformations. Data curation is critical for scientific data digitization, sharing, integration, and use (Dou et al. 2012; Webopedia.com n.d.). *Examples*: Data warehouses, Data marts, Data Management Plan tools (DMPTool)[b]
Data visualization (and interaction)	This term refers to the presentation of data in a pictorial or graphical format (e.g., creating tables, images, diagrams and other intuitive ways to understand data). Interactive data visualization goes a step further: moving beyond the display of static graphics and spreadsheets to using computers and mobile devices to drill down into charts and graphs for more details, and interactively (and immediately) changing what data you see and how it is processed (Vitolo et al. 2015; Webopedia.com n.d.). *Examples*: Poly Maps, NodeBox, FF Chartwell, SAS visual Analytics, Google Map

[a]Distributed version control systems have been designed to ease the traceability of changes, in documents, codes, plain text data sets and more recently geospatial contents.
[b]DMP tools create ready-to-use data management plans for specific funding agencies to meet funder requirements for data management plans, get step-by-step instructions and guidance for your data and learn about resources and services available at your institution to help fulfill the data management requirements of your grant.

OSINT analytic tools provide frameworks for data mining techniques to analyse data, visualize patterns and offer analytical models to recognize and react to identify patterns. These tools should combine/unify indispensable features and contain integrated algorithms and methods supporting the typical data mining techniques, entailing (but not limited to) classification, regression, association and item-set mining, similarity and correlation as well as neural networks (Harvey 2012). Such analytics tools are software products which provide predictive and prescriptive analytics applications, some running on big open sources computing platforms, commonly parallel processing systems based on clusters of commodity servers, scalable distributed storage and technologies such as Hadoop and NoSQL databases. The tools are designed to empower users rapidly to analyse large amounts of data (Loshin 2015). The most predominant tools and techniques for OSINT collection and storage are summaries in Table 14.1.

14.4.2 Detection and Prevention of Cyber Threats

Techniques to make use of open sources involve a number of specific disciplines including statistics, data mining, machine learning, neural networks, social network analysis, signal processing, pattern recognition, optimization methods and visualization approaches (Chen and Zhang 2014; also Chapters in Part 2 of this book).

Gottschalk et al. (2011) presented a four-stage growth model for Knowledge Discovery to support investigations and the prevention of white-collar[3] crime in business organizations (Gottschalk 2010). The four stages are labelled:

1. Investigator-to-technology
2. Investigator-to-investigator
3. Investigator-to-information
4. Investigator-to-application

Through the proper exercise of knowledge, such processes can assist in problem solving. This four-part system attempts to validate the conclusions by finding evidence to support them. In law enforcement this is an important system feature as evidence determines whether a person is charged or not for a crime (Gottschalk et al. 2011) and the extent to which proceedings against them will succeed (see Chaps. 17 and 18).

Lindelauf et al. (2011) investigated the structural position of covert criminal networks using the *secrecy versus information trade-off characterization* of covert networks to identify criminal networks topologies. They applied this technique on evidence for the investigation of Jemaah Islamiyah's Bali bombing as well as heroin distribution networks in New York. Danowski (2011) developed a methodology combining text analysis and social network analysis for locating individuals in discussion forums, who have highly similar semantic networks based on watch-list members' observed message content or based on other standards such as radical content extracted from messages they disseminate on the internet. In the domain of countering cyber terrorism and inciting violence Danowski used a Pakistani discussion forum with diverse content to extract intelligence of illegal behaviour. Igbal et al. (2013) presented a unified data mining solution to address the problem of authorship analysis in anonymous textual communications such as spamming and spreading malware and to model the writing style of suspects in the context of cyber-criminal behaviour.

Brantingham (2011) offered a comprehensive computational framework for co-offending network mining, which combines formal data modelling with data mining of large crime and terrorism data sets "aimed towards identifying common and useful patterns". Petersen et al. (2011) proposed a node removal algorithm in the context of cyber-terrorism to remove key nodes of a terrorism network. Fallah (2010) proposed a puzzle-based strategy of game theory using the solution concept of the Nash Equilibrium to handle sophisticated DoS attack scenarios. Chonka et al. (2011) offered a solution through Cloud TraceBack (CTB) to find the source of DoS attacks and introduced the use of a back propagation neutral network, called Cloud

[3]White-collar crime is financial crime committed by upper class members of society for personal or organizational gain. White-collar criminals are individuals who tend to be wealthy, highly educated, and socially connected, and they are typically employed by and in legitimate organizations..

Table 14.2 Categorization of methods using open source data for cyber-criminal investigations

	Domain (Which)	Author (Who)	Methodology description (How)
Data mining	Criminal networks	Iqbal et al. (2012)	Proposing a framework that consists of *three modules*. 1. *click miner*, 2. *topic miner* and 3. *information visualizer*. It is a unified framework of *data mining and natural language* processing techniques to collect data from chat logs for intuitive and interpretable evidence that facilitates the investigative process for crime investigation. Available from: Online Messages (Chat Logs) extracted from Social Networks
	Activity boom in cyber cafes, and anomaly detection	Ansari et al. (2007)	Describing a typical fuzzy intrusion detection scenario for *information mining application in real time* that investigates vulnerabilities of computer networks Available from: Data available via ISPs
	Malware activities detection using fast-flux services networks (FFSN)	Wu et al. (2010)	Investigating detection solutions of Fast-flux domains by using Data Mining techniques (Linear Regression) to detect the FFSN[a] and analysing the feature attributes Available from: Data in two classes: white and black lists. The white list includes more than 60 thousands benign domain names; the black list has about 100 FFSNs domain names detected by http://dnsbl.abuse.ch
	Cyber terrorism resilience	Koester and Schmidt (2009)	Providing a supporting framework via FCA (Factor Concept Analysis) to find and fill information gaps in Web Information Retrieval and Web Intelligence for cyberterrorism resilience Available from: Small terrorist data sets based on 2002, 2005, London, Madrid
Text Mining	Counter Cyber Terrorism	Srihari (2009)	Using Unapparent Information Revelation (UIR) method to propose a new framework for different interpretation. A generalization of this taskinvolves query terms representing general concepts (e.g. indictment, foreign policy)
	Intrusion Detection System	Adeva and Atxa (2007)	Proposing detection attempts of either gaining unauthorised access or misusing a web application and introducing an intrusion detection software component based on text-mining techniques using "Arnas" system
Social Network Analysis	Cyber terrorism (detecting terrorist networks)	Chen et al. (2011)	Providing a novel graph-based algorithm that generates networks to identify hidden links between nodes in a network with current information available to investigators

(continued)

Table 14.2 (continued)

	Terrorist network fighting	Kock Wiil et al. (2011)	Offering a novel method to analyse the importance of links and to identify key entities in the terrorist (covert) networks using *Crime Fighter Assistant*[b] Available from: Open sources: 9/11 attacks (2001), Bali night club bombing (2002), Madrid bombings (2004), and 7/7 London bombings (2005)
	Network attacks (intrusion detection)	He and Karabatis (2012)	Using an *Automatic Semantic Network* with two layers: first mode and second mode networks. The first mode network identifies *relevant attacks based on similarity measures*; the second mode network is *modified based on the first mode and adjusts it by adding domain expertise* Available from: Selected data from the KDD CUP 99 data set made available at the Third International Knowledge Discovery and Data Mining Tools Competition[c]
Optimization methods (based on game theory)	Preventing DDoS attacks	Spyridopoulos et al. (2013)	Making *a two-player, one-shot, non-cooperative, zero-sum game* in which the attacker's purpose is to find the optimal configuration parameters for the attack in order to cause maximum service disruption with the minimum cost. This model attempts to explore the interaction between an attacker and a defender during a DDoS attack scenario Available from: A series of experiments based on the Network Simulator (ns-2) using the dumbbell network topology
	Trust management and DoS attacks	Li et al. (2009)	Proposing a defence technique using two trust management systems (Key Note and Trust Builder) and credential caching. In their two player zero-sum game model, the attacker tries to deprive as much resources as possible, while the defender tries to identify the attacker as quickly as possible Available from: KeyNote (open-source library for the KeyNote trust management system) as an example to demonstrate that a DoS attack can easily paralyze a trust management server
	Cyber terrorism	Matusitz (2009)	A model combining *game theory and social network theory* to model how cyber-terrorism works to analyse the battle between computer security experts and cyberterrorists; all players wish the outcome to be as positive or rewarding as possible

(continued)

Table 14.2 (continued)

Related works for conceptual frameworks	Cyber-crime investigation	Katos and Bendar (2008)	Presenting an information system to capture the information provided by the different members during a cyber-crime investigation adopting elements of the Strategic Systems Thinking Framework (SST). SST consists of three main aspects: 1. *intra-analysis*, 2. *inter analysis* and 3. *value-analysis*
	Computer hacking	Kshetri (2005)	Proposing a conceptual framework based on factors and motivations, which encourage and energize the cyber offenders' behaviour: 1. Characteristics of the source nation 2. Motivation of attack 3. Profile of target organization (types of attack)
	Preventing white collar crime	Gottschalk (2011)	Developing an organizing framework for knowledge management systems in policing financial crime containing four stages to investigation and prevention financial crimes: 1. Officer to technology systems 2. Officer to officer systems 3. Officer to information systems 4. Officer to application systems
	Detecting cyber-crime in financial sector	Lagazio et al. (2015)	Proposing a multi-level approach that aims at mapping the interaction of both interdependent and differentiated factors with focusing on system dynamics theory in the financial sector. The factors together can facilitate or prevent cyber-crime, while increasing and/or decreasing its economic and social costs.
	Capturing and analysing military intelligence to prevent crises	Song (2011)	Proposing a military intelligence early warning mechanism based on open sources with four modules (1. collection module, 2. early-warning intelligence processing, 3. early warning intelligence analysis, 4. preventive actions) to help the collection, tracking, monitoring and analysis of crisis signals used by operation commanders and intelligence personnel to support preventive actions

[a]Creates a fully qualified domain name to have hundreds (or thousands) IP addresses assigned to it
[b]A knowledge management tool for terrorist network analysis
[c]This training dataset was originally prepared and managed by MIT Lincoln Labs

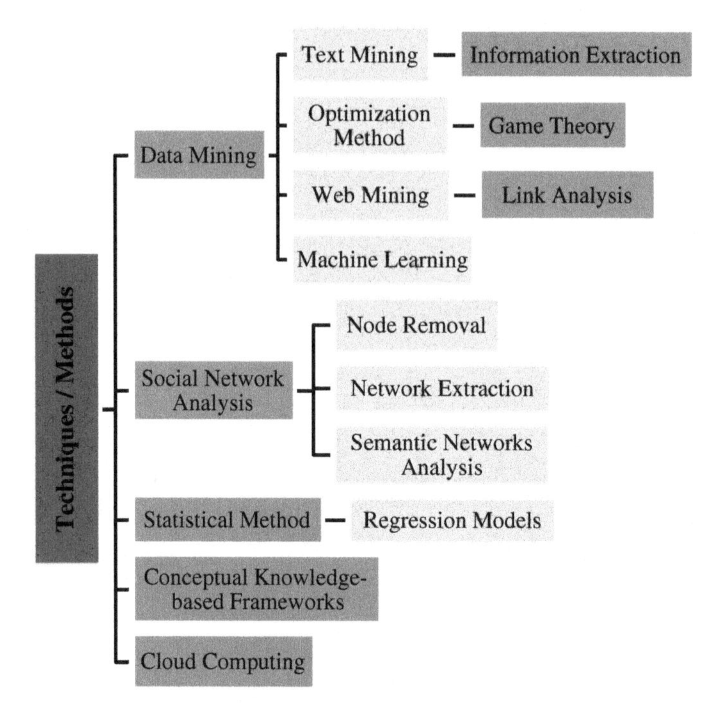

Fig. 14.3 Categorization of cyber-crime investigation methods and models

Protector, which was trained to detect and filter against such attack traffic. Mukhopadhyay et al. (2013) suggested a Copula-aided Bayesian Belief Network (CBBN) to assess and to quantify cyber-risk and cyber vulnerability assessment (CVA).

In summary, the field of computational criminology includes a wide range of computational techniques to identify:

1. Patterns and emerging trends
2. Crime generators and crime attractors
3. Terrorist, organized crime and gang social and spatial networks
4. Co-offending networks

Current models and methods are summarized Table 14.2 according to providing cyber-crime types (which), author (who), methodology (how) and open sources used for testing.

While many approaches seem to be helpful for cyber-crime investigation, existing literature suggests that social network analysis (SNA), data mining, text analysis, correlational studies and optimization methods specifically with focus on big data analysis of open sources are the most practical techniques to aid

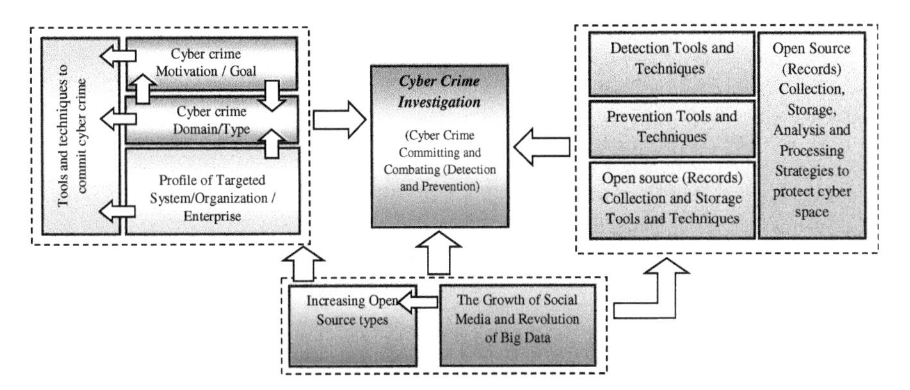

Fig. 14.4 Cybercrime investigation framework

practitioners and security and forensic agencies. Currently available techniques can be categorized in a schematic diagram such as Fig. 14.3.

14.5 Conclusions

The impact of cyber-crime has necessitated intelligence and law enforcement agencies across the world to tackle cyber threats. All sectors are now facing similar dilemmas of how to best mitigate against cyber-crime and how to promote security effectively to people and organizations (Jahankhani et al. 2014; Staniforth 2014). Extracting unique and high value intelligence by harvesting public records to create a comprehensive profile of certain targets is emerging rapidly as an important means for the intelligence community (Bradbury 2011; Steele 2006). As the amount of available open sources rapidly increases, countering cyber-crime increasingly depends upon advanced software tools and techniques to collect and process the information in an effective and efficient manner (Kock Wiil et al. 2011).

This chapter reviewed current efforts of employing open source data for cyber-criminal investigations. Figure 14.4 provides a summary of the findings in the form of an integrative Cybercrime Investigation Framework.

References

Adeva JJG, Atxa JMP (2007) Intrusion detection in web applications using text mining. Eng Appl Artif Intell 20:555–566

Agarwal VK, Garg SK, Kapil M, Sinha D (2014) Cyber crime investigations in India: rendering knowledge from the past to address the future. ICT and critical infrastructure: proceedings of the 48th annual convention of CSI, vol 2, Springer International Publishing Switzerland, pp. 593–600. doi:10.1007/978-3-319-03095-1_64

Ames DP, Horsburgh JS, Cao Y, Kadlec J, Whiteaker T, Valentine D (2012) Hydro desktop: web services-based software for hydrologic data discovery, download, visualization, and analysis. Environ Model Software 37:146–156

Ansari AQ, Patki T, Patki AB, Kumar V (2007) Integrating fuzzy logic and data mining: impact on cyber security. Fourth international conference on fuzzy systems and knowledge discovery (FSKD 2007). IEEE Computer Society

Appel EJ (2011) Behavior and technology, Internet Searches for Vetting, Investigations, and Open-Source Intelligence. Taylor and Fransic Group, pp. 3–17. ISBN 978-1-4398-2751-2

Boncella RJ (2003) Competitive intelligence and the web. Commun AIS 12:327–340

Bradbury D (2011) In plain view: open source intelligence. Comput Fraud Secur 5–9

Brantingham PL (2011) Computational Criminology. 2011 European intelligence and security informatic conference. IEEE Computer Society. doi:10.1109/EISIC.2011.79

Burden K, Palmer C (2003) Internet crime: cyber crime—A new breed of criminal? Comput Law Secur Rep 19(3):222–227

Buneman P, Khanna S, Chiew Tan W (2000) Data provenance: some basic issues. University of pennsylvania scholarly commons. Retrieved fromhttp://repository.upenn.edu/cgi/viewcontent.cgi?article=1210&context=cis_papers

Burwell HP (2004) Online competitive intelligence: increase your profits using cyber-intelligence. Facts on Demand Press, Tempe, AZ

Chauhan S, Panda K (2015) Open source intelligence and advanced social media search. Hacking web intelligence open source intelligence and web reconnaissance concepts and techniques. Elsevier, pp. 15–32. ISBN: 978-0-12-801867-5

Chauhan S, Panda K (2015) Understanding browsers and beyond. Hacking web intelligence open source intelligence and web reconnaissance concepts and techniques. Elsevier, pp. 33–52. ISBN: 978-0-12-801867-5

Chen A, Gao Sh, Karampelas P, Alhajj R, Rokne J (2011) Finding hidden links in terrorist networks by checking indirect links of different sub-networks. In: Kock Wiil U (ed) Counterterrorism and open source intelligence. Springer Vienna, pp. 143–158. doi:10.1007/978-3-7091-0388-3_8

Chen H, Chiang RHL, Storey VC (2012) Business intelligence and analytics: from big data to big impact. Bus Intell Res 36(4):1–24

Chen LP, Zhang CY (2014) Data-intensive applications, challenges, techniques and technologies: A survey on Big Data. Inform Sci 314–347

Chertoff M, Simon T (2015) The impact of the dark web on internet governance and cyber security. Global Commission on Internet Governance. No. 6

Chonka A, Xiang Y, Zhou W, Bonti A (2011) Cloud security defence to protect cloud computing against HTTP-DoS and XML-DoS attacks. J Netw Comput Appl 34:1097–1107

Clark RM (2004) Intelligence analysis: a target-centric approach. CQ Press, Washington, DC

Danowski JA (2011) Counterterrorism mining for individuals semantically-similar to watchlist members. In: Kock Wiil U (ed) Counterterrorism and open source intelligence. Springer Berlin Heidelberg, pp. 223–247. doi:10.1007/978-3-7091-0388-3_12

Dou L, Cao G, Morris PJ, Morris RA, Ludäscher B, Macklin JA, Hanken J (2012) Kurator: a Kepler package for data curation workflows. International Conference on Computational Science, ICCS 2012, Procedia Computer Science, vol 9, pp. 1614–1619. doi:10.1016/j.procs.2012.04.177

Enbody R, Soodo A (2014) Intelligence gathering. Elsevier Inc, Targeted cyber attacks. ISBN 9780128006047

Fallah M (2010). A puzzle-based defence strategy against flooding attacks using game theory. IEEE Trans Dependable Secure Comput 7:5–19

FlashPoint (2015) Illuminating The Deep & Dark Web: the next Frontier in Comprehensive IT Security. FlashPoint

Fleisher C (2008) OSINT: its implications for business/competitive intelligence analysis and analysts. Inteligencia Y Seguridad 4:115–141
Ghel R (2014) Power/freedom on the dark web: A digital ethnography of the Dark Web Social Network. New media and society
Google 2014 Learn about Sitemaps. ps://support.google.com/webmasters/answer/156184?hl=en
Gottschalk P (2010) White-collar crome: detection, prevention and strategy in business enterprises. Universal-Publishers, Boca Raton, Florida, USA. ISBN-10: 1599428393, ISBN-13: 9781599428390
Gottschalk P, Filstad C, Glomseth R, Solli-Sæther H (2011) Information management for investigation and prevention of white-collar crime. Int J Inf Manage 31:226–233
Govil J, Govil J (2007) Ramifications of cyber crime and suggestive preventive measures. Electro/information technology. Chicago, pp 610–615. IEEE. doi:10.1109/EIT.2007.4374526
Gregory M, Glance D (2013) Cyber-crime, cyber security and cyber warfare. Security and networked society. Springer, pp 51–95. ISBN: 978-3-319-02389-2
Harvey C (2012) 50 top open source tools for big data. Retrieved 01 July 2015, from http://www.datamation.com/data-center/50-top-open-source-tools-for-big-data-1(2,3).html
He P, Karabatis G (2012) Using semantic networks to counter cyber threats. IEEE. doi:10.1109/ISI.2012.6284294
Hobbs Ch, Morgan M, Salisbury D (2014) Open source intelligence in the twenty-first century. Palgrave, pp. 1–6. ISBN 978-0-230-00216-6
Hoque N, Bhuyan H, Baishya RC, Bhattacharyya DK, Kalita JKV (2014) Network attacks: taxonomy, tools and systems. J Netw Comput Appl 40:307–324. doi:10.1016/j.jnca.2013.08.001
Igbal F, Fung BCM, Debbabi M (2012) Mining criminal networks from chat log. 2012 IEEE/WIC/ACM international conferences on web intelligence and intelligent agent technology. Macau, pp. 332–337. IEEE. doi:10.1109/WI-IAT.2012.68
Iqbal F, Binsalleeh H, Fung BCM, Debbabi M (2013) A unified data mining solution for authorship analysis in anonymous textual communications. Inf Sci 231:98–112
Jahankhani H, Al-Nemrat A, Hosseinian-Far A (2014) Cybercrime classification and characteristics. In: Akhgar B, Staniforth A, Bosco F (eds.) Cyber crime and cyber terrorism investigators' handbook. Elsevier Inc., pp. 149–164. doi:10.1016/B978-0-12-800743-3.00012-8
Kang MJ (2012) Intelligence in the internet age: the emergence and evolution of Open Source Intelligence (OSINT). Comput Hum Behav 28:673–682. doi:10.1016/j.chb.2011.11.014
Kim W, Jeong OR, Kim Ch, So J (2011) The dark side of the Internet: attacks, costs and responses. Inform Syst 36:675–705
Kapow Software (2013) http://www.kofax.com/go/kapow/wp-building-your-osint-capability. Retrieved from http://www.kofax.com: http://www.kofax.com/go/kapow/wp-building-your-osint-capability
Katos V, Bednar PM (2008) A cyber-crime investigation framework. Comput Stand Interfaces 30:223–228. doi:10.1016/j.csi.2007.10.003
Koops BJ, Hoepman JH, Leenes R (2013) Open-source intelligence and privacy by design. Computer Law and Security Review. 2(9):676–688
Kshetri N (2005) Pattern of global cyber war and crime: a conceptual framework. J Int Manage 11:541–562
Koester B, Schmidt SB (2009) Information superiority via formal concept analysis. In. Argamon S, Howard N (eds) Computational methods for counterterrorism. Springer, pp. 143–171. doi:10.1007/978-3-642-01141-2_9
Kock Wiil U, Gniadek J, Memon N (2011) Retraction note to: a novel method to analyze the importance of links in terrorist networks. In: Wiil UK (ed) Counterterrorism and open source intelligence. Springer Vienna, p. E1. doi:10.1007/978-3-7091-0388-3_22
Lagazio M, Sherif N, Cushman M (2015) A multi-level approach to understanding the impact of cyber crime on the financial sector. Comput Secur 45:58–74

Li J, Li N, Wang X, Yu T (2009) Denial of service attacks and defenses in decentralized trust management. Int J Inf Secur 8:89–101. Springer
Lindelauf R, Borm P, Hamers H (2011) Understanding terrorist network topologies and their resilience against disruption. In: Kock Wiil U (ed.) Counterterrorism and open source intelligence. Springer, Vienna, pp 61–72. doi:10.1007/978-3-7091-0388-3_5
Loshin D (2015) How big data analytics tools can help your organization. Retrieved from http://searchbusinessanalytics.techtarget.com/feature/How-big-data-analytics-tools-can-help-your-organization
Matusitz J (2009) A postmodern theory of cyberterrorism: game theory. Inform Secur J: Glob Perspect 18:273–281. Taylor and Francis. doi:10.1080/19393550903200474
Mukhopadhyay A, Chatterjee S, Saha D, Mahanti A, Sadhukhan SK (2013) Cyber-risk decision models: To insure IT or not? Decis Support Syst 56:11–26. Retrieved from http://dx.doi.org/10.1016/j.dss.2013.04.004
Nykodym N, Taylor R, Vilela J (2005) Criminal profiling and insider cyber crime. Digital Invest 2:261–267. Elsevier
Omand D, Miller C, Bartlett J (2014) Towards the discipline of social media intelligence (2014). In: Hobbs, Morgan, Salisbury (eds.) Open source intelligence in the twenty-first century. Palgrave, 24–44. ISBN 978-0-230-00216-6
Petersen RR, Rhodes CJ, Kock Wiil U (2011) Node removal in criminal networks. 2011 European intelligence and security informatics conference. IEEE Computer Society, pp. 360–365.
PWC cyber security (2015) https://www.pwc.com/us/en/increasing-it-effectiveness/publications/assets/2015-us-cybercrime-survey.pdf. Retrieved from http://www.pwc.com/cybersecurity
Simmons C, Ellis C, Shiva S, Dasgupta D, Wu Q (2014) AVOIDIT: a cyber attack taxonomy. Annual symposium on information assurance. Office of Naval Research (ONR).
Song J (2011) The analysis of military intelligence early warning based on open source intelligence. Int Conf Intell Secur Inform (ISI). p. 226. IEEE
Spyridopoulos T, Karanikas G, Tryfonas T, Oikonomou G (2013) A game theoric defence framework against DoS/DDoS cyber attacks. Comput Secur 38:39–50
Staniforth A (2014) Police investigation processes: practical tools and techniques for tackling cyber crime. In: Akhgar B (ed.) Cyber crime and cyber terrorism investigator's handbook. Elsevier, pp. 31–42
Srihari RK (2009) Unapparent information revelation: text mining for counterterrorism. In: Argamon S, Howard N (eds) Computational methods for counterterrorism. Springer, Berlin Heidelberg, pp 67–87
Steele RD (2006) Open source intelligence. In Johnson LK (ed.) Strategic intelligence: understanding the hidden side of government (intelligence and the quest for security). Praeger, pp. 95–116
Sui D, Cavarlee J, Rudesill D (2015) The deep web and the darknet: a look inside the internet's massive black box. Wilson Center, Washington
Szomszor M, Moreau L (2003) Recording and reasoning over data provenance in web and grid services. On the move to meaningful internet systems, pp. 603–620.
Tilmes C, Yesha Ye, Halem M (2010) Distinguishing provenance equivalence of earth science data. Int Conf Comput Sci (ICCS). p. 1–9
Vitolo C, Elkhatib Y, Reusser D, Macleod CJA, Buytaert W (2015) Web technologies for environmental Big Data. Environ Model Softw 63:185–198
Wall DS (2005) The internet as a conduit for criminal activity. In: Pattavina A (ed) Information technology and the criminal justice system. Sage Publications, USA. ISBN 0-7619-3019-1
Wall DS (2007) Hunting shooting, and phishing: new cybercrime challenges for cybercanadians in the 21st century. The ECCLES centre for american studies
Wall DS (2008) Hunting shooting, and phishing: new cybercrime challenges for cyber canadians in the 21st Century. The Eccles Centre for American Studies. www.bl.uk/ecclescentre. The British Library Publication
Wang SJ (2007) Measures of retaining digital evidence to prosecute computer-based cyber-crimes. Comput Stand Interfaces 29:216–223. Elsevier

Webopedia.com. (n.d.). Webopedia.com
Wu J, Zhang L, Qu S (2010) A comparative study for fast-flux service networks detection. Netw Comput Adv Inf Manage (NCM). pp 346–350. IEEE
Yuan T, Chen P (2012) Data mining applications in E-Government information security, 2012 international workshop on information and electronics engineering (IWIEE). Proc Eng 29:235–240

Chapter 15
Combatting Cybercrime and Sexual Exploitation of Children: An Open Source Toolkit

Elisavet Charalambous, Dimitrios Kavallieros, Ben Brewster, George Leventakis, Nikolaos Koutras and George Papalexandratos

Abstract This chapter presents the UINFC2 "*Engaging Users in preventing and fighting Cybercrime*" software platform, showcasing how software tools designed to detect, collect, analyse, categorise and correlate information that is publically available online, can be used to enable and enhance the reporting, detection and removal capabilities of law enforcement and hotlines in response to cybercrimes and crimes associated with the sexual exploitation of children. It further discusses the social, economic and wider impact of cybercrime on a European and global scale, highlighting a number of challenges it poses to modern society before moving on to discuss the specific challenge posed by the proliferation of online child exploitation material and discussing the functionalities of the UINFC2 system as a response mechanism.

15.1 Introduction

Alongside the threat posed by traditional forms of organised crime and terrorism, cybercrime has developed into a primary security challenge across the EU as a result of the potentially massive financial implications resulting from security breaches and the inherently cross border nature of attacks. The formation and prominence of the European Cybercrime Centre (EC3) and other dedicated cybercrime units across national and regional police forces in recent years further demonstrates the scale and significance of the challenge that cybercrime poses to European society as a whole. In addition to penetrating the everyday lives of

E. Charalambous · N. Koutras
Advanced Integrated Technology Solutions & Services Ltd, Egkomi, Cyprus

D. Kavallieros · G. Leventakis · G. Papalexandratos
Center for Security Studies (KEMEA), Hellenic Ministry of Interior and Administrative Reconstruction, Athens, Greece

B. Brewster (✉)
CENTRIC/Sheffield Hallam University, Sheffield, UK
e-mail: B.Brewster@shu.ac.uk

B. Akhgar et al. (eds.), *Open Source Intelligence Investigation*,
Advanced Sciences and Technologies for Security Applications,
DOI 10.1007/978-3-319-47671-1_15

citizens, the day-to-day operations of business and the management of critical national infrastructure, the expansion in the use of the internet as an enabler, and in some cases a dependent vector, for crime has also continued at a rapid pace. Existing forms of serious and organized crime including the trade and supply of weapons and drugs, and financial crime have moved online whilst the internet has become the de facto mechanism for the dissemination of illegal and indecent content related to the sexual exploitation of children (European Commission 2015).

Child Sexual Exploitation (CSE) is so defined by directive 2011/93/EU of the European Parliament on combating the sexual abuse and sexual exploitation of children, and the creation and dissemination of child pornography (European Parliament 2011). Under this definition, CSE consolidates around 20 criminal offences including, but not limited to, the direct sexual abuse of children such as engaging in sexual activities with a person under the age of legal consent, the sexual exploitation and coercion of children to engage in prostitution or the creation of child pornography, the possession, access or distribution of child pornography and the online solicitation of children for the purposes of engaging in sexual acts. Europol distil this definition further, referring to CSE as a crime within which perpetrators engage in sexual activity or sexual contact with a minor, thereby violating established legal and moral codes with respect to sexual behaviour (Europol 2015). In subsequent sections of this chapter we will outline in more detail the scale and nature of the problem of online CSE before focusing primarily on article 5 of directive 2011/93/EU and offences related to child pornography, particularly in the development of a novel system to detect and assist in the removal of illegal content using natural language processing and web crawling techniques—The UINFC2 Platform. Although the directive itself refers to '*child pornography*' this suitability of this term in particular is somewhat disputed as 'pornography' itself is used to define consented sexual activity between adults whereas children do not, and legally cannot, give consent and are such the victims of crime (Kettleborough 2015). Instead, we refer to content that depicts the exploitation of children as 'child sexual abuse images' in order truly reflect the nature and gravitas of the crimes that are being discussed (INTERPOL n.d.).

15.2 The Extended Impact of Cybercrime

With the continued growth of internet use and its penetration across different aspects of people's daily lives, both professionally and socially, the internet has become a hotbed for crime that not only relies on the particulars of the online environment (cyber-dependant crime), but also for further enabling existing forms of criminality, such as trafficking and the trade in illegal goods and services, crimes for which the internet acts as a means to extend the scope and reach of such activity (cyber-enabled crime). Online crime, whether 'dependant' or further 'enabled' by the opportunities afforded through the internet provides those acting online with tools that enable increased levels of anonymity, invisibility and scope. Combining these factors with the limitations that affect our ability to prevent and respond to cybercrime, such as the lack of uniform, consistent and truly international legal jurisdiction, the internet has become the preferred, if not the de facto, means for facilitating and enabling a broad spectrum of malicious activity.

The relative criminal phenomenon that is cybercrime, in all its various guises, continues to evolve alongside changes in technology and the sustained and vociferous proliferation in internet adoption across Europe and the wider world, underpinned by widespread access to connected mobile devices. In response, organisations such as Europol continue to outline cooperation between key stakeholders, such as law enforcement, reporting bodies and content hosts, as a key capacity building mechanism in developing European society's response (Europol 2015). For obvious reasons, concrete statistics regarding the true nature and scale of CSE are hard to come by. In a similar vein to Human Trafficking, a crime which in some instances overlaps and shares traits with CSE, children are often trafficked with the underlying objective of being sexually exploited. The true extent to which CSE has spread is largely unknown. What we do know is that across Europe, the United States and Australasia, internet adoption, as of 2013, had reached 73.2 % and has continued to grow since (National Crime Agency 2013). Such a profound statistic has introduced the very real threat of CSE, as a large number of underage children are exposed to the internet. Increasingly, this exposure is facilitated by mobile devices (e.g. smartphones and tablets). In the UK, CEOP (National Crime Agency 2013) estimated that in 2013 smartphone ownership among 5–15 year olds had increased by 20 % year over year, whilst on average the amount of time they spend online also continues to increase. These figures coincide with societal and cultural trends that see online activity featuring ever more prominently in our social lives as well as impacting on how we consume multimedia content for both educational and entertainment purposes. The headline statistic however is that global reports of child sexual abuse images and videos have risen by 417 % since 2013 (Internet Watch Foundation 2016), a figure even more significant considering it does not take into account the unreported and unknown 'dark figure'.

Cybercrime has the potential to impact the full spectrum of societal stakeholders, with attacks having potentially devastating direct and indirect financial implications on the individuals, private organisations and governments affected. A survey conducted by PriceWaterhouseCoopers (2016) indicates that although cybercrime is statistically the second most reported type of crime, only 40 % of the organisations surveyed professed to having a dedicated cyber incident response plan, while McAfee (2014) indicates that globally the economic losses stemming from cybercrime are estimated to represent 0.8 % of the world's Gross Domestic Product (GDP). Although all industrial sectors are impacted by cybercrime, the financial implications on sectors such as financial services, utilities and energy providers, technology and heavy industry rank amongst the highest (Ponemon Institute 2015).

Unfortunately, the impact of cybercrime is not solely limited to that of an economic nature. Cases involving the creation and dissemination of child sexual abuse images and child trafficking can have a profound physical and psychological impact on victims. Child Sexual Abuse Material (CSAM) refers to visual material depicting sexually explicit activities involving a child under the 18 years of age. Research on CSAM has shown that it is rather impossible to have a precise indication of the amount of material which is distributed or hosted online (Carr 2011), however the most recent numbers emerging from Europol's EC3 (Europol 2015) are alarming. According to the report, more than 1.4 million CSAM images were hosted and hidden using Dark Web services such as Tor (The Onion Router), in the process estimating that globally more 300,000 unique users had accessed them over the course of 2015.

Due to a number of high profile, recent, and newly exposed historical CSE cases, the issue has become a priority across UK policing, with the proliferation of the sharing of indecent images online outlined as a particular issue in national threat assessments. In 2012 it was estimated that around 50,000 individuals were involved in the downloading and distribution of illegal online child sexual exploitation content in the UK alone (National Crime Agency 2013). While in 2014, INHOPE, an international reporting network made up of primarily European members, reported the presence of illegal content at more than 89,000 URLs, an almost two-fold increase on figures from 2013 and almost three times the number recorded in 2012 (INHOPE 2014).

15.3 Tools for Law Enforcement

It is becoming increasingly difficult to discuss new and emerging challenges in the security domain without coming across research advocating the use of 'big' and open-source data in relation to a wide range of contemporary initiatives and research projects proposing their use to enhance the ability of decision makers to take action. Other chapters in this book have discussed the use of data from social media to enhance first responder's ability to respond in crisis situations (Gibson et al. 2014), while other research has proposed and developed software solutions to enhance strategic situational awareness in response to organized crime threats (Brewster et al. 2014).

A number of existing approaches have focused on the use of images and image hash data to detect the presence of illegal CSE content online. Fronting this is INTERPOL's International Child Sexual Exploitation (ICSE) image database, a tool that allows specialist investigators to share and compare data about known victims, places and abusers (INTERPOL 2016). Large enterprise solutions, such as Microsoft's PhotoDNA platform have also implemented powerful tools that enable the conversion of images into hashes, and subsequently compared against databases of known data, such as the US National Center for Missing and Exploited Children (NCMEC), the ICSE and other authorities (Microsoft n.d.). While the majority of efforts, so as those mentioned previously, have focused on images, increasingly there are projects that aim to make use of natural language processing and the identification of keywords, phrases and other contextual information on the web and contained within the metadata associated with media content (Rutgaizer et al. 2012; Westlake et al. 2012).

Among the challenges that law enforcement practitioners face in the fight against child exploitation, one can identify the lack of legal tools and cooperation, the use of advanced hiding techniques and improving measures to manage the ever increasing amounts of forensic data as some of the most prominent that we face today. These challenges are further complicated and enhanced by the transnational, trans-jurisdictional nature of cybercrime in general, as in many instances the involved victims, offenders and communication vectors exist across different countries and jurisdictional boundaries, providing barriers to identification, investigation and prosecution. Unfortunately, this does not take place within a stagnant environment, instead tools and behaviours change alongside social and technological trends. For example, current tendencies toward using encrypted communication and privacy enhancing services such as ‘Tor’, in order to avoid known surveillance methods used by law enforcement and intelligence agencies, continue to challenge existing crime prevention and response techniques. Analysts and officers are increasingly faced with vast amounts of data to resolve, and as hard disk capacities regularly exceed several terabytes, the time and resources needed to investigate and forensically analyse data has risen steeply.

Despite the prominence of these challenges, positive steps are being taken at regional, national and now international levels to improve the rate of cybercrime and CSE reporting at both an individual citizen level. Provision of online reporting mechanisms, such as those provided by organisations like Inhope and proposals by international law enforcement (i.e. Europol and Interpol) to facilitate the fusion of information from across member states, and beyond, considering information from open-sources and other closed mediums demonstrate the value in current research efforts into stakeholder engagement in cybercrime and online child exploitation reporting approaches, such as those undertaken by the UINFC2 project. This value is demonstrated especially when considering the challenges and barriers that still exist to the effective reporting and coordination of information, such as low levels of citizen and private sector engagement because of awareness levels, privacy and reputational fears, lack of consistency in the information being acquired in different member states, differences in the way information is being used and stored, and low levels of information sharing between nations and the agencies within them (see Chap. 10 for further details on OSINT tools).

15.4 The Role of OSINT

Publicly accessible information capable of generating knowledge is commonly referred to as Open Source Intelligence (OSINT). OSINT, in its many guises has been discussed and defined in earlier chapters (e.g., Chaps. 12–14). Here, we will take forward the loose definition of using and analysing data that is freely available online in order to aid investigators in developing new, or expanding the scope of existing, insights relevant to efforts proactively targeting the removal of illegal content, or in the development of intelligence products to assist in other processes.

Spanning the era of social media, the average internet user maintains profiles across multiple platforms. This, in conjunction with the fact that a large number of social media platforms and networks are available, each targeting a different audience and using different communication models, contain information about and from hundreds of millions of people, providing data that may be used, alongside traditional methods and closed intelligence sources, to combat crime. As a result, the desire for tools capable of gathering publicly accessible information from open websites, forums and social media platforms has continued at a vociferous pace. Such tools aid the gathering, correlation and analysis of information such as images, videos and textual descriptions as well as the metadata associated with the aforementioned media file types.

Metadata (i.e. information that assorts the payload of a file or a document) can be used to reveal information of great value to businesses looking to learn more about their customers, or by law enforcement looking to learn more about nominals and other persons of interest. In its simplest form, analysis of this sort of content might be used to detect market trends or to reveal behavioural and purchasing habits of customers for marketing purposes. Metadata can also be used to tell the story behind one's actions as well as the characteristics of their personal preferences. Owing to its open nature the utilisation of this potentially revealing data, is not limited to law enforcement, with phishing and social engineering attacks now regularly making use of information individuals freely post online to exploit human security vulnerabilities. Every online action, even those not involving the conscious sharing of information, leaves behind a trace which may be used for a number of malicious purposes. Website visitation leaves the users trace on the hosting server, while sending an email may embed information related to the user's IP address, the transport path from the sender to the receiver, information about the emailing client used along with its version or even the user's login id; information most would think twice before consciously sharing.

The internet itself is a network of networks, connecting millions of computing devices and performing information interchange throughout the world (Kurose 2005). Routing information is not only stored on network components but also on the end devices (source and sink). Multiple, existing, open-source tools can represent jigsaw pieces which, when put together, can form a snapshot of our internet persona. Computer Forensics science deals with the preservation, identification, extraction and documentation of computer evidence. The latest statistic showed that 5973 TB of data was processed in Fiscal Year (FY) 2013, a 40 % increase on FY 2011 (U.S. Department of Justice 2013). Many studies continue to investigate new, improved and novel methods to be used in data and network forensics (Arthur and Venter 2004; Geiger 2005; Lalla and Flowerday 2010; Meghanathan et al. 2010), however, there is still the need for advanced OSINT forensic tools to develop investigators capacity and capability to find and interrogate specific information pertinent to investigations (see also Chaps. 6 and 7).

Currently, this capacity is lagging behind that of offenders, meaning that law enforcement struggles to keep pace with the proliferation of crime and the means used by the individuals and groups committing it. Towards enhancing this, the UINFC2 platform has been developed and foresees use by law enforcement and hotlines for the efficient and effective identification of criminal activity in an effort to minimize the impact of cybercrime and the proactive detection of illegal networks of activity.

15.5 The UINFC2 Approach

The primary aim of the UINFC2 platform is to utilize information available in reports made by citizens as well as publicly available intelligence, through the means of web crawling, to assist report handling and management by hotlines and law enforcement. The project assists citizen reporting with the implementation of a cybercrime taxonomy, linking the different areas of cybercrime with easy to understand questions designed to extract the information needed to remove illicit content and pursue those perpetrating crime.

The UINFC2 platform supports three different types of internal users, administrators, law enforcement and hotlines. Due to the different nature of the organisations, LEAs and hotlines use different installations of the platform isolated from each other and equipped with different features appropriate to imposed legislation and investigative powers. The platform itself implements two distinct components, the first is aimed at citizens to improve the reporting of incidents whilst the second is oriented for use by LEAs and Hotlines. The platform is embedded with the knowledge of domain experts realised through functionality allowing for the use of keywords, phrases and more sophisticated text mining/analysis methods, and graphical and data visualization representations. These components are discussed in more detail in subsequent sections of this chapter.

15.5.1 Citizen Reporting Form

The frameworks for reporting cybercrime around Europe do not yet take a standardized approach, while in each country hotlines and law enforcement provide their own channels for cybercrime reporting; the structure and nature of the information they request varies from service to service. Moreover, it has been noted that existing reporting procedures often only accommodate the reporting of incidents under three or four very generic branches of criminal activity, sometimes leaving investigators with insufficient or inconclusive information creating the additional requirement to follow up the report to request additional information.

Towards this direction, the UINFC2 platform implements its own reporting procedure supporting a number of features to enhance the reporting procedure

UINFC2 Reporting Form File Report

The UINFC2 Reporting form is designed and implemented to be used ONLY for the purposes of the UINFC2 research project. Thus, data will be accessible and will only be used by authorized personnel (members of the consortium and pilot users). Data will solely be used for the purposes of the project and throughout the lifecycle of the project. Law Enforcement Agencies and Hotlines will not receive this report, therefore please be advised to use the links bellow if you wish to report an incident.

Reporting Form

Type of Incident Child Sexual Abuse Material

Description **Child Sexual Abuse Material (CSAM)** refers to visual material depicting sexually explicit activities involving a child under the 18 years of age.

I believe that my report is related to material whose inappropriate use could put in danger an underage person. However, the material does not indicate misacts.

Yes No

Technology Source Website

Website*

Image URL

Additional Information Provide us with any information you believe we would find useful during investigation

Fig. 15.1 Citizens reporting form

whilst also increasing the accuracy and completeness of the data gathered. The implemented taxonomy organizes cybercrime into seven easily distinguishable categories presented in a tree structure. Each condition is represented by a question and leaf nodes represent specific aspects of cybercrime. As a result, the sue of the taxonomy in this procedure enables the capture of more accurate, detailed information about the specific act being reported, without creating the need for the citizen to articulate specific details about what they are reporting. A snapshot of the UINFC2 reporting form is shown in Fig. 15.1. Upon platform setup, the administrator is capable of selecting which report fields are mandatory/optional; ensuring compliance with existing national standards while the flexibility of the mechanism provided means that only relevant fields and information is displayed.

Despite the fact that the UINFC2 project implements its own reporting form, for the previously mentioned reasons, the backend of the system allows for the integration of reports based on existing standards made through existing reporting channels (regional/national LEAs and hotlines) so that the analytical aspects of the system may be used against information provided through channels external to the platform itself.

15.5.2 LEA/HOTLINE UINFC2 Platform

Considering the vast, and continually expanding, amount of data that is available and publicly accessible online, the need for tools that enable the efficient processing and retrieval of suspicious content is increasingly important. The UINFC2 platform not only implements mechanics that are capable of extracting possibly suspicious content (a web crawler) out of masses of data, but also incorporates tools for discerning possible leads in criminal activity. This is done mainly through networking tools and image metadata extraction. Despite the fact that these types of data are often neglected, they can be of investigative importance. The analysis of information found through DNS resolution and image metadata can provide means for investigators to identify links to criminal activity and gather key information related to potential offenders' identities and whereabouts. The information, such as the legal owner of a given web domain, can be of significance in supporting the investigative process (Allsup et al. 2015).

The utility and benefit of the platform is illustrated in the following example. In the system two reports were submitted, one related to *child sexual abuse material* and another related to *hacking*. The report was made through the platform's 'Citizen Reporting Form' and in both reports the suspicious web-site was reported as "http://www.uinfc2.eu/wp/en/". For illustrative purposes it is assumed that the same LEA/Hotline user received both reports. Once the reports are received, the user is able to crawl the suspicious web-site, in the process crawling other URLs connected to the seed URL. Considering that two different types of incident were reported, the user will investigate the web-page for known keywords and phrases related to the specifics of each incident, such as "*cybercrime*", "*DDOS*", "*Trojan horse sale*", "*12yob*", "*9yog*", "*XSS code*" and "*Hacktivism*". The interface used to facilitate this process is shown in Fig. 15.2.

Once the web-crawling on the suspicious link is finished, in this particular instance ***361*** instances of the aforementioned keywords/phrases were identified in total. Furthermore, the UINFC2 platform can visually imprint the connection between web-pages as well as the density and diversity of suspicious content with regards to the provided keywords (see Figs. 15.3 and 15.4). The first graph (Fig. 15.3) depicts the "path" of the crawled content based on the depth (sub-links) the crawler reached. This assists the investigator to quickly identify connections between the seed URL and sub-links of high interest, enabling the dismissal of entry points which do not bring results, and the depth and seeds used to guide the depth of the crawl to be revised based upon the results in order to focus and reduce the scope of any actions taken as a result of the information provided.

The second graph (Fig. 15.4) shows the connection between crawled content. Higher levels of density indicate homogeneity in the content while sparse levels indicate low levels of heterogeneity and therefore dissimilarity in the content. Crawling of the reported web-source resulted in three highly dense clusters of websites and a smaller one indicating disparity in the volume of identified keywords. This implies or better leads the investigator into identifying a new group of

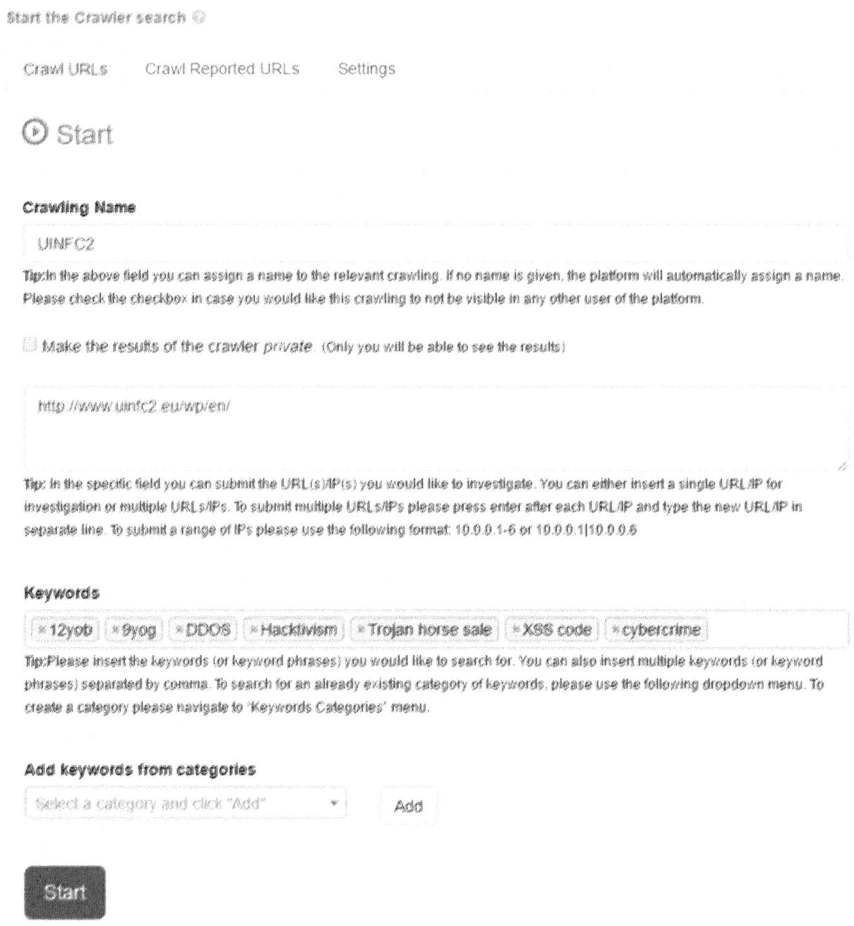

Fig. 15.2 Web crawling mechanism

suspicious websites initially hidden from the original source, possibly previously unexplored. In both graphs the colour of the nodes depicts how many different keywords were identified (green implies that no keywords were identified while the colour (from red to black), the bigger the number of keywords detected). The size of the nodes is analogous to the frequency of the identified keywords (larger size means higher frequency).

In the example provided earlier in Fig. 15.2, the investigator provided the keywords/phrases directly into the system by hand, but the UINFC2 platform can also categorise content in a two-tier system composed of base categories which then break down into subcategories so that core sets of keywords and phrases can be recycled across multiple investigations. A base category can be seen as general crime area encompassing a number of more specific crime areas referred to as

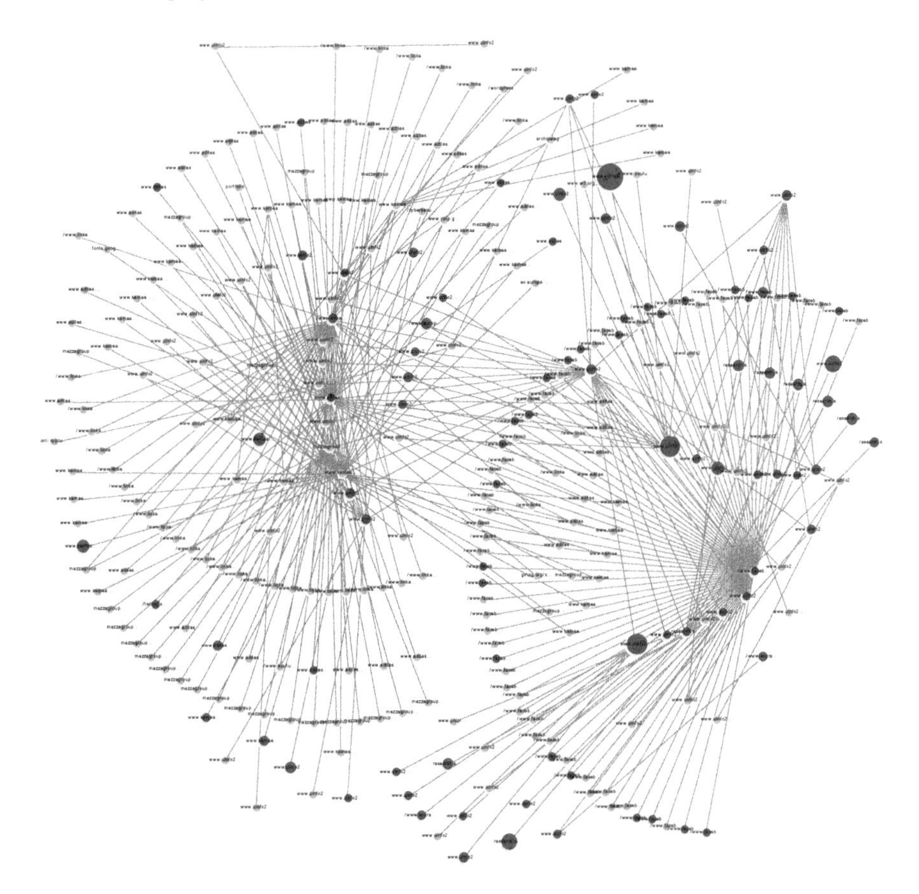

Fig. 15.3 Crawler path

subcategories. In UINFC2, a base category may also include keywords that not necessarily belong to a category while a keyword might belong to multiple categories at the same time (multi-label classification). The investigator can choose to crawl the reported web-site using the categories of the UINFC2 platform instead of providing them before the crawling starts (Fig. 15.5).

For our example the base categories are *Hacking* and *Child Sexual Abuse* respectively. Hacking breaks down into two subcategories, *Hacking Attacks* and *Hacking for sale*, respectively, while Child Sexual Abuse contains the *Child exploitation* subcategory. "*DDOS*" and "*XSS code*" are included in the *Hacking Attacks* subcategory, the "*Hacktivism*" and "*Trojan horse sale*" belongs to *Hacking for sale* subcategory and "*12yob*" and "*9yog*" are included into the *Child Exploitation* while "*cybercrime*" belongs to all categories.

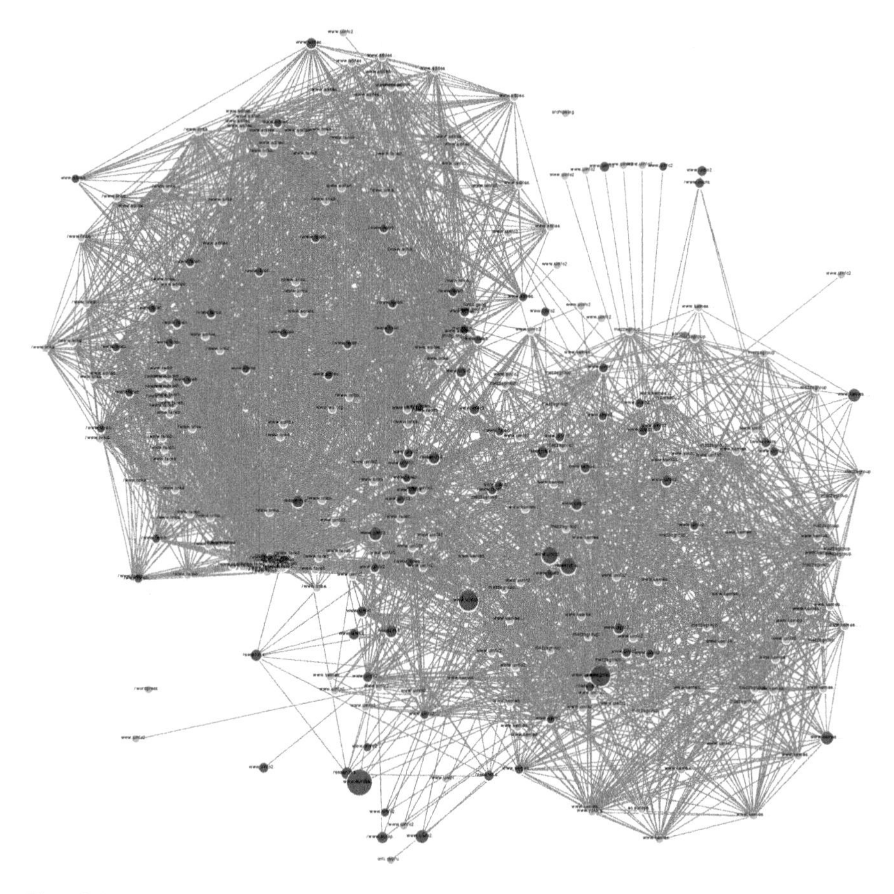

Fig. 15.4 Crawler links

It is worth noting that the administrator of the platform is responsible for creating base keyword categories (e.g. Child Sexual Abuse and Hacking) while the user can create subcategories under the base categories (parental relationship) as well as keywords for each subcategory. Moreover, each user is entitled to decide whether their categories are to be treated as private or public, by the system, and to decide if the category will be editable by other users. Public categories—and their content—are visible to other users bearing the same user role whereas private categories are only visible to their owner. As a result of this structure, users can crawl a website using one or more keyword categories specifically constructed for each respective aspect of cybercrime.

Another important functionality of the platform is the automated classification/categorization of crawled links and submitted citizen reports. Based

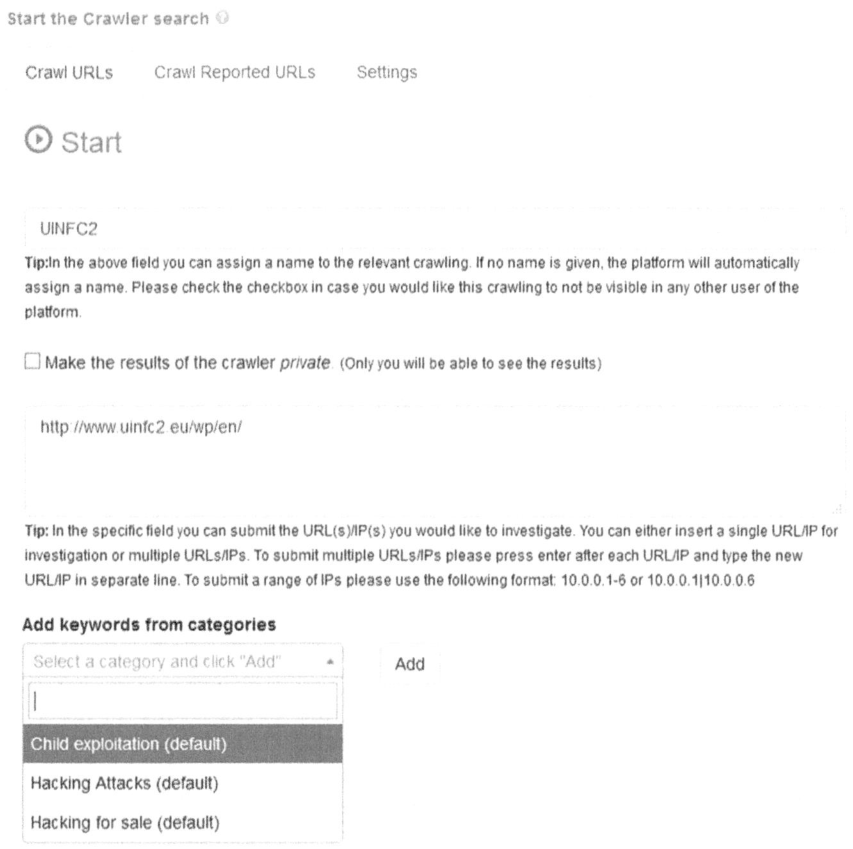

Fig. 15.5 UINFC2 crawler interface

on the embedded knowledge and the structure of system categories, the platform generates a possible class label allowing the user to know whether the internal structure of a reported link confirms the associate content type (specified by the user). This increases the investigator's insight on the case before even needing to examine the preliminaries of the report. Once crawling has concluded the results for suspicious content, the platform also provides a first estimation on the category of criminal activity; this is done based on keywords found in system categories such as Hacking and Child Sexual Abuse Material. The platform operator may then determine whether the contents of the link are legal or illegal by setting the appropriate selector.

The platform also includes logging functionality, creating a file that tracks step-by-step the actions taken by the user during an investigation, from the initial report right through the keywords and URLs used to conduct a crawl and the statistics, graphs and output data produced as a result. Such functionality is designed to aid the investigator in tracing the actions taken leading to a specific conclusion being made about a specific report, web page or other piece of information—providing a tangible record of the steps taken throughout the investigation. Such information can be used to aid with issues related to the admissibility of any information used as part of a larger investigative process.

The network tools which have been integrated into the UINFC2 platform enable investigators to reveal important information about the host servers (name, IP etc.) as well as information regarding the people responsible for the site domain. To this extent, the reported website (http://www.uinfc2.eu/wp/en/) was resolved by the investigator in order to gain insights like the website's owner, hosting country and more (Fig. 15.6). Further analysis with networking forensic tools could potentially reveal email address information linking key contacts in revealing perpetrators.

LEA/Hotline users may also gain insight with the generated intelligence analytics and charts. The provided statistics and charts of the crawled content includes, but is not limited to; the popular keywords (over time and per country) based on their frequency in total (from all crawling) while at the same time the user can identify the keywords that were identified more times based on the geo location of the crawled web-source (information extracted from the DNS resolving). Furthermore, the charts offer graphical representation of the aforementioned information in addition to a chart depicting the number of crawled URLs per country and a heat map which depicts the frequency of the identified suspicious keywords, as shown in Fig. 15.7.

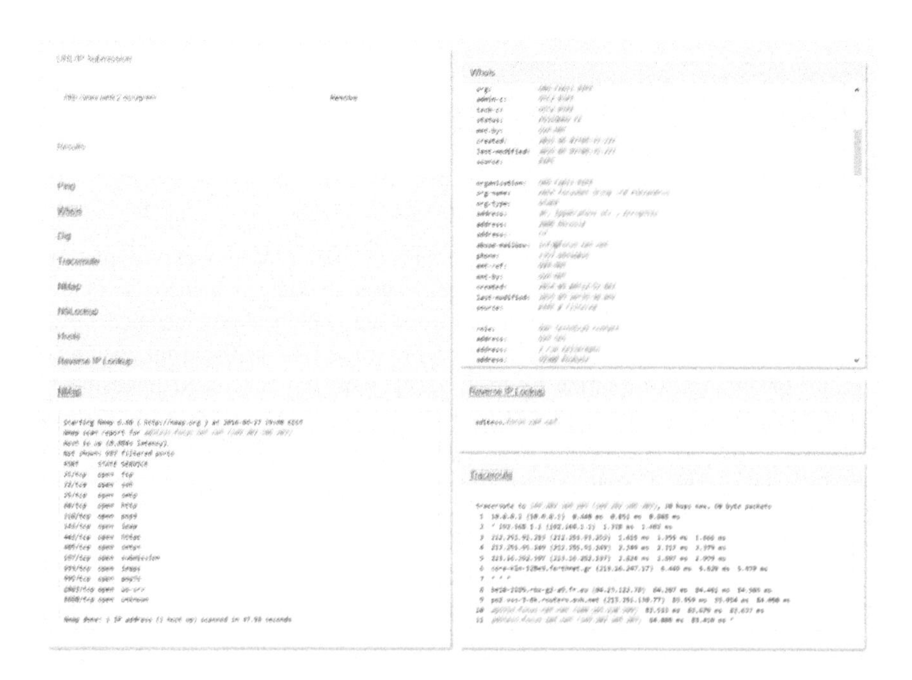

Fig. 15.6 Results of UINFC2 OSINT networking tools

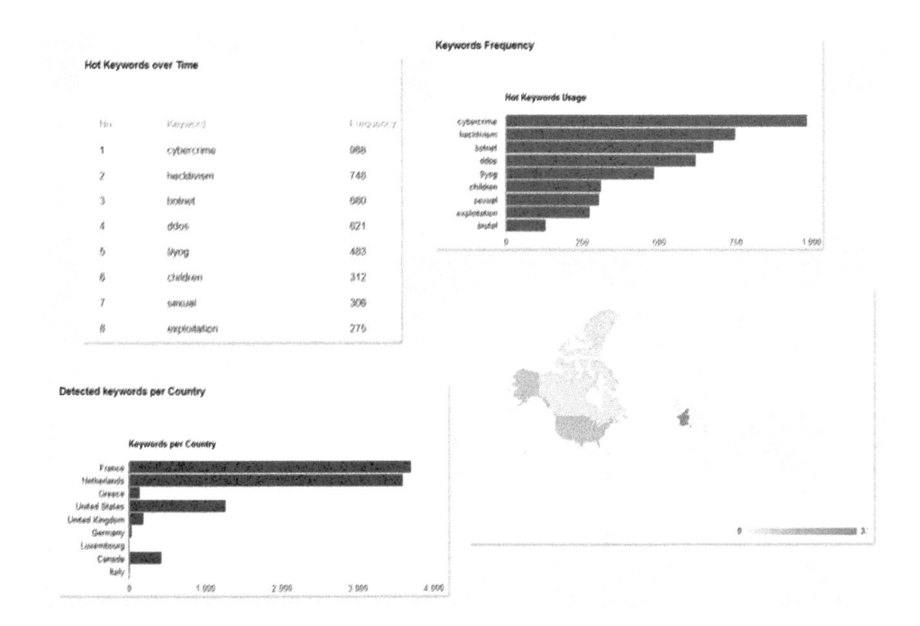

Fig. 15.7 UINFC2 statistics and charts

15.6 Concluding Remarks

The rise of cybercrime presents a great danger to public safety and wellbeing, posing continued challenges to LEAs as the tools used by and vectors through which individuals and groups commit crime continue to evolve. The right to anonymity, the wide range of available applications/devices used for internet access along with other challenges associated with international jurisdiction, technological change, training, education and awareness continue to escalate the severity of the challenge faced by those charged with safeguarding society and its citizens. The continued proliferation of connected devices, from traditional computers to mobile devices and the internet of things means young and vulnerable individuals are exposed to the internet like never before, creating new vectors for exploitation and abuse.

In response, law enforcement agencies and reporting authorities are increasingly looking to harness the power of technological tools to assist in the management and analysis of data to expedite and enrich existing investigative processes. Towards this objective, the UINFC2 platform combines a number of open-source tools to enable the analysis of information from the web and reports, narrowing the gap between citizens and law enforcement to provide means for incidents and content to be more comprehensively and accurately reported and subsequently analysed to detect related activities, and to detect suspicious or illegal material through keyword detection and text mining practices.

Acknowledgments The research leading to these results was funded by the European Commission's Prevention of and Fight against Crime, Illegal Use of Internet (INT) under grant agreement number [HOME/2013/ISEC/AG/INT/4000005215].

"Co-funded by the Prevention of and Fight against Crime Programme of the European Union"

This project has been funded with the support from the European Commission. This publication reflects the views only of the authors, and the European Commission cannot be held responsible for any use which may be made of the information contained therein.

References

Allsup R, Thomas E, Monk B, Frank R, Bouchard M (2015) Networking in child exploitation: assessing disruption strategies using registrant information. In: Proceedings of the 2015 IEEE/ACM international conference on advances in social networks analysis and mining 2015, ACM, pp. 400–407

Arthur KK, Venter HS (2004) An investigation into computer forensic tools. In: ISSA, pp. 1–11

Brewster B, Andrews S, Polovina S, Hirsch L, Akhgar B (2014) Environmental scanning and knowledge representation for the detection of organised crime threats. In: International conference on conceptual structures. Springer International Publishing, pp. 275–280

Carr J (2011) The internet dimension of sexual violence against children. In: Council of Europe, protecting children from sexual violence-A comprehensive approach, pp. 281–282

European Commission (2015) The European agenda on security. Retrieved from http://ec.europa.eu/dgs/home-affairs/e-library/documents/basic-documents/docs/eu_agenda_on_security_en.pdf

European Parliament (2011) DIRECTIVE 2011/92/EU of the European parliament and of the council of 13 December 2011 on combating the sexual abuse and sexual exploitation of children and child pornography

Europol (2015) The internet organised crime threat assessment

Geiger M (2005) Evaluating commercial counter-forensic tools. In: DFRWS

Gibson H, Andrews S, Domdouzis K, Hirsch L, Akhgar B (2014) Combining big social media data and FCA for crisis Response. In: Utility and cloud computing (UCC), 2014 IEEE/ACM 7th international conference on (pp. 690–695). IEEE

INHOPE (2014) Facts figures and trends—the fight against online child sexual abuse in perspective. Retrieved 1 Aug 2016, from http://www.inhope.org/tns/resources/statistics-and-infographics/statistics-and-infographics-2014.aspx

Internet Watch Foundation (2016) Annual Report 2015

INTERPOL (2016) Victim identification/crimes against children/crime areas/internet/home—INTERPOL. Retrieved 1 August 2016, from http://www.interpol.int/Crime-areas/Crimes-against-children/Victim-identification

INTERPOL (n.d.) Crimes against children: appropriate terminology. Retrieved from http://www.interpol.int/Crime-areas/Crimes-against-children/Appropriate-terminology

Kettleborough D (2015) What's wrong with 'child pornography'? Retrieved from http://wp.me/p2RS15-9f

Kurose JF (2005) Computer networking: A top-down approach featuring the internet, 3/E. Pearson Education India

Lalla H, Flowerday S (2010) Towards a standardised digital forensic process: E-mail forensics. In: ISSA

McAfee (2014) Estimating the global cost of cyber crime [June 2014], pp. 1–22
Meghanathan N, Allam SR, Moore LA (2010) Tools and techniques for network forensics. arXiv preprint arXiv:1004.0570
Microsoft (n.d.) PhotoDNA cloud service. Retrieved 1 Aug 2016, from https://www.microsoft.com/en-us/photodna
National Crime Agency (2013) Threat assessment of child sexual exploitation and abuse
Ponemon Institute (2015) 2015 cost of cyber crime study: global
PriceWaterhouseCoopers (2016) Global economic crime survey 2016
Rutgaizer M, Shavitt Y, Vertman O, Zilberman N (2012) Detecting pedophile activity in bittorrent networks. In: International conference on passive and active network measurement. Springer Berlin Heidelberg, pp. 106–115
U.S. Department of Justice (2013) Regional computer forensics laboratory annual report for fiscal year 2013. Retrieved from https://www.rcfl.gov/downloads/documents/fiscal-year-2013
Westlake B, Bouchard M, Frank R (2012) Comparing methods for detecting child exploitation content online. Proc.—2012 Eur Intell Secur Inform Conf EISIC 2012:156–163

Chapter 16
Identifying Illegal Cartel Activities from Open Sources

Pál Vadász, András Benczúr, Géza Füzesi and Sándor Munk

Abstract In a truly free marketplace, business entities compete with each other to appeal and to satisfy the purchasing needs of their customers. This elegant and efficient process can only succeed when competitors set their prices independently. When collusion occurs among competitors, prices rise, quality is often compromised and the public at large loses. In all developed countries around the world, price fixing, bid rigging and other forms of collusion are illegal and prosecuted through judicial systems. The relevance of OSINT for this form of activity is two-fold: as covertly conducted activity between parties, market manipulation and price fixing is particularly difficult to detect and prove while, at the same time, it is particularly susceptible to automated information discovery which can be vital for law enforcement agencies. However, finding even weak threads of evidentiary material requires extensive human and financial resources. This chapter proposes an automated methodology for text and data analysis, which aims to save both professional time and cost by equipping investigators with the means to detect questionable behavioural patterns thus triggering a more intimate review. This is followed by working examples of how OSINT characteristics and techniques come together for law enforcement purposes.

András Benczúr: Support from the "Momentum—Big Data" grant of the Hungarian Academy of Sciences.

P. Vadász (✉) · S. Munk
National University of Public Service, Budapest, Hungary
e-mail: pal.vadasz@uni-nke.hu

A. Benczúr
Institute for Computer Science and Control of the Hungarian Academy of Sciences (MTA SZTAKI), Budapest, Hungary

G. Füzesi
Hungarian Competition Authority, Budapest, Hungary

B. Akhgar et al. (eds.), *Open Source Intelligence Investigation*,
Advanced Sciences and Technologies for Security Applications,
DOI 10.1007/978-3-319-47671-1_16

16.1 Introduction

Consumers worldwide have come to expect the benefits of open, fair and honest competition in the marketplace. This means the availability of the best products and services at the lowest prices. In a truly free marketplace, business entities compete with each other to appeal and to satisfy the purchasing needs of their customers. This elegant and efficient process can only succeed when competitors set their prices independently. When collusion occurs among competitors, prices rise, quality is often compromised and the public at large loses. In all developed countries around the world, price fixing, bid rigging and other forms of collusion are illegal and prosecuted through judicial systems.

The concept of "transparency" is an essential element of all democratic governments for many reasons—with few more important than in the arena of public procurement. The purchasing of goods and services by government bodies is an especially vulnerable area for corruption, discrimination, and the distortion of fair marketplace competition and optimum pricing.

It is beyond the scope of this paper to analyse in detail the economic impact of cartel activities. However, an example of the aggregated cartels' economic impact in 12 developing countries (Brazil, Chile, Colombia, Indonesia, South Africa, Mexico, Pakistan, Peru, Russia, South Korea, Ukraine and Zambia) in terms of affected sales related to Gross Domestic Product (GDP) taken in average during 2002, revealed that impact varied from 0.01 to 3.74 %, with a maximum of 6.38 % (South Africa) in 2002. In terms of cartels' excess revenues, the actual harm reached 1 % of GDP for South Korea in 2004 and South Africa in 2002. Analysis shows that a cartel can decrease the production level by about 15 % on the given market (Ivaldi et al. 2014).

Throughout the professional and scientific literature, there is a multitude of publications related to healthy competition, reducing the costs of corruption in public procurement, fighting against collusion, cartels and bid rigging. Following 2005 in particular, a number of well documented articles appeared in scientific journals (Hinloopen and Martin 2006; Marshall and Marx 2007; Morgan 2009; Padhi and Mohapatra 2011; Werden et al. 2011; Cosnita-Langlais and Tropeano 2013), in scholarly books and research papers (Harrington 2006, 2008; Hüschelrath and Veith 2011; Morozov and Podzolkina 2013; Abrantes-Metz 2013; Wensinck et al. 2013), and in official government documents (Danger et al. 2009a, b; OECD 2012).

In economics, a cartel is defined as an agreement between competing firms to control prices or to exclude entry of a new competitor in a market.

Bid rigging (or collusive tendering), one form of cartel, occurs when businesses, that would otherwise be expected to compete, secretly conspire to raise prices or lower the quality of goods or services acquired through a bidding process. Bid rigging can be particularly harmful in public procurement. If bidders agree among themselves to unfairly compete in the procurement process, the public is denied a fair price. If suppliers engage in bid rigging, taxpayers' money is wasted as governments pay more than an appropriate price (Danger et al. 2009a, b). From a Hungarian point of view, in addition to national authorities, the two main international actors in the fight against bid rigging are the European Union (EU) and the Organisation for Economic Co-operation and Development (OECD).

In 2003, the European Commission announced a comprehensive EU policy against corruption (COM 2003). This communication noted that public procurement represents about 15 % of the GDP in the EU and underscored the importance of the community directives on public procurement to ensure transparency and non-discriminatory access to procurement opportunities. Moreover, it emphasized that collusive behaviour between companies participating in a tender may constitute a direct infringement of the prohibition laid down in Article 81 of the EC Treaty. In 2015, a detailed study was jointly prepared by PricewaterhouseCoopers and Ecorys for the EU Commission (Wensinck et al. 2013) and which presented a methodology to estimate the costs of corruption in public procurement. The study further provided information and tools to improve the public procurement rules and practices in and among the EU member states.

In 2009, the OECD published two policy documents with the title banner "Competition—The Key to Productivity and Growth". The first document, "Detecting bid rigging in public procurement" (Danger et al. 2009a), outlined a methodology to assist governments improve public procurement by thwarting bid rigging. The second document (Danger et al. 2009b) presented additional detailed concepts and descriptions. It also contained two checklists, one for aiding the design of the procurement process to reduce risks of bid rigging, and the other, a model for detecting bid rigging in public procurement. In 2012, the guidelines were reissued as an annex to an OECD Recommendation (OECD 2012).

At the present time, approximately 100 of the 163 World Trade Organization (WTO) member countries, including some 50 developing and transition countries, have adopted competition laws. Not all countries, however, are WTO members. Further, there are no "competition rules" that exist at the WTO level. In spite of the ever expanding globalization of trade, the regulation of competition predominantly exists at a state level. It is true that, at a national scale, competition rules are found in all EU and European Free Trade Association (EFTA) member states.

It is also true that in a number of major international free trade agreements [such as the North American Free Trade Agreement (NAFTA) and the Central European Free Trade Agreement CEFTA)] competition law regulations are included in varying width and depth. Attempts have been by made the General Agreement on Traffics and Trade (GATT), under the auspices of the WTO, to secure broader and strengthened competition rules, but these efforts (Singapore 1996, Doha 2001, Cancun 2003) have not been successful. While no imminent change can be expected in the foreseeable future, a practice of "soft" cooperation is becoming evident. Along with the OECD, the International Competition Network (ICN) is enabling a "loose" collaboration among legal experts, economists and academicians from the member competition authorities (WTO 2016).

16.2 The Principles

The basic definition of the cartel as a crime is defined in this section. The forms of cartels are listed, together with significant patterns for later machine-readable recognition.

16.2.1 The Definition of a Cartel

In economics, a cartel is an agreement between competing firms to control prices or to exclude entry of a new competitor in a market. It is a formal organisation of sellers or buyers that agree to fix selling prices, purchase prices or reduce production using a variety of tactics (Bishop and Walker 2010). The aim of such collusion is to increase individual members' profits by reducing competition.

There are several forms of cartel, such as:

- Setting minimum or target prices, i.e., price fixing
- Reducing total industry output, i.e., fixing market shares
- Allocating customers
- Allocating territories
- Bid rigging
- Illegally sharing information
- Coordination of the conditions of sale or
- A combination of the above

In this section, special focus is given to the category of bid rigging, a form of fraud in which a commercial contract is promised to one party—even though several other entities also "appear" to present a bid. This form of collusion of bidding companies (the winner and the pseudo losers) is illegal in most countries. It is a form of price fixing and market allocation, and is often practised where contracts are determined by a call-for-bids, for example, in the case of government construction contracts, Information and Communications Technology (ICT) procurement, etc. (Danger et al. 2009a).

Since, a cartel like all corruption is covertly conducted automated information discovery can be vital for law enforcement agencies[1]. Finding even weak threads of evidentiary material requires extensive human and financial resources. Our proposed automated methodology for text and data analysis saves both professional time and cost by equipping investigators with the means to detect questionable behavioural patterns thus triggering a more intimate review. All of this is done by taking advantage of open source intelligence.

[1]In some countries cartel detection may be the task of the General Accounting Office or a Comptroller's Office. In Hungary, it is the Cartel Office which can well be considered a LEA.

It is important to highlight that in our methodology all sources used are online, and we firmly believe—should be by law—remain unhindered, freely and publicly available. By definition, open source intelligence (OSINT) is publicly available. The availability of relevant information resources, whether free or paid, varies country by country. Even supposedly "free" content may have restrictions applied through compulsory identification and registration, password protection, or by mandatory human challenge-response tests (i.e., "CAPTCHA") preventing automated or robotic accessing and indexing for further dissemination.

One may well argue that the paucity of freely accessible public procurement data in any state is reflective of the degree of fear of unwanted disclosure by corrupt, or potentially corrupt, government procurement officials. In other words, free public access to such data can be viewed as a directly proportional measure as to the level of democracy in a given society.

16.2.2 The Sources of Information

In this section, the sources of information for the automatic evaluation of cartel cases are described.

Across the literature of the field, it is commonly accepted that methods of detecting bid rigging can be built using a combination of different types of information applied to one of two approaches or techniques—*reactive techniques* or *proactive techniques.*

Reactive techniques triggering an investigation is an approach based on an external event, such as receipt of a complaint (from a competitor, a customer, an agency, or an employee), possession of external information (from whistle-blowers, or informants), or information obtained from leniency applicants (Anti-cartel Enforcement Manual 2015; Hüschelrath 2010).

In present times, authorities responsible for protecting the integrity of open and fair competition rely heavily on reactive detection techniques, particularly on leniency/amnesty programmes. It is not prudent or sufficient, obviously, to rely on one single tool, or technique.

The second approach, applying ***proactive techniques***, to which our focus belongs, is detection characterized by initiative; it is not reliant on an external triggering event. An abundance of proactive techniques are presented across the literature field (Anti-cartel Enforcement Manual 2015; Hüschelrath 2010; Abrantes-Metz 2013; Harrington 2006; Hüschelrath and Veith 2011; OECD 2013). These include: analysing past cartel and competition cases; monitoring the market, industry, press and internet reporting; cooperation with competition and other national and foreign investigative agencies; and quantitative devices to highlight suspicious patterns, among other strategic screens.

Cartel screens have no generally accepted definition, rather, they are commonly identified as "*economic tools designed to analyse observable economic data and information, such as information on various product and market characteristics, data on costs, prices, market shares, various aspects of firm behaviour,* etc. *and flag markets which may either have been affected by collusion, or which may be more susceptible to collusion*" (OECD 2013). In known bid rigging cases, several patterns and screens are typically present and a comprehensive (but not necessarily exhaustive) list can be found in the OECD checklist (Bishop and Walker 2010).

Screening methods can be partitioned into structural and behavioural groupings. The first, structural, is based on data about an industry and identifies markets where it is more likely that a cartel will form. In contrast, behavioural methods employ data about the conduct or actions of the firms that, in itself, provides evidence that a cartel has formed. Behavioural screens are usually based on publicly available datasets, e.g. public procurement data (Anti-cartel Enforcement Manual 2015; Harrington 2006).

One way of detecting of cartels is through quantitative analysis—conducting calculation of elementary and complex indicators. These indicators can both include and signal many types of collusive behavioural patterns, and if used in combination of traditional investigative methods, can provide more efficient use of limited resources.

Such indicators have been discussed at length in numerous publications (Abrantes-Metz 2013; Harrington 2006; Tóth et al. 2015; Wensinck et al. 2013) and can be applied toward achieving different practicable solutions.

A detailed study done by Hungarian researchers (Tóth et al. 2015) defined and classified the major types of collusive bidding, developed a set of elementary and complex indicators expected to signal collusion in public procurement, and demonstrated how such indicators can be deployed to fit diverse market realities—all combined into a coherent framework.

An important phase of cartel detection from publicly available datasets is extracting relevant data from these datasets. In today's practice, public procurement data is available mainly in structured or semi-structured formats (HTML, XML, JSON, CSV). So to enable evaluation, data extraction should be done by specially programmed solutions (Fóra 2014; Tóth et al. 2015). The future in this topic should also be based on linked open data paradigm, where competent authorities publish their open government data in a more usable format (Vojtech et al. 2014).

In 2012, an unpublished case study derived from publicly available datasets was prepared for the Hungarian Cartel Office. The study tracked actual government procurement with data acquired from public sources (OSINT). But for one step, the process was completely automated. (The datasets syphoned from the procurement database were not automatically compared with the Public Company Registry records as this required separate programming for the purposes of the case study.) It should be noted that even this partially automated effort revealed substantial suspicious nodes and lines that had not been previously detected by investigators through human intelligence and other means.

Unfortunately, despite the tantalizing promise suggested by the results of this case study, the Hungarian Cartel Office, for whatever reason, chose not to pursue the project. Consequently, work on the algorithm has not been continued.

In the following section, we outline further discussion on the matters of data capture and automation.

16.2.2.1 Government Procurement Records

Unless a specific procurement is ascribed relief by the Hungarian National Security Committee of Parliament from the obligation of confidentiality stipulated by the Public Procurement Law, all details of the transaction must be posted on the internet. There is one "quasi" exception, namely, the individual calls of frame agreements, which can be accessed by designated users only and which are protected by passwords and CAPTCHA challenge-response features.

After registration, the database can be accessed without constraints by search engines.

Relevant main fields of the Procurement Database are as follows:

- Name of the organisation running the procurement
- Basic contract data
- Date of publication
- CPV[2] code
- Name of the winning company
- Short and detailed description of the subject of procurement
- Short description of the contract
- The winner price
- The type of bidding
- Name of the losing companies
- The loser prices
- Subcontractors' data
- Procurement consultants' data
- Reasons for eventual failure of the public procurement process

16.2.2.2 Company Registry

The Hungarian Company Registry contains all relevant information on commercial organisations that are, or had been, active, in Hungary since 1991. Unfortunately, the database cannot be openly accessed by public search engines, with the exception of designated contract partners. To our knowledge, at present, it is not possible for a non-established player to purchase the database. While access to individual records is possible, the content is CAPTCHA-protected. Commercial providers offer additional company information, such as financial data or the network of owners, etc., visualised.

16.2.2.3 Legal Databases

Court decisions on occasions provide enlightening glimpses into the unlawful behaviour of companies or key executives operating behind the scenes. In fact, it is

[2]Common Procurement Vocabulary, see: http://simap.ted.europa.eu/web/simap/cpv (27.06.2016).

not uncommon to discover that names of key executives reappear at different organisations or to uncovered instances of individuals possessing ***identical*** backgrounds repeatedly turn up at multiple companies like a chameleon. Legal databases can be valuable tools toward identifying possible fraudulent undertakings.

16.2.2.4 Other Open-Source Intelligence (OSINT) sources

News accounts and social media (platforms such as forums, blogs, Facebook, Twitter, etc., can serve to spotlight hints of internal dissatisfaction within a company, or otherwise help publicly raise instances of "whistleblowing" or similar potentially important information.

16.2.3 Cartel Patterns

In this section, various methods of bid rigging are discussed. These methods can serve as patterns that potentially result in the recognition of illegal activities. The following methods listed are derived from the directives of the EU and OECD (Hölzer 2014; Danger et al. 2009b) as well as from private interviews.

The patterns below, cross-matched with other records, suggest an indication of a suspicious cartel case.

- *Winner steps back*. The probability of bid rigging is high if a winner withdraws its bid and lets the second best (and substantially more expensive) bidder obtain the deal.
- *Invalid bids*. Some or all bidders but one submit invalid bids, so that the agreed entity wins.
- *High prices*. All or most of the tenders are significantly higher than in previous procedures, higher than the companies' list prices or higher than a reasonable estimate of the costs.
- *Coordinated high prices*. All but one of the tenders are extraordinarily high, well above market standard.
- *Market coordination based on geographic allocation*. A company submitting tenders that are significantly higher in some geographic areas than others, without any obvious reason, such as differences in costs, may suggest it is involved in a bid-rigging cartel.
- *Suspiciously similar prices*. If several companies have submitted tenders with identical or suspiciously similar prices, it may indicate that they have agreed to share the contract.
- *Suspected boycott*. If no tenders are received, there may be a coordinated boycott with the intention of influencing the conditions of the contract.
- *Suspiciously few tenders*. Too few companies submit tenders; this may indicate a market-sharing cartel.

- *Suspiciously similar tenders.* If tenders refer to industry agreements that affect the price, the companies may have agreed, for example, to apply common price lists, delayed payment fees or other sales conditions for the sector.
- *Suspicious subcontracting arrangements.* If the company that won the contract assigns or subcontracts part of the contract to a competitor that submitted a higher priced bid in the same procedure, this may suggest a bid-rigging cartel.
- *Suspected joint tenders.* A joint tender submitted by more companies than necessary to perform the assignment may be illegal.
- *Cyclical winning.* Coordinated bidding series with different bidders.

Publications and the research literature, generally speaking, proffer the assumption that individual companies or collective entities belonging to a specific interest group committing cartel fraud are independent from one another and remain static, non-changing organizations throughout the period while under investigation. This is a faulty assumption since people, companies and organisations engage in less than obvious relationships which are indecipherable based upon procurement data.

In order to discern hidden ties, one must thoroughly investigate a multitude of sources beyond simply an official government Procurement Bulletin. An obvious investigatory tip-off, for example, may stem from a situation where records reveal that several companies are owned by the same individual. In such a case, the probability of maleficence is high.

Though there can be numerous variations of hidden control, some typical examples include:

- Seemingly different owners, but one of them is related to the other
- The owner's wife is another owner, but she is listed by her maiden name
- Different owner companies, but they possess the same address under an offshore trust
- Companies seemingly have nothing to do with one another, but use the same bank branch, the same auditor and/or their internet webpages are registered under the same provider
- Common directors, supervisory board members, managing officers or other senior executives
- "Dummy" or "straw man" figurehead is somehow related or connected to the actual owner. Sometimes also referred to as a "hollow suit," this individual might be a classmate, a neighbour, a golf club fellow, or an acquaintance of the "true" owner that perhaps was made through a random encounter including a social media website. The "front man" might have previously been named in the press or in court document.

Not realising the existence of any of these hidden relationships, an investigator might be satisfied with the integrity of the bidding and contract award process. Over a prolonged lifetime, numerous fraudulent procurement contracts might easily be successful perpetuated.

The most obvious example of a cartel case possibly "flying under the radar screen" of investigators occurs when financial compensation is transferred through subcontracting or informal payments by a company related to the "winner" to an entity or entities connected to the "losers." If, however, the hidden relationship between the companies at the surface and those corresponding beneath in the shadows are not discovered, the fraud remains undetected.

No "silver bullet" for exposing this behaviour is offered in our discourse. The criminal methodological possibilities are, without exaggeration, countless, while the techniques for conveying illicit cash payments are practically impossible to discover by data or text analytic tools.

It is important to emphasize that a procurement record raising suspicion does not necessarily establish cartel crime. Expert human analysis providing final judgement cannot be completely substituted by any robot or artificial intelligence (AI) technology.

16.2.4 Security Models

In order to devise a system to raise suspicion of cartel fraud, we rely on security models that were originally designed to implement policies to disallow certain behaviours or to only allow certain others. Automated fraud detection methods can be based upon the design of these security models.

Early security models only allowed the specification of positive authorizations, which were later extended by negative and mixed rules (Bertino et al. 1997). Positive, negative and mixed security models correspond to positive and negative security parameters in rule-based systems. Negative rules list known disallowed patterns, drawn from episodes observed in the past, while positive rules allow only known behaviour and disallow any deviation.

In the next section, we will describe how the positive and negative security models may be used to derive automated fraud detection systems. These security models also correspond to training models in machine learning.

16.2.4.1 Negative Security Models and Supervised Learning

Negative security modelling developed using past security risk events; consequently, it can only detect present security incursions that have similar precedents. Most experts agree that approach is only able to detect a small fraction of hazardous events, particularly against the backdrop of the complicated procedures of large modern companies.

Negative security models correspond to supervised machine learning. A supervised model is trained based upon a set of positive and negative scenarios that is it provided. As a general critique to supervised fraud detection, the system may only identify known fraudulent episodes.

Supervised machine learning has very powerful, measurable methods. This advantage of measurability has led to an explosion of continually enhanced methodologies, which will be described in Sect. 16.4.

16.2.4.2 Positive Security Models and Unsupervised Learning

Under a positive security model, all events that deviate from the patterns of typical activities may be considered as fraudulent. The principle advantages of a positive security model is that it needs no predefined set of known fraudulent events and that it may detect previously unknown types of malicious activities.

Unsupervised learning is the branch of machine intelligence that may assimilate and build upon its knowledge base without a set of predefined labelled events. Among unsupervised methods include anomaly detection, a class of methods that examines all available data and identifies "outliers," which deviate from the data set universe.

Despite of the promise of being able to detect previously unknown types of fraud, unsupervised methods have two main disadvantages that limit their use in practice. First, it is difficult to evaluate their performance if no labelled data is available. In this case, there is no information to compare the performance of two models one against the other. Second, the notion of an outlier cannot be well defined. The largest contract values, or sectors with only very few procurement procedures may well be identified as outliers, however, they may not correspond to fraud.

Over the past decade, machine learning research has turned mostly to supervised models leaving humans interact by creating training instances. "Active" learning (Tong and Koller 2001) is a promising direction when a human in the loop can contribute toward gradually improving the classifier by providing the most important labelled cases. In our examination, we focus on supervised machine learning—combined with manually identifying new fraud cases as described in Sect. 16.3.4, labelled Feature engineering.

16.3 Data Acquisition from Open Sources

16.3.1 The Architecture

Figure 16.1 represents the architecture of the entire proposed application. There are four basic sources:

- The procurement database
- The Company Registry
- The legal text bases
- Optional external databases such as press, forum, social media, etc.

The fusion centre prepares a data table of the available numeric and categorical data for each bidding action in the form needed for the predictive analytics methods. The actions (events or cases) form the rows or the data table, while columns are called variables, attributes or features. An example of data is shown in Table 16.1.

16.3.2 Entity Extraction

Public procurement, press or Company Registry information is normally stored in text databases in PDF format and can be retrieved through RSS or otherwise with crawlers. It is not an easy task to retrieve specific information from an unstructured form in plain text file and then transform the data into a structured database. Specific fields such as price, Common Procurement Vocabulary (CPV) code or the name of a person or organisation, etc., must be recognised in the continuous text flow and the content of the field retrieved. In order to achieve this, text-mining technology (grammar-based methods and statistical models) must be applied. This process is called Named Entity Extraction (NER).

16.3.3 Filtering Out Suspicious Items in the Fusion Centre

Having retrieved all relevant information from the procurement database, the Company Registry, the legal database and public sources, and by applying the cartel patterns as noted above, the algorithm can automatically filter out those procurement records that, by definition, possess suspicious traits. The task is fairly straightforward inasmuch as each and every record falling within the searched timeframe is matched against the given patterns. The suspicious ones are kept for further processing whereas the rest are simply dropped. It is important to emphasize that this filtering is an evolving process. New cartel cases identified by the investigators should be added as patterns for future processing. It is also worth considering the establishment of an international database to store emerging new patterns and in recognition that cartel methods are not country or language specific.

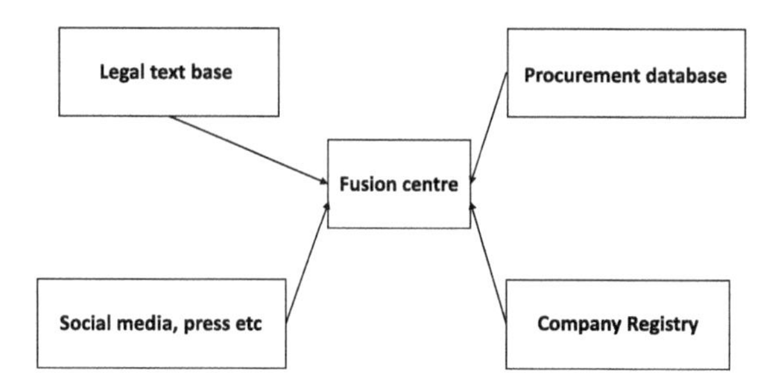

Fig. 16.1 The architecture of the open source data acquisition

Table 16.1 An example procurement data set for predictive analytics

ID	Bid prize	Second bid prize	Reserve prize	Times winner withdrawn	Percent of wins in past	Percent of reserve prize	Distance of location	…	Cartel fraud (label)	M1 (logistic regression)	M2 (depth 2 tree)	M3 (boosted tree)
1	1000	1100	1200	0	0.2	…	…	…	No	0.17	No	3
2	1000	1200	1100	1	0.5	…	…	…	Yes	0.49	No	−1
3	2000	3100	3200	1	0.1	…	…	…	No	0.00	No	1
4	2000	2200	2100	0	0.3	…	…	…	No	0.39	No	3
5	3000	3100	3200	2	0.6	…	…	…	Yes	0.84	Yes	−3
6	3000	3200	3100	0	0.5	…	…	…	No	0.54	No	1
7	3000	4100	4200	1	0.2	…	…	…	No	0.00	No	1
8	3000	3200	3100	3	0.3	…	…	…	Yes	0.96	Yes	−1
9	4000	5200	5100	0	0.3	…	…	…	No	0.00	No	3
10	4000	4200	4100	0	0.7	…	…	…	Yes	0.69	Yes	−1

The last four columns contain the true label and the predictions of two sample models.

While this paper describes simple database matching, the potential of applying machine learning algorithms to identify hidden relationships holds exciting promises for further exposure of sophisticatedly constructed cartels today in and the future.

Quite often, a few fraudulent cases can very easily be identified by considering certain extreme values in the procurement database. For example, an outlying number of identical prices can easily be observed in a price histogram (Funderburk 1974) leading to the identification of "boilerplate" or "templatic" competition documents as an indicator of collusion. In fact, a similar approach was used to identify the perpetuators behind an episode of social media spam by spotting precisely equally large engagements for messages (Pálovics et al. 2014).

Bootstrapping is a technique that starts first begins by identifying certain fraudulent cases by manual investigation of the attribute distributions, and then adding these cases as positive training instances to identify additional, and perhaps, less obvious and better concealed instances of fraud.

16.3.4 Feature Engineering

The goal of feature engineering is to provide a sufficiently rich attribute table for the machine learning methods. A portion of the existing literature on cartel fraud detection relies on traditional statistics of the procurement process. For example, one particular model (Padhi and Mohapatra 2011) considers bid price to reserve price ratios and identifies cartel cases by this single value alone. This approach only performs linear regression for the bid to reserve price and statistically analyse the regression coefficients, however the findings are inconclusive for a predictive model. This model also requires manual outlier investigation by thoroughly examining cluster distributions.

Other numeric attributes may be tagged as the number of participants in the bid, the number and percent of successful and lost bids of the applicant, and derived statistics—such as deviance from sectorial and regional averages. The cases when the winner withdrew from a bid appear less frequent, and hence, different normalization is required for the count of such episodes.

In Sect. 16.2.3, labelled Cartel patterns, we outlined a number of collusive bidding patterns, each of which has a corresponding set of features that can be computed and used by machine learning techniques. The techniques and the corresponding features are summarized in Table 16.2. Additional features can be generated, for example, based upon the distance of the corporate headquarters from project location, or the percentage of organisation's free capacity, and/or the experience of the contractor involved in comparable projects. Note the definition of the last type of features may be problematic to incorporate if subcontractors are further involved.

Table 16.2 Collusive patterns and corresponding feature engineering technologies

Collusive pattern	Numeric or nominal features
Same company wins most of the time	Number and fraction of wins, by region, sector, etc.
Bid rotation	Time series descriptors of winners
Few or no new participants; bid participants are well aware of competitors	Histograms of participation distribution
Bidding does not erode target prize	Bid to reserve prize statistics
Participants withdraw	Withdraw count by bid and participant

16.3.5 Fitted Parameters of Economic Models

Fitted parameters of economic models may also serve as features. Several similar economic models are surveyed in Detecting Cartels (Harrington 2008) and price time series observations in Behavioural Screening and the Detection of Cartels (Harrington 2006). Porter and Zona propose statistics over the rank order of competitors' bids, which can also directly be used as a feature by a machine learning algorithm (Porter and Zona 1992).

We propose a different approach using both statistics and fitted parameters. Instead of investigating the possible values and thresholds one by one, all the raw and derived variables (e.g. prices and ratios of prices) are computed as features for supervised classification. Classification methods will then be able to automatically select relevant variables, their threshold, and even potential rules that reveal connections between further values.

16.3.6 Network Science and Visualization

Network science (Lewis 2011) is a relatively recent research area for analysing the topology, stability and emergence of interconnected pairs of events. In the context of the procurement process, it is possible to analyse the complexity of data comprising authorities, auctions, participants separating winners among competitors. Subcontractors, and relationship of ownership, board members, key officers, etc., may also be considered.

In order to define features, the various distances in the graph are used to characterize the bids by their pairwise distance, or by their distance from known fraudulent cases. The simplest distance represents the number of interactions (common participation in auctions, ownership, membership, etc.) between the two cases. Advanced measures such as SimRank (Jeh and Widom 2002) may rely on random walk distances over the network.

The product of the search and matching is a delineation of points which reveals entities that possess suspicious business relationships, including ties to individuals who have some kind of role with or connection to these companies. The linkages of graphs are previous procurement bidding instances of entities competing with (won against, lost to) each other (through direct or indirect ownership, or through relatives). The nodes are companies and people. To better comprehend and fully analyse this network, such a visualisation is extremely helpful and enlightening.

Figure 16.2 is a snapshot of a Sentinel application. The centre of the cartel network can be enlarged for a more thorough analysis. A Sentinel application enables the user to view the history of events following a timescale. Obviously other visualisation tools such as Analyst Notebook from i2/IBM may be employed as well.

In this enlarged, detailed view of a principal cartel network, the ties between the nodes are noted as "accomplice of," whereas the bidding companies and the customers are designated as unique nodes. The darkness of the node reflects its degree of involvement in the cartel. The dark grey rectangle represents the corpus of the octopus, or in System Network Architecture (SNA) terms, the "Alpha User".

16.4 Machine Learning Methodologies

Different names have been assigned to the discipline of producing qualitative predictions derived from existing data. Traditionally, predictive methods belong to statistics, which was already using decision trees (Safavian and Landgrebe 1998) and logistic regression (Cox 1958) from early times.

Toward the end of the 1990s, the field of study known as of "data mining" arose bringing attention to the size of the data (Han and Kamber 2001). Data mining had an equal focus on supervised and unsupervised methods, however, not all data mining techniques were later proved to perform well in practice.

At roughly the same time, in the area of machine learning, new techniques were coming into prominence, most notably support vector machines (SVM) and boosting. SVM proved to work very well for text classification (Joachims 1998) and, more recently, for time series (Daróczy et al. 2015). Boosting is the preferred technique used in most data analysis challenge solution projects (Chen and Guestrin 2016).

More recently, a class of learning algorithms using very large artificial neural networks of specialized structure known as deep learning is generating widespread interest for its success in variety of areas including image (Krizhevsky et al. 2012) and audio (Lee et al. 2009) classification. Deep learning has also performed well in recommender systems (Wang et al. 2015), however, as of present time, it appears to be less suited for the application needs of this project.

Anomaly detection (Chandola et al. 2009) can be based on most of the classifiers by turning them to one-class grouping selections. Anomaly detection methods, however, suffer from the general problem of unsupervised methods.

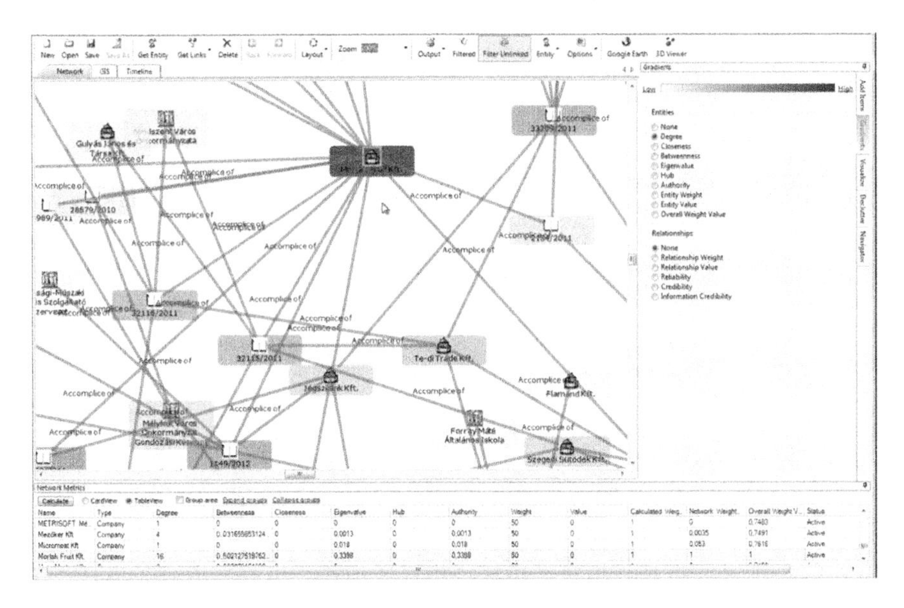

Fig. 16.2 Detailed view of a successfully investigated network in Hungary. In the *centre* is the hub of the network

16.4.1 Evaluation of Predictive Methods

We start the overview of the machine learning techniques applicable for fraud detection by describing the measures to evaluate the quality of a predictive method.

We distinguish between two types of evaluation metrics, the first based on a binary normal (negative) versus fraud (positive) prediction, and the other according to a ranking of the events tied to their expected risk of being fraudulent. We note that a ranked list of events can be turned into a binary prediction by specifying a threshold such that events above the threshold are classified fraudulent and those beneath deemed normal.

Binary predictions can be evaluated based on the so-called "class confusion matrix" (Powers 2011). This matrix has four elements: true positive (TP), true negative (TN), false positive (FP) and, false negative (FN), for events as shown in Table 16.3.

Employing the class confusion matrix, we may define a large number of quality metrics, including:

$$\text{accuracy} = (\text{TP} + \text{TN})/\text{total number of cases},$$

$$\text{precision} = \text{TP}/(\text{TP} + \text{FP}),$$

Table 16.3 Class confusion matrix and quality metric values based on binary prediction of model M1 > 0.5 in Table 16.1

	Actual cartel fraud: yes	Actual cartel fraud: no
Predicted by M1 as fraud: yes	True positive (TP): 3 (IDs 5, 8, 10)	False positive (FP): 1 (ID 6)
Predicted by M1 as fraud: no	False negative (FN): 1 (ID 2)	True negative (TN): 5 (IDs 1, 3, 4, 7, 9)

In the example, Accuracy = (3 + 5)/10 = 0.8 Precision = 3/(3 + 1) = 0.75; Recall = 3/(3 + 1) = 0.75; False positive rate = 1/(1 + 5) = 0.16

$$\text{recall, or true positive rate} = \text{TP}/(\text{TP} + \text{FN}),$$

$$\text{false positive rate} = \text{FP}/(\text{FP} + \text{TN}).$$

For ranked list evaluation, the key notion is the Receiver Operating Characteristic (ROC) curve (Swets 1996), which plots the true positive rate against the false positive rate. The Area Under the ROC Curve (AUC) metric has value 1 for a perfect prediction and 0.5 for a random forecast. A commonly used variant is $\text{Gini} = 2 \cdot \text{AUC} - 1$.

Utilizing the above metrics and a set of labelled events, we train a model on the labelled instances and then evaluate its performance over the same instances. This so-called "re-substitution method" has a drawback: it overemphasizes performance on known instances and may reveal little of the outcome of yet unseen new events.

There are several options to separate the evaluation from the training data. It is possible to select a random proportion of the labelled data for "training" and retain the rest for "testing." Alternatively, historic data may be used for training and the more recent for testing (Tan et al. 2013).

The advantage of a separate testing set is that it may not only detect if the model is insufficiently simple, but also highlight if irrelevant attributes of the training data not generalized for unseen cases is overused. In technical phrasing a model "underfits" if there is another approach that performs better on both training and testing data. Conversely, a model "overfits" if another method shows weak performance over the training data but has better outcome with the testing data.

16.4.2 Logistic Regression

Simply stated, a logistic regression model is a monotonic transformation of the linear combination of the variables (Cox 1958). The advantage of the linear model is that it explains the importance of the variables by the coefficients as weights.

Logistic regression has a drawback: it is prone to overfitting correlating variables.

In the example of Table 16.1 the correct model would be

$$\text{score} = \text{expit}\,(-0.2 \cdot \text{Times winner withdrawn} - 0.8 \cdot \text{Percent of wins in past} + 0.5)$$

of accuracy 100 %. While we may easily obtain models that overfit the large correlating numeric attributes such as model M1 in Table 16.1:

$$\text{score} = \text{expit}\,(-0.0136 \cdot \text{bid prize} + 0.004 \cdot \text{second bid prize} + 0.009 \cdot \text{Reserve prize} - 1.0 \cdot \text{Times winner withdrawn} - 0.15 \cdot \text{Percent of wins in past})$$

with accuracy only 60 %. Variable selection, normalization and various optimization procedures are crucial for the quality of logistic regression.

16.4.3 Decision Trees

Decision trees (Safavian and Landgrebe 1998) are inductively built by splitting the actual set of events along the variable that separates the fraudulent events from the accepted normal cases.

A sample decision tree of depth 2 is seen in Fig. 16.3, evaluated in the last column as taken from Table 16.1.

The disadvantage of decision trees is that they tend to overfit deeper down in the tree, since the decisions are made based on decreasingly fewer events.

In considering the bottom left node of the tree in Fig. 16.3, note that a single cartel case could not be separated from normal transactions. Further, neither of the variables "Times winner withdrawn" nor "Percent of wins in past" can separate this case. Therefore, two more layers would be required to handle only the remaining one single cartel case.

16.4.4 Boosting

"Boosting" (Freund and Schapire 1997) has as its principal concept the element of training simple classifiers, for example, small decision trees, by gradually improving the prediction quality in iteration cycles. The main advantage compared to large decision trees is that boosting obtains training gained over the entire data and not just a subset in all the iterations.

In Fig. 16.4, a sample boosted tree model is shown for the data set in Table 16.1. The positive or negative sign of the boosted tree model has accuracy 100 % over the sample data.

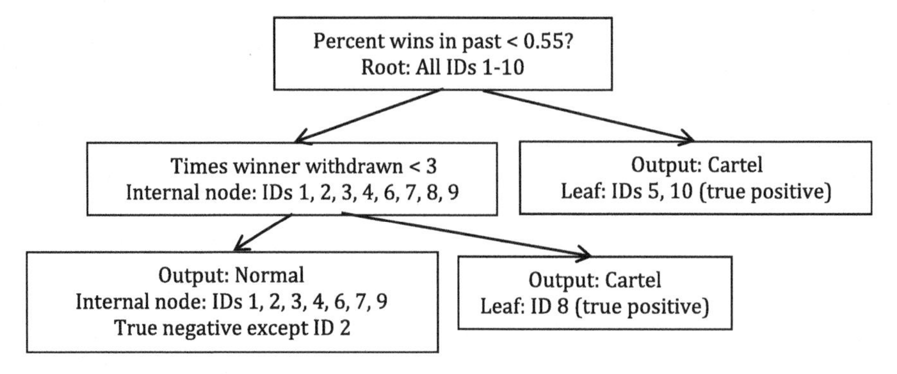

Fig. 16.3 Depth two decision tree example over data of Table 16.1

16.5 Conclusion and Further Work

In order to identify cartels, the following methodology is used in the following sequence:

1. Potential data sources are identified
2. Information extraction method is determined (crawler and text mining)
3. Network is created and visualized
4. Machine learning techniques are used
5. Results are analysed

Fig. 16.4 Boosted tree example over data of Table 16.1

```
Tree 1
    if Times winner withdrawn < 1:
   score = -1
  else:
   score = 1
Tree 2
  if Percent wins in past < 0.4:
   score = score - 1
  else:
   score = score + 1
Tree 3
  if Times winner withdrawn < 2:
   score = score - 1
  else:
       score = score + 1
```

The essence of the process is to distil the relevant information from an otherwise vast universe of data that can be analysed by Social Network Analysis (SNA) and graph theory tools. The SNA plays a key role here, so it is used as a framework for investigating cartels. Structuring information from unstructured data, web mining and text mining have also been used in several areas (Nemeslaki and Pocsarovszky 2011; Nicholls 2011).

As our main findings, we identified key variables composed of past participation and wins, competition prizes, locations and networked properties of the companies. In our proof of concept, we were able to identify cartel cases yet unknown to the Hungarian Cartel Office. Further, we highlighted the significance of strongly relevant OSINT data other than procurement.

In future work, we plan to conduct a systematic analysis of the Hungarian procurement-related OSINT data to quantitatively measure the performance of the machine learning methods. We envisage participation in a joint EU-funded project with the view of applying the method used for the Hungarian project toward other European countries.

References

Abrantes-Metz RM (2013) Proactive vs. reactive anti-cartel policy: the role of empirical screens. Social Science Research Network. http://ssrn.com/abstract=2284740, 15 June 2016

Anti-cartel Enforcement Manual (2015) International competition network. http://www.internationalcompetitionnetwork.org/working-groups/current/cartel/manual.aspx, 15 June 2016

Bertino E, Samarati P, Jajodia S (1997) An extended authorization model for relational databases. IEEE Trans Knowl Data Eng 9(1):85–101

Bishop S, Walker M (2010) The economics of EC competition law. Sweet and Maxwell, Andover, UK, p 171

Chandola V, Banerjee A, Kumar V (2009) Anomaly detection: a survey. ACM Comput Surv (CSUR) 41(3):15

Chen T, Guestrin C (2016) Xgboost: a scalable tree boosting system. arXiv preprint arXiv:1603.02754

COM (2003) 317, Communication from the Commission to the Council, the European Parliament, and the European Economic and Social Committee on a Comprehensive EU Policy against Corruption. Commission of the European Communities, Brussels, 2003

Cosnita-Langlais A, Tropeano JP (2013) Fight cartels or control mergers? On the optimal allocation of enforcement efforts within competition policy. Int Rev Law Econ 34:34–40

Cox Dr (1958) The regression analysis of binary sequences (with discussion). J Roy Stat Soc B 20:215–242

Danger K, Capobianco A (2009a) Guidelines for fighting bid rigging in public procurement, OECD. https://www.oecd.org/competition/cartels/42851044.pdf, 13 Mar 2016

Danger K, Capobianco A (2009b) Detecting bid rigging in public procurement, OECD. www.oecd.org/competition/cartels/42594486.pdf, 30 Jan 2016

Daróczy B, Vaderna P, Benczúr, A (2015) Machine learning based session drop prediction in LTE networks and its SON aspects. In: 2015 IEEE 81st vehicular technology conference (VTC Spring). IEEE

Fóra G (2014) Közbeszerzési eljárások dokumentumainak adatfeldolgozása 1998–2004 (Data extraction from public procurement documents 1998–2004). 8th Report. Korrupciókutató Központ, Budapest

Freund Y, Schapire RE (1997) A decision-theoretic generalization of on-line learning and an application to boosting. J Comput Syst Sci 55(1):119–139

Funderburk DR (1974) Price fixing in the liquid-asphalt industry: economic analysis versus the 'hot document'. Antitrust Law Econ Rev 7:61–74

Han J, Kamber M (2001) Data mining: concepts and techniques. Morgan Kaufmann

Harrington JE (2006) Behavioral screening and detection of cartels. In: Ehlermann CD, Atanasiu I (eds) European competition law annual 2006: enforcement of prohibition of cartels. Hart Publishing, Oxford, Portland

Harrington JE (2008) Detecting cartels. In: Buccirossi P (ed) Handbook in antitrust economy. MIT Press

Hinloopen J, Martin S (2006) The economics of cartels, cartel policy, and collusion: introduction to the special issue. Int J Ind Organ 24(6):1079–1082

Hölzer H (2014) Checklist of the Swedish Competition Authority in Harmonization of public procurement system in Ukraine with EU standards, Report on the development of diagnostic tools and Guidelines for the AMCU for identifying and preventing bid-rigging cases, Annex 3., Crown Agents Ltd., 2 Oct 2014. http://eupublicprocurement.org.ua/wp-content/uploads/2015/01/Report-on-Bid-Rigging-Diagnostics-ENG.pdf, 10 Jan 2016

Hüschelrath K (2010) How are cartels detected? The increasing use of proactive methods to establish antitrust infringements. J Eur Compet Law Pract 1(6)

Hüschelrath K, Veith T (2011) Cartel detection in procurement markets. Discussion Paper No. 11-066. Centre for European Economic Research, Mannheim

Ivaldi M, Khimich A, Jenny F (2014) Measuring the economic effects of cartels in developing countries, published in the framework of the CEPR PEDL program. http://unctad.org/en/PublicationsLibrary/ditcclpmisc2014d2_en.pdf, 7 Feb 2016

Jeh G, Widom J (2002) SimRank: a measure of structural-context similarity. In: Proceedings of the eighth ACM SIGKDD international conference on Knowledge discovery and data mining. ACM

Joachims T (1998) Text categorization with support vector machines: learning with many relevant features. In: European conference on machine learning. Springer, Berlin

Krizhevsky A, Sutskever I, Hinton GE (2012). Imagenet classification with deep convolutional neural networks. In: Advances in neural information processing systems

Lee H, Pham P, Largman Y, Ng AY (2009) Unsupervised feature learning for audio classification using convolutional deep belief networks. In: Advances in neural information processing systems. pp 1096–1104

Lewis TG (2011) Network science: theory and applications. Wiley

Marshall RC, Marx LM (2007) Bidder collusion. J Econ Theory 133(1):374–402

Morgan EJ (2009) Controlling cartels—implications of the EU policy reforms. Eur Manage J 27 (1):1–12

Morozov I, Podkolzina E (2013) Collusion detection in procurement actions. Working Paper. National Research University, Higher School of Economics (Russia), Moscow

Nemeslaki A, Pocsarovszky K (2011) Web crawler research methodology. In: 22nd European Regional ITS Conference, Budapest 2011: Innovative ICT Applications—Emerging Regulatory, Economic and Policy Issues 52173, International Telecommunications Society (ITS). http://econstor.eu/bitstream/10419/52173/1/67255853X.pdf/, 7 Feb 2016

Nicholls SN (2011) Detecting, mitigating & fighting bid rigging in public procurement, fair trading commission. http://www.ftc.gov.bb/library/2011-02-07_ftc_guidelines_checklist_procurement.pdf, 30 Jan 2016

OECD (2012) Recommendation of the OECD council on fighting bid rigging in public procurement. OECD Council, Paris, 2012

OECD (2013) Ex officio cartel investigations and the use of screens to detect cartels. Policy Roundtables. OECD, 2013

Padhi SS, Mohapatra PKJ (2011) Detection of collusion in government procurement auctions. J Purchasing Supply Manage 17(4):207–221

Pálovics R, Ayala-Gómez F, Csikota B, Daróczy B, Kocsis L, Spadacene D, Benczúr AA (2014, Oct). RecSys Challenge 2014: an ensemble of binary classifiers and matrix factorization. In: Proceedings of the 2014 Recommender Systems Challenge. ACM, p. 13

Porter RH, Zona JD (1992) Detection of bid rigging in procurement auctions, No. w4013, National Bureau of Economic Research

Powers DMW (2011) Evaluation: from precision, recall and F-measure to ROC, informedness, markedness and correlation (PDF). J Mach Learn Technol 2(1):37–63

Safavian SR, Landgrebe D (1998, May/June) A survey of decision tree classifier methodology. IEEE Trans Syst Man Cybern 22:660–674

Swets JA (1996) Signal detection theory and ROC analysis in psychology and diagnostics: collected papers, Lawrence Erlbaum Associates, Mahwah, NJ

Tan PN, Steinbach M, Kumar V (2013) Data mining cluster analysis: basic concepts and algorithms

Tong S, Koller D (2001, Nov) Support vector machine active learning with applications to text classification. J Mach Learn Res 2:45–66

Tóth B, Fazekas M, Czibik Á, Tóth JI (2015) Toolkit for detecting collusive bidding in public procurement. With examples from Hungary. Working Paper series: CRC-WP/2014:02, Version 2.0—Corruption Research Center, Budapest

Vojtěch S, Mynarz J, Węcel K, Klímek J, Knap T, Nečaský M (2014) Linked open data for public procurement. In: Auer S, Bryl V, Tramp S (eds) Linked open data. Creating knowledge out of interlinked data. Lecture Notes in Computer Science 8661. Springer, Heidelberg, pp. 196–213

Wang H, Wang N, Yeung DY (2015) Collaborative deep learning for recommender systems. In: Proceedings of the 21th ACM SIGKDD international conference on knowledge discovery and data mining. ACM

Wensinck W, De Vet JM et al (2013) Identifying and reducing corruption in public procurement in the EU. PricewaterhouseCoopers, Ecorys, Brussels

Werden GJ, Hammond SD, Barnett BA (2011) Deterrence and detection of cartels: using all the tools and sanctions. Antitrust Bull 56(2):207–234

WTO official site (2016) https://www.wto.org/english/thewto_e/minist_e/min03_e/brief_e/brief08_e.htm, 16 July 2016

Part IV
Legal Considerations

Chapter 17
Legal Considerations for Using Open Source Intelligence in the Context of Cybercrime and Cyberterrorism

Alison Lyle

Abstract The aim of this chapter is to raise awareness of some of the legal issues around open source investigations of cybercrime and cyberterrorism. The issues raised relate to different stages of the investigation process and highlight instances within each where various types of legislation may affect the activities carried out, or the progress of a case. Rather than attempt to provide an in-depth legal analysis, the author uses UK and European law to illustrate difficulties that may be encountered or factors that may need to be considered in this particular type of investigation. General issues of data protection and human rights are outlined and addressed and more specific topics such as lawful actions, disclosure and digital evidence are explored. The chapter also considers the reasons behind the legal issues and suggests some ways that problems may be overcome.

17.1 Introduction

This chapter outlines some of the main areas that illustrate the legal issues around cybercrime and cyberterrorism investigations using open source intelligence. Although the areas addressed present issues that may arise in all jurisdictions, a UK perspective is adopted in order to explore them and illustrate the effect that legal restraints can have on this type of investigation. It is hoped that this approach will serve to highlight issues that can then be applied to different legal systems. As well as UK law, legislative instruments, policies, procedures and guidelines at European and international level will be considered so that a more holistic view can be presented.

Employed by West Yorkshire Police, UK.

A. Lyle (✉)
Wakefield, UK
e-mail: Alison.Lyle@westyorkshire.pnn.police.uk

B. Akhgar et al. (eds.), *Open Source Intelligence Investigation*,
Advanced Sciences and Technologies for Security Applications,
DOI 10.1007/978-3-319-47671-1_17

One of the difficulties in relation to cybercrime is that there are no legal definitions or specific legislation at the time of writing, so the laws that apply to traditional crimes need to be adapted to be relevant to the cyber environment and the characteristics of offences committed therein. Because the laws were not developed to take account of the particular nature of cybercrimes or open source intelligence investigations, this can be a source of difficulty (Sampson 2015). Another consideration is whether the judges, courts and perhaps juries are appropriately equipped, in terms of knowledge, expertise and technology; that is, whether criminal justice systems themselves are adequate to be able to achieve justice in this area.

Cybercrime and cyberterrorism are usually referred to together, due to the 'cyber' element of the modi operandi that are common to both. In real terms, both categories of offending often involve the same actions but what distinguishes them is the motive or intent of those carrying them out; that is, the *actus reus* will be the same, but the *mens rea* will be different. For example, online fraud could be carried out by a teenager who has a dream of owning a sports car; the same action carried out by members of a terrorist organisation with the aim of furthering their cause would be seen and treated in a different way. Similarly, denial of service attacks can be hackers testing their skills or sinister and threatening acts of terrorists.

These distinctions are important and significant in a criminal justice system. If the motive and intent of the crime is known from the outset, the investigation and prosecution will follow very different courses. Apart from the case being brought under different legislation, the more serious crimes will have a wider range of investigative opportunities and resources made available to them. This is partly necessary because a more serious crime will be heard in a higher court which, in the UK, means that the evidence has to be prepared and presented in a more thorough way. However, many of the issues presented here will apply to both.

17.2 Citizens' Perceptions and Human Rights

Instances of cybercrime and cyberterrorism are rapidly increasing, and present serious threats to both individuals and societies. They threaten and violate human rights and cause a great deal of harm to many, from citizens to critical infrastructures. Because of the ubiquitous nature of the cyber environment where these crimes are carried out, detecting and investigating them involves law enforcement authorities accessing that same environment; this has the potential to impact on many more people than would be the case with traditional criminal investigations. Therefore, a dilemma exists: protecting citizens at the expense of their rights, or limiting actions to protect human rights? This debate has been long and heated and is the subject of intense feelings. Perhaps it is well illustrated by the case in 2016 between the FBI and Apple[1] which was resolved out of court, thereby not setting any useful precedent. However, the issues raised are important; Apple stood firm against FBI requests to access an iPhone, stating that sacrificing privacy, security and data protection puts people at greater risk. New surveillance powers, so useful for the fight against cybercrime and cyberterrorism, are often fiercely opposed by action groups but there are significant numbers of people equally concerned about cybercrime.[2] The subject of competing interests is not a new one and the difficulties are highlighted with the use of open source intelligence.

The Council of Europe has produced a guide for internet users, to clarify understanding of existing human right particularly in this context. In relation to public authorities, it is expressly stated that any interference should not be arbitrary, should pursue a legitimate aim in accordance with the European Convention on Human Rights[3] (ECHR), such as the protection of national security or public order, public health or morals and must comply with human rights laws.

17.3 Investigatory Powers

The special powers provided, through legislation, to police investigators allow certain privileges, in respect of access and actions, that would otherwise be considered intrusive or in violation of fundamental human rights such as privacy and data protection. The legislation recognises the importance of balancing these conditional rights against those relating to public or national security. Police powers are made at national level and take into account nations' particular practices, policies, cultures and priorities. This is crucial, but in relation to investigating online crimes, to which territorial boundaries are often irrelevant, the differences can be unhelpful and may provide frustrating obstacles.

When defining and categorising cybercrime and cyberterrorism in order to deal with it in a criminal justice system, it is generally the case that existing criminal laws have been applied or adapted rather than new ones enacted. In relation to investigations however, due to the difficulties of obtaining digital evidence and the large-scale problem that cybercrime and cyberterrorism is, many countries have enacted specific laws to enable law enforcement authorities (LEAs) to access and seize digital information and evidence. Many of these have granted investigators

[1]Hern A, Apple declares victory in battle with FBI, but the war continues (The Guardian, online report, 29 March 2016). Available at: https://www.theguardian.com/technology/2016/mar/29/apple-victory-battle-fbi-but-war-continues-san-bernardino-iphone?utm_source=esp&utm_medium=Email&utm_campaign=GU+Today+main+NEW+H&utm_term=164296&subid=13209147&CMP=EMCNEWEML6619I2. Accessed 29 March 2016.

[2]PA Consulting 'Cybercrime Tipping Point: public perceptions' (Online survey report, 2015). Available at: http://www.paconsulting.com/our-thinking/cybercrime-2015/. Accessed 5 May 2016

[3]Council of Europe, *European Convention for the Protection of Human Rights and Fundamental Freedoms, as amended by Protocols Nos 11 and 14*, 4 November 1950, CETS 5.

wide powers which override normal perceptions of privacy and have met with strong opposition. As this area of legislation and the associated case law develops, the degree to which the capabilities will be exercised remains to be seen.

17.3.1 Existing and Proposed Powers

In the UK, the current legislation which controls police powers in relation to carrying out surveillance and accessing digital communications is the Regulation of Investigatory Powers Act 2000 (RIPA). If correct procedures have not been followed, or the appropriate level of authorisation obtained, the defence in a case brought before a court may claim abuse of process and have evidence excluded or the proceedings stayed. Directed or intrusive surveillance of any kind, including during online investigations, should be carried out under a RIPA authority.

In 2015 as a response to the Snowden revelations, the Intelligence and Security Committee (ISC) of the UK Parliament published a report[4] on the powers of the UK intelligence and security agencies in relation to covert interception and acquisition of communications data. The report calls for increased transparency and greater recognition of privacy protections to be provided by a single Act of Parliament. Although this relates specifically to the intelligence and security agencies, it illustrates the concerns should be taken account of at all levels. These concerns were again raised in April 2016 in respect of UK surveillance laws, including the new Investigatory Powers Bill, which were scrutinised at European level after widespread condemnation of excessive powers and indiscriminate surveillance of digital personal data, amongst other things. The European Court of Justice will deliver its decision by June 2016, which will set clearer guidance in respect of lawful interception of digital personal data.

A new Intelligence and Security Services Bill in the Netherlands proposes new powers which will enable intelligence and security services to carry out surveillance and collection of personal data on a large scale. An example of the proposed powers would be the ability to intercept communications, using a particular chat app, between cities and countries. This would go some way to resolving issues around cross-border evidence gathering but has caused strong reaction from privacy groups.[5]

---

[4]Intelligence and Security Committee of Parliament 'Privacy and Security: A modern and transparent legal framework' (Online report, HMSO, 12 March 2015) https://b1cba9b3-a-5e6631fd-s-sites.googlegroups.com/a/independent.gov.uk/isc/files/20150312_ISC_P%2BS%2BRpt%28web%29.pdf?attachauth=ANoY7cqVUAbEn1Wlwy7jBr_BmXr5OO72-sBuBienMmDWi5kUa5hxQOafIjfEREk8eMJIL5ssJov7xru8L-Omuu2mtMnCZigEKs2t55QqorpbNX9k5oI_q3p8-xG3cpB7yHC-RWXX8wGGhWnhDKVkorvJjgeSLgDwpzH3p2ktFSi79_8RuQEJfmdLmi-LPk2iWZF6zMhiZmA2EFvfTft-b2rHRKOyzxeqwhfr5ekhEaE2zSSk5W4dAlACpqjJM5qUJ9kMBvFkn-2d&attredirects=0. Accessed 8 May 2016.

[5]Siedsma T and Austin E, 'Dutch dragnet surveillance bill leaked' (Online report, EDRi, 4 May 2016) https://edri.org/. Accessed 4 May 2016.

The introduction of a new surveillance law in France, following the Charlie Hebdo attacks, allows wide and indiscriminate surveillance of internet and mobile phone traffic and was widely criticised in all quarters, including the UN Human Rights Committee, Amnesty International and others.[6]

17.3.2 (Un)Lawful Practices

It has been acknowledged[7] that legal issues around open source intelligence investigations are uncertain in many areas, due to the lack of case law dealing specifically with these practices. However, an awareness needs to be maintained of the risks that this type of investigation may pose. Carrying out online investigations often involves officers accessing and downloading or copying material which may be used as evidence; while this is an essential part of such enquiries, some acts may result in the commission of an offence.

Officers must be cautious when carrying out online investigative work using a fake profile on a social media platform, which could fall into the category of committing an offence in the UK under the Computer Misuse Act 1990.[8] It is arguable whether creating a fake profile in order to gain access to a user's account would amount to 'unauthorised' under the Act, but with ambiguous case law in this area,[9] caution is advised. There are also clear risks if any material so obtained is to be relied on as evidence (see this chapter). Any directed surveillance online would currently require a RIPA authority in the UK; whether such action would be considered as such may be open to question.

Practices such as obtaining screenshots of online sources or carrying out enquiries using search engines for publicly available information present little problem but routine, prolonged monitoring of social media accounts may amount to illegal surveillance, if the appropriate legal authority was not in place. There is a fine line between carrying out specific, focused, short-term observations and engaging in activities which could be said to be a violation of privacy under Article 8 ECHR.

In the case where an online investigation involved child sexual offences, officers may, in the course of the investigation, commit an offence under the Sexual Offences Act 2003, which prohibits the taking or making of indecent photographs of a child.

---

[6]Siedsma T and Austin E, 'France: New surveillance law a major blow to human rights' (Online report, Amnesty International, 24 July 2015) https://www.amnesty.org/en/latest/news/2015/07/france-new-surveillance-law-a-major-blow-to-human-rights/. Accessed 6 May 2015.

[7]Hill M QC and Davis W, 'Social Media and Criminal Justice'(Winter 2013) Issue 4, Criminal Bar Quarterly p 5. Available at: https://www.criminalbar.com/files/download.php?m=documents&f=141106211508-CBQWinter2013.pdf. Accessed 19 April 2016.

[8]As amended by the Serious Crimes Act 2015.

[9]DPP v. Bignell [1998] 1 Cr. App.R. 1 and R. v. Bow Street Metropolitan Stipendiary Magistrate ex parte United States (No. 2) [2000] 2 AC 216).

However, in the UK, there is a Memorandum of Understanding Agreement between the Crown Prosecution Service (CPS) and the Association of Chief Police Officers (2016)[10] (ACPO) that state no action will be taken against officers or professionals acting to report or investigate such an offence. Care must be taken in respect of similar enquiries being carried out in other jurisdictions Jurisdiction (2016), where agreements may differ or not exist. In the UK there is always an overriding consideration of whether a prosecution is in the public interest; nonetheless, care must be taken to avoid committing an offence in order to prove one.

17.4 Data Protection

Open source investigations use public, web-based sources as a resource to assist with detecting, investigating and prosecuting cybercrime. It is an invaluable and well-used method, providing information that would otherwise be unavailable or very difficult to obtain. In many instances, crimes are solved using open source information that would otherwise leave victims without justice. However, in order to protect the privacy and personal data of individuals, data protection and human rights laws will usually impose restrictions on activities which involve any processing of personal data.

17.4.1 The Legislation

The main current piece of legislation for data protection at EU level is the Data Protection Directive[11] and although policing purposes do not come within its remit, the principles will still apply. In particular, there must be a legal basis and the processing (which includes collecting information) must be necessary in a democratic society and proportionate to the specific aim pursued. Even where the legitimate aim of the prevention or detection of crime has been accepted, the court will still consider the necessity and proportionality principles.[12] These principles will also be considered in cases involving the violation of rights under the Charter of Fundamental Rights of the European Union[13] (the Charter) which affords pro-

[10]Replaced on 1 April 2015 by The National Police Chief's Council (NPCC).

[11]Directive 95/46/EC of 24 October 1995 on the protection of individuals with regard to the processing of personal data and on the free movement of such data [1995] OJ L 281/31, as amended.

[12]See: S and Marper v United Kingdom, Appl. No. 30562/04 (ECtHR 4 December 2008); Z v Finland, Appl. No. 2209/93 (ECtHR 25 February 1997).

[13]Charter of Fundamental Rights of the European Union [2010] OJ C 83/02.

tection to rights of privacy[14] and data protection.[15] There is currently no EU legislation covering data protection issues relating to police duties in domestic situations; the Framework Decision[16] addresses processing of personal data for purposes of police and judicial cooperation but only applies to data transmitted between Member States. It provides minimum standards to be maintained.

Additionally, privacy and data protection are protected by the Council of Europe, which is separate to the European Union, but the laws from each compliment and influence each other. The main legal instrument here is the European Convention on Human Rights (ECHR) which has influenced all data protection law at European level. The European Court of Human Rights (ECtHR) has consistently held that the processing of personal data interferes with Article 8 ECHR, which protects privacy, and such interferences must have a legal foundation and be justified.[17] Due to the nature of many open source investigations, it may often be the case that personal data of non-suspects is held, which is an area of concern. However, anonymised data does not come within the remit of data protection laws so if this can be carried out, ideally at the earliest opportunity, requirements will not apply.

Another piece of legislation, which is based on the ECHR, is Convention 108[18] which is an internationally binding legal instrument that addresses data protection and includes police activities. The same principles apply as for the Data Protection Directive, which is based upon this agreement.

In May 2016, two new significant pieces of legislation were introduced; the General Data Protection Regulation[19] (GDPR) and the Police Directive,[20] both of which will apply to open source investigations from 2018 and will replace the current Data Protection Directive. The GDPR will impose obligations on protection of personal data that is processed in a general way by investigators and police forces, and the Police Directive will apply to the protection of data during investigations. Even though no longer a member of the European Union, UK forces will

---

[14]Ibid Article 7.

[15]Ibid Article 8.

[16]Council Framework Decision 2008/977/JHA of 27 November 2008.

[17]ECtHR. Leander v. Sweden. No. 9248/81, 26 March 1987; ECtHR. M.M. v. the United Kingdom, No. 24029/07. 13 November 2012.

[18]Council of Europe *Convention for the Protection of Individuals with regard to Automatic Processing of Personal Data,* CETS 108 1981.

[19]Regulation (EU) 2016/679 of the European Parliament and of the Council of 27 April 2016 on the protection of natural persons with regard to the processing of personal data and on the free movement of such data, and repealing Directive 95/46/EC (General Data Protection Regulation).

[20]Directive (EU) 2016/680 of the European Parliament and of the Council of 27 April 2016 on the protection of natural persons with regard to the processing of personal data by competent authorities for the purposes of the prevention, investigation, detection or prosecution of criminal offences or the execution of criminal penalties, and on the free movement of such data, and repealing Council Framework Decision 2008/977/JHA.

still have to comply with these standards in order to exchange and transfer data when collaborating with EU forces.

17.4.2 Further Considerations

There are police powers in all nations that provide a legal basis for the interference with conditional rights such as privacy, freedom of expression and the freedom to impart and receive information. European Member States were required to do this to give effect to the Data Protection Directive, but when the GDPR takes effect in 2018 this will be directly applicable to all Member States, thereby creating greater legal harmony. National laws will still be needed to implement the Police Directive however, which is designed to both strengthen citizens' rights and reduce the burden on public authorities to allow them to carry out their duties effectively.

Open source intelligence is highly valuable in respect of helping to solve crimes. The difference between this use as a valuable resource and its use as a method of surveillance really comes down to the principles of necessity and proportionality derived from data protection legislation which, in turn, is influenced by human rights instruments. Because the data protection laws are based on, and compliment, fundamental human rights, they are of particular importance. Carrying out cyber-crime investigations using open source intelligence will inevitably engage these rules and care must be taken to comply with requirements. An additional consideration is public perception of the use of personal data from open sources, which is a sensitive topic in the wake of the Snowden revelations.

Retaining data from open sources is linked to data protection and was controlled by a European Directive[21] until it was declared invalid in the joined cases of Seitlinger and Digital Rights Ireland[22] in 2014. The EU stated that data retention laws should be provided at national level and the UK introduced the Data Retention and Investigatory Powers Act 2014; however, this too was declared unlawful by the UK High Court in 2015.[23] Both pieces of legislation were deemed to be inconsistent with Articles 7 and 8 of the Charter, which protect privacy and data protection rights, and did not comply with the principle of proportionality. This

[21]Directive 2006/24/EC of 15 March 2006 on the retention of data generated or processed in connection with the provision of publicly available electronic communications services or of public communications networks and amending Directive 2002/58/EC [2006] OJ L 105.

[22]Joined Cases C-293/12 and C-594/12 *Digital Rights Ireland Ltd v. Minister for Communications, Marine and Natural Resources, Minister for Justice, Equality and Law Reform, Commissioner of the Garda Siochana, Ireland, The Attorney General, intervener: Irish Human Rights Commission and Karntner Landesregierung, Michael Seitlinger, Christof Tschohl and others* [2014].

[23]Mason Hayes and Curran 'Stopping the 'DRIP' of data—UK Court finds revised data retention legislation unlawful' (Online report, Institute of Paralegals, July 2015) http://www.lexology.com/library/detail.aspx?g=c2b1a42b-5fa9-435f-a201-c0458e35fbe6. Accessed March 2016.

emphasises the importance of incorporating such principles in the context of open source investigations, where data from open sources may be collected and retained.

17.5 Data Acquisition

It has been shown that data protection and privacy laws impose obligations and duties on LEAs collecting, processing and retaining personal information from open sources. Those same laws providing the same protections can also make obtaining digital evidence, in the form of personal data, more difficult.

It can frequently be the case that evidence for cybercrimes is held on servers in other countries. In such situations, the laws in the host country will take precedence and sometimes complex and lengthy procedures will ensue. This can result in delays in the investigation, with the potential for damaging consequences such as a known offender being able to continue committing offences during the period. Such situations can arise when LEAs have discovered an online crime using open source intelligence and have evidence, but need to link that evidence with an identified individual in order to bring a prosecution (see Chaps. 6 and 16).

It is not unusual that the host country is the United States, which can cause delays. The length of the delay may depend on which company is holding the data required. A company such as Apple take their customers' rights to privacy and data protection very seriously and often put this as a higher priority than assisting the fight against cybercrime. Most investigations do not involve a request for access to a phone, as in the high profile case involving the FBI,[24] and generally require personal data to support evidence. If Apple refuse to provide this, then a preservation order must be put on the material held and the process of formally obtaining it through the US legal system begun. Other countries, such as Canada and other organisations such as Facebook, can frequently cause similar difficulties.

17.6 Rules of Evidence

Rules surrounding evidence exist largely in the interests of ensuring a fair trial. These are rules and procedures which must be followed for evidence to be admitted at court and used to support the prosecution case in criminal proceedings. These are discussed in greater detail in this chapter. In brief the first stage relates to the identification and seizure of material which may or may not become evidence. In the area of cybercrime, most countries have enacted laws which allow police officers to obtain digital evidence which is thought to relate to a criminal

[24]Summary of case and issues available at: http://www.nytimes.com/indexes/2016/03/29/technology/bits/index.html?nlid=70241408. Apple iPhone Standoff Ends, for Now (2016)

investigation, or to preserve it until processes to release it are completed. The storage, examination and presentation of evidence must be highly controlled so that continuity and integrity can be demonstrated.

During open source investigations, it can frequently be the case that evidence of a crime has been discovered and seized but the offender needs to be identified. This is, of course, a crucial piece of evidence without which no charges can be brought.

17.6.1 Seizing Digital Evidence

In the UK, the Association of Chief Police Officers[25] (ACPO) provides guidance on obtaining, examining and presenting digital evidence. These guidelines help to create a national approach in these matters. Whilst the same rules and laws that apply to documentary evidence also apply to digital evidence, emphasis is placed on integrity and continuity which need special attention due to the ease with which they could be tampered with or automatically changed. Any access to original data must only be made by a person competent to do so.

Another issue specifically relating to digital evidence is the requirement for it to be examined and analysed by an expert. Usually, the original format would not be suitable to present to a jury, therefore additional processes relating to this special requirement must be followed.

When seizing digital evidence, officers should be selective and consider whether it is relevant to the investigation. Seizing a family computer, for example, may not take the investigation further forward but has the potential to violate rights of privacy under human rights legislation, and cause problems for the disclosure officer acting under the Criminal Procedure and Investigations Act 1996 (CPIA) (Criminal Procedure and Investigations Act Code of Practice 2016).[26]

Of particular relevance to investigators seizing evidence from open sources is the consideration of leaving a 'footprint' when accessing sites, which will alert website owners to law enforcement presence or interest. It is also important that records are kept of all actions taken when capturing evidence of this type.

17.7 Unused Material

Disclosure of material gathered or created during the course of an investigation is regulated and guided in the UK by the CPIA and by the Code[27] issued under that Act, as well as the Attorney-General's guidelines and judicial protocol. The

[25]Replaced on 1 April 2015 by The National Police Chief's Council (NPCC).

[26]As amended.

[27]The Criminal Procedure and Investigations Act 1996 (Code of Practice) Order 2015.

Disclosure Manual (Disclosure Manual 2016)[28] also applies in the UK. Any breaches can result in the collapse of a trial or the discontinuance of a case, so it is vital to comply. Disclosure of material capable of undermining the prosecution case or assisting the defence is crucial in the interests of a fair criminal justice process so is well regulated and will usually be scrutinised by defence.

The duty to retain, record and reveal all material relating to an investigation applies from the beginning of the investigation, but of particular importance to many open source investigations is that the beginning of an investigation can include a period of surveillance before the cybercrime was committed. In such cases, the material generated, rather than the surveillance operation itself, would be disclosable.

Due to the special characteristics of cybercrime, it may frequently be the case that techniques and tactics are used which need to be protected. If the legal basis for the action was a RIPA authority, this would need to be disclosed on a schedule listing sensitive information so that details are not passed to the defence.

Another factor to be considered and is of particular relevance to open source investigations, is that much of the material generated or retained may contain personal or sensitive information of individuals not connected to the investigation. This would particularly apply to social media accounts which contain details of other people. Care must be taken to store information securely and only redacted versions put on disclosure schedules. If redaction would affect the meaning of the material, then an unedited version should be placed on the sensitive schedule.

If the material or information is highly sensitive, this needs to be brought to the attention of the prosecutor at the earliest opportunity so that a decision can be made about whether a Public Interest Immunity application is required.

17.8 Different Jurisdictions

It is part of the nature of cybercrime and cyberterrorism that different jurisdictions are involved in the offences and investigations. Using open source intelligence also means carrying out investigations in borderless environments. This can cause problems on several fronts due to the different national laws of the countries involved. There is often the difficulty of who is to investigate and where the prosecution is to be brought. As has been referred to, there are also difficulties in relation to data acquisition between some countries.

Difference in national legislation can be the result of various influencing factors. The priority given to certain types of crime can vary as a result of, for example, major national incidents, cultures and beliefs or case law. The resulting legislation is, in turn, capable of influencing accepted norms in societies and thus of the perceived seriousness of different types of criminal behaviour. This variation in

[28]An agreed document between the Association of Chief Police Officers and the Crown Prosecution Service to ensure compliance with CPIA 1996.

perceptions and approaches can impact on cooperation and collaboration between investigators and other stakeholders in different countries.

As well as different laws creating unevenness in approaches to investigations they can also affect the response to requests for assistance, resulting in delays. Another issue raised by different legislation might be where a country with wide powers requests assistance from a country where LEAs have fewer powers and may not be able to access the information sought. This might particularly be the case during the present time when surveillance and seizure powers across European Member States are in the stages of development and enactment.

17.9 Overcoming Problems

As has been widely acknowledged, it is essential that there is collaborative working in order to deal effectively with cybercrime and cyberterrorism. Cooperation not only between different LEAs on national, European and international levels, but between LEAs and other sectors, such as private industry, citizens and non-governmental organisations.

17.9.1 Europol

In April 2016, MEPs endorsed the new EU Regulation to enable the EU police agency Europol (Europol: MEPs endorse new powers to step up EU police counter-terrorism Drive 2016) to operate faster and more effectively in the fight against cross-border crimes and terrorist threats, which includes cybercrime and cyberterrorism. With a strong emphasis on data protection, the new legislation which is endorsed by Civil Liberties Committee MEPs, will enhance Europol's effectiveness; an example of this is their ability to contact Facebook directly to order the takedown of pages supporting terrorism. There are rules for specialist cybercrime units such as the Internet Referral Unit and obligations for Member States to provide relevant data to Europol so that their activities are fully informed. The new rules will be effective from 1 April 2017.

17.9.2 Joint Investigation Teams

The Joint Investigation Team (JIT) network was established in 2005, implementing the 'The Hague' programme[29] and allows cooperation and collaboration of

[29]The Hague Programme: strengthening freedom, security and justice in the European Union (OJ C 53, 3.3.2005, p. 1), and the Council and Commission action plan implementing the Hague

investigation teams. Each Member State has a National Expert as a point of contact. A JIT includes judges, prosecutors and LEAs and is established for a defined period for a defined purpose. This arrangement allows the direct gathering and exchange of information and evidence without using formal Mutual Legal Assistance channels.

17.9.3 Eurojust

Eurojust facilitates cooperation in relation to investigations of serious cross-border crimes. Dealing with large, complex cases usually involving two or more Member States, the organisation can be used as final arbiter where national prosecutors fail to reach an agreement on where the prosecution should take place. Usually, the prosecution should be based in the jurisdiction where the majority of either the criminality or the greatest loss occurred. In other instances, the case will be assessed on its particular details and various factors relating to legal proceedings will be considered. In this way, the most desirable outcome can be identified whilst using or avoiding different legislation. Eurojust and the European Judicial Network facilitate cooperative working across Europe and beyond.

17.9.4 CEPOL

CEPOL is the European Police College, an agency of the European Union that works to achieve greater European police cooperation by providing training and development for officers from across Europe. A new Regulation[30] is expected to be in force in July 2016, which gives it increased importance in addressing the European dimension of serious and organised crime and terrorism, amongst other things. They have the task of assessing, defining and implementing training requirements for a wide group of law enforcement officials across Europe and beyond. This will further the aim of achieving greater unity in the fight against cybercrime and cyberterrorism, and may reduce the difficulties faced when investigating across two or more jurisdictions.

(Footnote 29 continued)

Programme on strengthening freedom, security and justice in the European Union (OJ C 198, 12.8.2005, p. 1).

[30]Regulation (EU) 2015/2219 of the European Parliament and of the Council of 25 November 2015 on the European Union Agency for Law Enforcement Training (CEPOL).

17.9.5 Interpol

The world's largest international police organisation aims to unite police forces across the globe. Providing support, secure data channels and expertise, Interpol facilitate cooperative, effective investigations into cybercrime. The Interpol Global Complex for Innovation brings together expertise from law enforcement and private industry to carry out research into emerging cybercrimes and the tools and methods to combat it.

17.10 Summary

Failure to comply with legal requirements, as well as policies and good practice guidelines, may mean that valuable evidence is inadmissible and an opportunity to deal with cybercrime is missed. It may also impact on the effectiveness of future investigations, if tactics have been unknowingly revealed. Professional integrity, public confidence, justice for victims and individual officer safety are all at risk of being compromised too. The risk of jeopardising any of these things is too great to ignore.

1. The legislation and guidance applicable to investigations must be understood and applied, where necessary. Rather than a 'one size fits all' approach, each case should be assessed on its individual merits.

Examples:

- Consider whether the activity requires legal authority, i.e. Regulation of Investigatory Powers Act 2000.
- Data protection rules and principles will apply to personal data (see below).
- Consider whether the activity is likely to infringe privacy rights, i.e. even though personal information is openly available, collecting it for law enforcement reasons may still violate human rights.
- Consider whether the activity will inadvertently breach criminal laws (see below)
- Ensure that relevant guidelines and policies are adhered to. All those involved in the investigation must be aware of these.

2. It is possible that the act of carrying out open source investigations can inadvertently trigger various legal issues, awareness of this possibility must be maintained.

Examples:

- Using a fake profile to access social media accounts may be an offence under the Computer Misuse Act 1990.
- Downloading material as evidence in child sexual abuse cases, for example, may constitute an offence under the Sexual Offences Act 2003.
- Several investigators accessing the same personal account or details of an individual may be defined as prolonged or directed surveillance when considered as a whole, which would require legal authority.
- Establishing a personal contact with a subject using a fake profile in order to access personal information, for example becoming a 'friend' on social media, may be defined or challenged as being a 'relationship', thereby requiring legal authority.

3. Much open source investigation work will interfere with privacy rights protected by Article 8, ECHR. If this is the case, then legal justification for this interference must be established.

Examples:

- Individuals posting personal information on open access platforms may still have a legitimate expectation of privacy, which will be infringed by some investigation activities, such as the collection and storage of it.
- Seizing a family computer as evidence may constitute a violation of privacy in relation to others not subject to the investigation. Processing personal data of non-suspects is a particular area of concern.
- Wide-ranging and indiscriminate collection of personal data from open sources is likely to be considered a violation of privacy, even though for legitimate law enforcement purposes.

4. The collection, processing and storage of personal data in relation to an investigation must comply with data protection laws and the data protection principles are of overriding importance.

Examples:

- Any processing of personal data must be for a specific purpose ('law enforcement' would not suffice), be the minimum necessary to achieve the specific purpose and not further processed for unrelated reasons.
- Even where a legitimate purpose had been established, a court would still consider whether the aim could have been achieved by less intrusive means and apply the necessity and proportionality principles.
- Adequate safeguards for the storage of personal data must be provided and data must not be kept for longer than is necessary.

- Personal data belonging to individuals not connected with the investigation should be anonymised or deleted.

5. The seizing, storage, examination and presentation of evidence must be highly controlled so that integrity and continuity can be demonstrated.

Examples:

- Examination of devices not seized under legislation such as the Police and Criminal Evidence Act 1984 (e.g. remotely) will require authority under Part III of the Police Act 1997.
- Seizing an excessive amount of material could result in collateral intrusion and also cause disclosure problems. Both of which are subject to legal constraints.
- Cyber experts should be contacted to access and preserve data wherever necessary; their specialist knowledge can often save valuable evidence that would otherwise be lost, in a way that will be admissible in court.
- All evidence must be stored according to data protection rules and evidential requirements. In the UK, the Information Commissioner's Office recently fined a police force £160,000 for leaving unencrypted DVD's containing sensitive material in an insecure location. Such action would also mean loss of integrity of evidence as it could be tampered with; this would leave it open to challenges by the defence.

6. Records of all activity must be kept, in the interests of statutory data protection and disclosure obligations.

Examples:

- In order to comply with legal disclosure obligations, records must be kept and any material generated in the course of an investigation retained, whether evidential or not. The beginning of an investigation can include surveillance to determine whether or not an offence has been committed.
- An audit trail must be available that shows all activity relating to online investigations and justifications or reasons recorded.

17.11 Conclusion

The legal issues outlined here are all capable of having an impact on investigators carrying out open source investigations of cybercrime and cyberterrorism; some have a positive impact and provide guidance and protection whilst others may cause difficulties and prolong investigations, with various consequences.

However, most of the legal issues can be said to be derived from human rights considerations. Apart from the national, European and international instruments that specifically protect fundamental rights and freedoms, the data protection laws are all derived from the ECHR, and are closely linked with the right to privacy. Similarly, the laws controlling evidential and disclosure matters are all designed to give effect to and protect the right to a fair trial and prevent miscarriages of justice.

It has also been shown that public perception is an important factor to consider; whilst it isn't a legal issue, the way that police activities are perceived on a general scale can have enormous impact. After several high profile cases, along with the Snowden revelations, public feeling is very high in terms of both public authorities and fundamental rights and freedoms. This can affect how case law develops which, in turn, can affect legislation. Indeed, the latest developments in European law are as a direct result of these very things.

Perhaps we are living in the digital age, where people live their lives, and commit crimes, in a cyber environment, but never have human beings been more united on a global scale in respect of demanding the protection of their human rights. We might differentiate between cybercrime and traditional crime, but essentially it comes down to real people causing damage or loss to other real people and the legal issues relate to this.

Although it is an area which has developed so rapidly that appropriate legislation failed to keep up, new laws are being developed and introduced that facilitate cooperative working, that help determine the extent and nature of police powers and that address conditional human rights and determine an acceptable degree of intrusion in return for increased security and online safety. It is this balance of rights that needs to be achieved and maintained and which should form the basis of all investigations.

References

Apple iPhone Standoff Ends, for Now (2016) New York Times, Bits daily report. http://www.nytimes.com/indexes/2016/03/29/technology/bits/index.html?nlid=70241408. Accessed 29 March 2016

Association of Chief Police Officers (2016) ACPO good practice guide for digital Evidence. http://www.digitaldetective.net/digital-forensicsdocuments/ACPO_Good_Practice_Guide_for_Digital_Evidence_v5.pdf. Accessed 28 March 2016

Bowcott O (2016) European court to consider legality of UK surveillance laws. The guardian online 11 April 2016. http://www.theguardian.com/world/2016/apr/11/european-court-to-considerlegality-of-uk-surveillance-laws. Accessed 8 May 2016

Criminal Procedure and Investigations Act Code of Practice (2016) Statutory guidance, UK Government, february 2016. https://www.gov.uk/government/publications/criminal-procedure-and-investigations-act-code-of-practice. Accessed 7 May 2016

Director of Public Prosecutions (2016) Guidelines on prosecuting cases involving communications sent via social media. Online guide, Crown Prosecution Service, June 2012. http://www.cps.gov.uk/legal/a_to_c/communications_sent_via_social_media/index.html. Accessed 8 May 2016

Disclosure Manual (2016) Online guide, Crown prosecution service. http://www.cps.gov.uk/legal/d_to_g/disclosure_manual/. Accessed 8 May 2016

Europol: MEPs endorse new powers to step up EU police counter-terrorism Drive (2016) European Parliament online press release, 28 April 2016. http://www.europarl.europa.eu/news/en/news-room/20160427IPR24964/Europol-MEPs-endorse-new-powers-to-step-up-EU-police-counter-terrorism-drive. Accessed 28 April 2016

Jurisdiction (2016) Online guide, Crown Prosecution Service. http://www.cps.gov.uk/legal/h_to_k/jurisdiction/#an12. Accessed 7 May 2016

Law Enforcement and Cloud Computing (2016) Linklaters online report, October 2011. http://www.linklaters.com/Insights/Law-Enforcement-Cloud-Computing/Pages/Index.aspx. Accessed 7 May 2016

Lovells H (2016) UK Parliamentary report calls for a new legal framework for UK secret intelligence agencies. Online article, Institute of Paralegals, 12 March 2015. http://www.lexology.com/library/detail.aspx?g=20574274-2074-42b9-b514-ac5a9930bbd2. Accessed 8 May 2016

Memorandum of Understanding between Crown Prosecution Service (CPS) and the Association of Chief Police Officers (ACPO) concerning Section 46 Sexual Offences Act 2003 (2016) Online article, Crown Prosecution Service. https://www.cps.gov.uk/publications/agencies/mouaccp.html. Accessed 8 May 2016

New CEPOL Regulation (2016) Published on the Official Journal of the European Union Online report, CEPOL February 2016. https://www.cepol.europa.eu/media/news/new-cepol-regulation-published-officialjournal-european-union. Accessed 8 May 2016

Sampson F (2015) The legal challenges of big data application in law enforcement. In: Akhgar et al (eds) Application of big data for national security—a practitioner's guide to emerging technologies, 2015. Elsevier, Oxford, UK

Chapter 18
Following the Breadcrumbs: Using Open Source Intelligence as Evidence in Criminal Proceedings

Fraser Sampson

Abstract Intelligence and evidence are fundamentally different and while evidence can always provide some degree of intelligence the reverse is not the case. If intelligence is to be relied on evidentially it will need to meet the same forensic standards and clear the same legal hurdles as any other form of evidence. Therefore LEAs need to be aware of these standards and hurdles at the outset and to ensure—so far as practicable—that they are in a position to address them. This chapter addresses some of the legal issues that arise if OSINT material is to be used in legal proceedings, particularly within countries that are signatories to the European Convention on Human Rights (ECHR).

Breadcrumbs[1]
noun

1. a series of connected pieces of information or evidence
2. a type of secondary navigation scheme that reveals the user's location in a website or Web application.

18.1 Introduction

The provenance, collation, interpretation, analysis and deployment of open source intelligence (OSINT) is becoming a highly topical and relevant area of policing. As has been considered in detail in earlier chapters OSINT can be considered as an element of a 'new age' in policing and as an adjunct to the 'longer arm of the law' (Chap. 3). In this chapter we are concerned with addressing some of the legal issues

[1]https://www.google.co.uk/#q=breadcrumbs+web+design (Accessed 12 June 2016).

F. Sampson (✉)
Office of the Police and Crime Commissioner for West Yorkshire, West Yorkshire, UK
e-mail: fraser_ospre@me.com

B. Akhgar et al. (eds.), *Open Source Intelligence Investigation*,
Advanced Sciences and Technologies for Security Applications,
DOI 10.1007/978-3-319-47671-1_18

that arise if OSINT material is to be used in legal proceedings, particularly within countries that are signatories to the European Convention on Human Rights (ECHR). While each jurisdiction will be governed by its own domestic laws there are some common elements around evidence and some overarching provisions within the ECHR that will apply to relevant proceedings in each of the 47 signatory States.[2] Both the generic principles of evidence and the ECHR are considered below.

The expansion of social media and Internet-based communication, together with its relevance for criminal investigation and national security, have been explored and discussed in the previous chapters. It is clear from the foregoing just how far Law Enforcement Agencies (LEA) have come to understand the power of these tools, not just as an adjunct to their own communications (Coptich and Fox 2010) but as a game-changing source of intelligence and investigation. The contribution of OSINT to inductive investigation has yet to be fully understood, still less harnessed, but the 'breadcrumbs' left by electronic data interactions by suspects, victims, witnesses and other persons of interests represent a phenomenological change in the intelligence world. Following those breadcrumbs—in both senses defined above—in order to find people, patterns, propensities or property is one thing; relying on the material to support a prosecution is another matter altogether. This chapter will consider some of the key elements in utilising OSINT material as evidence.

The developments in socio-digital behaviour have produced a whole new category of 'community' which can be seen as a virtual group which coalesces around a particular theme or event, groups which are evanescent in nature and probably unique in identity. Once the event/activity/interest that unites the members of the community diminishes, so does the digital community itself (for examples see Beguerisse-Díaz et al. 2014). Law Enforcement Agencies are increasingly requesting contributions from these digital communities and seeking material from citizens to investigate crime (see for example the request by police for 'dashcam' material in connection with the suspected attempt to abduct an RAF serviceman[3]).

18.2 What Is the Difference Between Intelligence and Evidence?

At its heart the principal difference between intelligence and evidence is purposive. The purpose of intelligence is wide ranging, almost undefined, and can cover an array of activities from supplying information on which to base an arrest (e.g., by giving rise to reasonable suspicion under the Police and Criminal Evidence Act 1984, s. 24) to the likely destination of a vulnerable person who has run away from home or understanding the lifestyle of someone suspected of benefiting from the

[2]http://www.coe.int/en/web/conventions/search-on-treaties/-/conventions/chartSignature/3—accessed 15 April 2016.

[3]http://www.bbc.co.uk/news/uk-england-norfolk-36853106. Accessed 26 July 2016.

proceeds of crime. Evidence, on the other hand, has one function: to assist a court or finder of fact to determine a matter that has come before it. Of course, if the matter coming before a court arose out the use of intelligence (for example a civil action against the police for wrongful arrest based on flawed information) then the two might overlap. Taking Staniforth's second category of intelligence (see Chaps. 2 and 3), the end user of OSINT material is essentially the organization producing or collating it while with *evidential material* the recipient will be the relevant tribunal. Generally a court will not be concerned with intelligence and in some cases in England and Wales will be prevented from considering it at all.[4] However, in some cases OSINT will potentially be helpful to parties either in a criminal prosecution or in some civil proceedings such as employment litigation, defamation or infringement of intellectual property. If OSINT is to be deployed and relied upon in criminal proceedings by LEAs there are some important practical considerations that need to borne in mind—and the earlier in the process of acquisition the better.

To illustrate those considerations consider a situation where investigators are inquiring into a robbery. Conducting OSINT research they find a photograph on a Facebook page that appears to have been taken at the time and in the location of the alleged offence. The photograph shows two people, one of whom is the registered user of the Facebook page. The photograph shows the two people, both male, standing in a park laughing and one of the males is holding up what looks like a handgun. Plainly this OSINT would be potentially relevant to the robbery inquiry for a whole range of reasons. In and of itself the material might be sufficient to put the two men at the scene of the offence and substantiate the grounds for their arrest. It might also be relevant in terms of search activity for a weapon and stolen property, for identification of suspects, associates, witnesses, clothing etc. But how far would the material be accepted by a court in a subsequent criminal trial? A good starting point in addressing that question would be the material's *relevance* and what *purpose* it would serve. The court would need, for example, to establish the facts in issue in the case and how far the Facebook material helped to prove any of those facts. If the men in the photograph admitted to having been present in that place and at that time but simply denied having been involved in the robbery, it would be of limited relevance. If, on the other hand, they denied having been present or even knowing each other, the material would be of greater relevance. If there was dispute about their whereabouts at the time and location it might be possible to show not only the content of the image but, if it had been created on a mobile device, where and when the image was made and transmitted. There might be a description of the offenders' clothing or other matters of their appearance, words used during the offence etc., some of which could be corroborated (or indeed contradicted) by the Facebook entry and any accompanying text. But unless the party relying on it can demonstrate the material's relevance to an issue in the proceedings it is likely to be inadmissible.[5]

---

[4]See for example ss. 17–19 of the Regulation of Investigatory Powers Act 2000.

[5]see e.g. *R v Blastland* (1986) AC 41.

18.3 Practical Issues

The requirement to demonstrate relevance *to a fact in issue* is a critical element in the rules of evidence within England and Wales and, as we shall see below, any other state that is a signatory to the ECHR. Then there will be issues of reliability. While a key concept in intelligence gathering, *reliability* has a very specific legal meaning when it comes to the rules of evidence. Before admitting the Facebook material the court would also want to know where the material came from, who made the photograph, who posted it on the page, how reliable the maker (if identified) is, how easily someone else could have made and posted the material, what the defendant has had to say about it and the integrity of the process by which it has reached the court. These considerations will not just affect the admissibility of the material but also the weight to be attached to it. The greater the likelihood that the material could have been altered or interfered with, the less weight it will carry even if it is held to be relevant.

A further and overriding consideration in a criminal trial will be the fairness of allowing the material to be adduced as evidence. In trials involving a jury it is often necessary for the judge to give specific directions about the evidence admitted, for what purpose(s) it can be considered (e.g. motive, identity, alibi etc.) and the limits of any inference that can be made from it. Generally material that has appeared in some open source with no reliable antecedents, with ready opportunities to interfere with/alter it and without anyone willing to testify to its provenance such material is unlikely to be of much use in criminal proceedings.

And a significant consideration where the material is being relied upon by an LEA will be the means by which it has been obtained. If the material has been obtained illegally or in breach of process (particularly if it has been obtained in breach of a defendant's rights under Art. 8 of the ECHR[6]) there will be further impediments to its being deployed as evidence.

18.4 Legal Framework

In most jurisdictions with developed legal systems the legal framework governing criminal proceedings will provide a defendant basic entitlements such as the right to a fair hearing before an impartial tribunal, a (qualified) right not to incriminate him/herself[7] and the right to challenge any witnesses testifying against him or her. In countries that are signatories to the ECHR these fundamental entitlements are set

[6]see *"Opinion on the status of illegally obtained evidence in criminal procedures in the Member States of the European Union*' 30 Nov 2003—Reference: CFR-CDF. opinion 3-2003.

[7]*Funke v. France*, 44/1997/828/1034; see also *O'Halloran and Francis v. the United Kingdom* (2007) ECHR 545; *Saunders v. the United Kingdom* (1997) 23 EHRR 313.

out in Art 6(1) and are likely to have parallels in other jurisdictions observing the rule of law. The legal framework is considered below.

The legal framework governing the acquisition and use of OSINT by LEAs in the UK is a mixture of European and domestic law, some of which creates particular challenges and dilemmas for LEAs (see Sampson 2015). As discussed above, the ECHR—and art 6(1) in particular—plays a central part in this framework; other jurisdictions beyond the 47 signatory states will have their own primary and secondary sources of protection for defendants in criminal proceedings.

18.5 European Convention on Human Rights

Article 6(1) of the European Convention on Human Rights provides that

Article 6—Right to a fair hearing

1. *In the determination of ... any criminal charge*[8] *against him, everyone is entitled to a fair and public hearing within a reasonable time by an independent and impartial tribunal established by law. ...*
2. *Everyone charged with a criminal offence shall be presumed innocent until proved guilty according to law*
3. *Everyone charged with a criminal offence has the following minimum rights*

    (a) *to be informed promptly, in a language which he understands and in detail, of the nature and cause of the accusation against him;*
    (b) *to have adequate time and facilities for the preparation of his defence;*
    (c) *to defend himself in person or through legal assistance of his own choosing or, if he has not sufficient means to pay for legal assistance, to be given it free when the interests of justice so require;*
    (d) *to examine or have examined witnesses against him and to obtain the attendance and examination of witnesses on his behalf under the same conditions as witnesses against him;*
    (e) ...

The admissibility of evidence is primarily a matter for regulation under national law[9] but Art. 6(1) requires that prosecuting authorities disclose *all material evidence in their possession for or against the accused.*[10] This duty of disclosure is

[8]Note that there is a 'civil limb' to the ECHR – see Art 6(1) and *Guide to Article 6 and the Right to a Fair Trial'* Council of Europe www.echr.coe.int (Case-law – Case-law analysis – Case-law guides). Accessed 12 April 2016.

[9]*Schenk v. Switzerland* (1988) ECHR 17; *Heglas v. the Czech Republic* (2007) ECHR 5564.

[10]*Rowe and Davis v. the United Kingdom* (2000) ECHR 91.

strengthened by domestic legislation[11] and is an important element in the evidential use of OSINT discussed in Chap. 17. Although the rules of evidence differ significantly 'civil law' jurisdictions (such as those countries whose legal systems evolved from the Napoleonic Code)[12] the effect of Art 6(1) and the broader entitlement to a fair hearing are very similar. As a general rule of fairness it can be safely assumed that the use of any OSINT material that is by its nature gravely prejudicial to the defendant is likely to be challenged and probably excluded.[13] The entitlement to a fair hearing also involves giving a defendant the proper opportunity to challenge and question a witness [per Art. 6(3)(d)] and that would include the maker of OSINT materials relied on against him or her.

Many, if not all, jurisdictions will have specific rules about hearsay evidence and its admissibility. In England and Wales hearsay is "a statement not made in oral evidence in the proceedings that is evidence of any matter stated"[14] and it is governed by statute[15] which provides fairly wide gateways through which hearsay evidence may be admitted. Clearly OSINT documents and material will, if used as proof of any matter stated within them,[16] fall within this definition and the statutory rules, together with relevant guidelines for prosecutors should be consulted.

In relation to Art. 6(3)(b) the "*facilities*" that the defendant must enjoy will include the opportunity to acquaint him or herself with the results of investigations carried out throughout the proceedings.[17] If the defendant is detained on remand pending trial those "*facilities*" may include "such conditions of detention that permit the person to read and write with a reasonable degree of concentration".[18] In order to facilitate the conduct of the defence, the defendant must not be hindered in obtaining copies of relevant documents and compiling and using any notes taken.[19] All these considerations *could* have particular significance when relying on OSINT from the Internet and all relevant materials used by the LEA will need to be made available or accessible to the defendant.[20]

---

[11]such as the Criminal Procedure and Investigations Act 1996 in England and Wales.

[12]see Law Society Gazette 11 April 2016, pp 13—15 London.

[13]For the general approach of the court in England and Wales see *Noor-Mohamed v R* [1949] ac 182.

[14]s.114 (1) Criminal Justice Act 2003.

[15]The Criminal Justice Act 2003.

[16]e.g., SMS messages—*R v Leonard* (2009) EWCA Crim 1251) but *cf R v Twist* [2011] EWCA Crim 1143.

[17]*Huseyn and Others v. Azerbaijan* (application nos. 35485/05, 45553/05, 35680/05 and 36085/05); *OAO Neftyanaya Kompaniya Yukos v. Russia* (2014) ECHR 906.

[18]*Mayzit v. Russia* application no. 42502/06; *Moiseyev v. Russia* (2011) 53 EHRR 9.

[19]*Rasmussen v. Poland* (application no. 38886/05).

[20]Questions are whether these entitlements will ever extend to being able to access relevant materials via the Internet, where the hard or downloaded copies are incomplete or insufficiently verifiable by the defendant?

18.6 Uses of OSINT as Evidence

Against that framework the potential evidential uses of OSINT are vast. For example the prosecution may want to use the defendant's use of certain expressions or idiosyncratic grammar to prove that she wrote a particular sentence in, say, a case of blackmail or harassment. Alternatively the state may wish to show that the defendant posted materials on social media showing that they were at a certain place at the time of an offence, that they were a member of a violent gang or that they were bragging openly about involvement in an incident.[21]

Of course some criminal offences (such as the making of threats or insulting comments[22] or posting 'revenge porn'[23]) might directly involve the use of 'open source' material such as that found on social media. In those cases the material will be directly relevant to the facts in issue. An example can be found in one case[24] where a juror posted a grossly inappropriate Facebook message during the trial of an alleged sex offender. It was held that this posting of the message amounted to a contempt of court as it had been calculated to interfere with the proper administration of justice. In that case the defendant had used his smart phone to send the message when travelling home on a bus[25] and the *message itself was direct evidence of the offence itself.* Alternatively such material might include a recording made by a witness on their mobile phone and posted on YouTube to prove the manner of an assault (kicking, stamping etc.) and the presence/absence of anyone else at the time, or the geo-locator of a phone to undermine evidence of alibi. However, much OSINT material is unlikely to be directly probative of an offence and is more likely to be relied on by way of background or contextual information or to corroborate/contradict a specific fact in issue. In addition it may be the *defendant* who wishes to rely on OSINT, for instance to show that unsolicited pictures had been submitted by a complainant on his Facebook page.[26] In such cases the same evidential principles will apply.

While these same principles can apply within the context of related civil proceedings by LEAs (such as applications for recovery of illegally obtained assets, injunctive relief or applications for confiscation orders) these are outside the scope of this book.

Finally, although OSINT is, by its nature, generally put into the public domain by others without the involvement of an LEA, investigators will need to be very cautious about any activity that me be regarded as encouraging or inciting the

---

[21]*Bucknor v R* (2010) EWCA Crim 1152.

[22]See for example http://www.theguardian.com/uk/2012/may/22/muamba-twitter-abuse-student-sorry—accessed 16 April 2016.

[23]see s. 33 Criminal Justice and Courts Act 2015.

[24]*Attorney General v. Davey* [2013] EWHC 2317 (Admin).

[25]*loc cit* at 6.

[26]*T v R* (2012) EWCA Crim 2358.

commission of an offence[27] and must not breach any laid down processes for accessing data.[28] As discussed above if the material has been has been obtained unlawfully there will be significant consequences and may even result in the dismissal of the entire case.[29]

18.7 Conclusion

Intelligence and evidence are fundamentally different and while evidence can always provide some degree of intelligence the reverse is not the case. If intelligence is to be relied on evidentially it will need to meet the same forensic standards and clear the same legal hurdles as any other form of evidence. Therefore LEAs need to be aware of these standards and hurdles at the outset and to ensure—so far as practicable—that they are in a position to address them.

References

Akhgar B, Saathoff G, Arabnia H, Hill R, Staniforth A, Bayerl PS (2013) Application of big data for national security: a practitioner's guide to emerging technologies. Elsevier, Waltham MA USA

Armstrong T, Zuckerberg M, Page L, Rottenberg E, Smith B, Costelo, D (2013) An Open Letter to Washington, 9 December

Beguerisse-Díaz M, Garduno-Hernandez G, Vangelov B, Yaliraki S, Barahona M. (2014) Interest communities and flow roles in directed networks: the Twitter network of the UK riots, Cornell University Library http://arxiv.org/abs/1311.6785

Blackman J (2008) Omniveillance, Google, Privacy in Public, and the Right to Your Digital Identity: a Tort for Recording and Disseminating an Individual's Image over the Internet, 49 Santa Clara L. Rev. 313

Bruns A, Burgess J (2012) #qldfloods and @QPSMedia: Crisis Communication on Twitter in the 2011 South-East Queensland Floods. Queensland University of Technology, ARC Centre of Excellence for Creative Industries and Innovation, Brisbane, Australia

Casilli A, Tubaro P (2012) Social media censorship in times of political unrest—a social simulation experiment with the UK riots. Bulletin de Methodologie Sociologique 115:5–20

Copitch G, Fox C (2010) Using social media as a means of improving public confidence. Safer Communities 9(2):42–48

Crowe A (2011) The social media manifesto: a comprehensive review of the impact of social media on emergency management. J Bus Continuity Emerg Planning 5(1):409–420

[27] *Khudobin v. Russia* (application no. 59696/00); *Texieira v Portugal* (application 44/1997/828/1034).

[28] See s. 1(1) of the Computer Misuse Act 1990 in England and Wales; *DPP v. Bignell* (1998) 1 Cr App R 1 and *R v. Bow Street Metropolitan Stipendiary Magistrate ex parte United States* (No.2) (2000) 2 AC 216).

[29] *El Haski v. Belgium (Application no. 649/08) Gäfgen v. Germany* (2010) ECHR 759.

Crump J (2011) What are the police doing on Twitter? Social media, the police and the public. Policy Internet 3(4):1–27

Denef S, Kaptein N, Bayerl PS, Ramirez L (2012) Best practice in police social media adaptation. Composite project

Earl J, Hurwitz H, Mesinas A, Tolan M, Arlotti A (2013) This protest will be Tweeted. Inform Commun Soc 16(4):459–478

Howard P, Agarwal S, Hussain M (2011) When Do States Disconnect their Digital Networks? Regime Responses to the Political Uses of Social Media 9 Aug 2011. (Online) http://ssrn.com/abstract=1907191. Accessed 25 Nov 2014

Kaplan A, Haenlein M (2010) Users of the world, unite! the challenges and opportunities of social media. Bus Horiz 53(1):59–68

Kavanaugh AL, Yang S, Li L, Sheetz S, Fox E (2011) Microblogging in crisis situations: Mass protests in Iran, Tunisia, Egypt. *CHI2011*, Vancouver, Canada, May 7–12 2011

Kokott J, Sobotta C (2013) The Distinction between Privacy and Data Protection in the Jurisprudence of the CJEU and the ECtHR. Int Data Privacy Law 3(4):222–228

Kotronaki L, Seferiades S (2012) Along the pathways of rage: the space-time of an uprising. In: Seferiades S, Johnston H (eds) Violent protest, contentious politics, and the neoliberal state. Ashgate, Surrey, pp 159–170

Lanier J (2013) Who Owns the Future?. Simon and Schuster NY USA

Liberty (2011) Liberty's Report on Legal Observing at the TUC March for the Alternative (Online) https://www.liberty-human-rights.org.uk/sites/default/files/libertys-report-on-legal-observing-at-the-tuc-march-for-the-alternative.pdf Accessed 22 Nov 2014

Lin D (1998) Extracting Collocations from Text Corpora. First workshop on computational terminology. Montreal, Canada, pp 57–63

Loveys K (2010) Come down from the roof please, officers tweeted, Mail Online 11 Nov 2010. (Online) http://www.dailymail.co.uk/news/article-1328586/TUITION-FEES-PROTEST-Met-chief-embarrassed-woeful-riot-preparation.html. Accessed 16 Mar 2011

McSeveny K, Waddington D (2011) Up close and personal: the interplay between information technology and human agency in the policing of the 2011 Sheffield Anti-Lib Dem Protest. In: Akhgar B, Yates S (eds) Intelligence management: knowledge driven frameworks for combating terrorism and organized crime. Springer, New York, pp 199–212

NETPOL Network for Police Monitoring (2011) Report on the Policing of Protest in London on 26 Mar 2011. (Online) https://netpol.org/wp-content/uploads/2012/07/3rd-edit-m26-report.pdf. Accessed 22 Nov 2014

NPIA (2010) Engage: digital and social media for the police service. National Policing Improvement Agency, London

Palen L, (2008) On line social media in crisis events Educause 3: 76-78. See also Baron G. Social media and crisis: a whole new game. http://www.youtube.com/watch?v=MFt7NXDhcmE

Papic M, Noonan S (2011) Social media as a tool for protest. *Security Weekly*, 3 Feb 2011. (Online) http://www.stratfor.com/weekly/20110202-social-media-tool-protest#axzz3LWjMNk4d. Accessed 10 Dec 2014

Poell T, Borra E (2011) Twitter, YouTube, and Flickr as Platforms of Alternative Journalism: The Social Media Account of the 2010 Toronto G20 protests. Journalism 13(6):695–713

Procter R, Crump J, Karstedt S, Voss A, Cantijoch M (2013) Reading the riots: what were the police doing on Twitter? Policing Soc: An Int J Res Policy 23(4):413–436

Russell A (2007) Digital communication networks and the journalistic field: the 2005 French Riots. Critical Stud Media Commu 24(4):285–302

Sampson F (2015) Cybercrime presentation Project Courage and CAMINO Cyber Security Workshop, Montpellier, France 9 April

Searls D (2012) The intention economy: when customers take charge. Harvard University Press Cambridge, USA

Seretan V, Nerima L, Wehrli E (2003) Extraction of multi-word collocations using syntactic bigram composition. Proceedings of International Conference on recent advances in NLP Issue: Harris 51. Publisher, Citeseer, pp 424–431

Smadja F (1993) Retrieving collocations from text: xtract. Computational Linguistics 19(1):143–177

Vieweg S, Palen L, Liu S, Hughes A, Sutton J (2008) Collective intelligence in disaster: an examination of the phenomenon in the aftermath of the 2007 virginia tech shootings. In: Proceedings of the information systems for crisis response and management conference (ISCRAM 2008)

Xiguang L, Jing W (2010) Web-based public diplomacy: the role of social media in the Iranian and Xinjiang Riots. J Int Commu 16(1):7–22

Printed by Printforce, the Netherlands